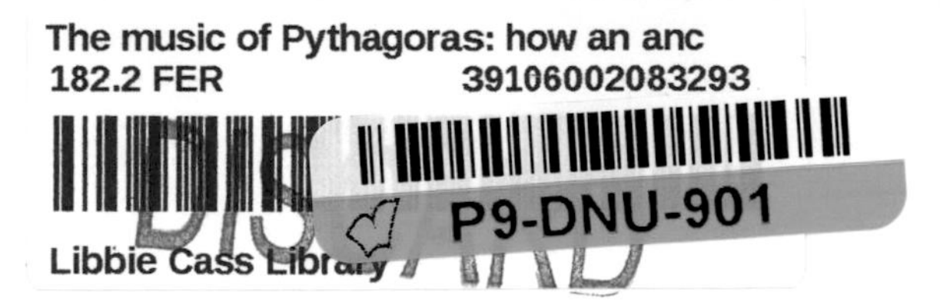

The Music of Pythagoras

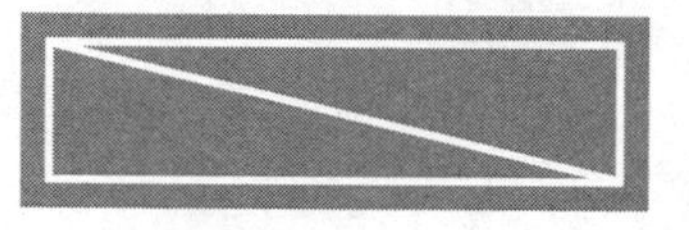

BY THE SAME AUTHOR

Tycho & Kepler:
The Unlikely Partnership That Forever Changed Our Understanding of the Heavens

Measuring the Universe: Our Historic Quest to Chart the Horizons of Space and Time

The Fire in the Equations:
Science, Religion, and the Search for God

Prisons of Light: Black Holes

Stephen Hawking: Quest for a Theory of Everything

The Music of Pythagoras

How an Ancient Brotherhood Cracked the Code of the Universe and Lit the Path from Antiquity to Outer Space

Kitty Ferguson

Walker & Company
New York

Published by Walker Publishing Company, Inc., New York
Distributed to the trade by Macmillan

Maps by Jeffrey Ward

All papers used by Walker & Company are natural, recyclable products made from wood grown in well-managed forests. The manufacturing processes conform to the environmental regulations of the country of origin.

LIBRARY OF CONGRESS CATALOGING-IN-PUBLICATION DATA
HAS BEEN APPLIED FOR.

ISBN-10: 0-8027-1631-8
ISBN-13: 978-0-8027-1631-6

Visit Walker & Company's Web site at www.walkerbooks.com

First U.S. edition 2008

1 3 5 7 9 10 8 6 4 2

Typeset by Westchester Book Group
Printed in the United States of America by Quebecor World Fairfield

To Serafina Clarke

Contents

Acknowledgments

I wish to thank all those friends who, during the years when I was researching and writing this book, have told me about ways—some of them odd and unexpected—that Pythagoras and the Pythagoreans have made an impact, or at least an appearance, in their own fields of study and interest. I also wish to thank my husband, Yale, for the help he has given me out of his own historical knowledge and library, his wonderful company on research journeys to Samos and Crotone, and his invaluable early critique of this book; Eleanor Robson, for her patient help in the area of Mesopotamian mathematics; John Barrow, for calling my attention to the "Sulba-Sûtras" and reconstructing the tunnel of Eupalinos on Samos for me out of a dinner napkin; the staff of the Museo Archeologico Nazionale di Crotone for their extraordinary helpfulness; and the librarians at the Chester Public Library, for their skill and willingness when I came to them with numerous unusual interlibrary loan requests.

Lifetimes and Other Significant Dates

CHAPTER 1

Pythagoras c. 570–500 B.C.
Thales fl. c. 585 B.C.
Anaximander 610–546 B.C.
Diogenes Laertius fl. c. A.D. 193–217
Porphyry c. A.D. 233–306
Iamblichus of Chalcis c. A.D. 260–330

CHAPTER 2

Babylonian exile of the Hebrews 598/7 and 587/6 to 538 B.C.
Rule of the Samian tyrant Polykrates 535–522 B.C.

CHAPTERS 3–6

Pythagoras' arrival in Croton 532/531 B.C.
Croton defeats and destroys Sybaris 510 B.C.
Death or disappearance of Pythagoras 500 B.C.
Second decimation of the Pythagoreans 454 B.C.

CHAPTER 7

Philolaus c. 474–399? B.C.
Parmenides 515 or 540–mid-5th century B.C.
Melissus early 5th century–late 5th century B.C.
Zeno of Elea c. 490–mid to late 5th century B.C.
Socrates c. 470–399 B.C.

CHAPTER 8

Plato 427–347 B.C.
Archytas 428–347 B.C.
Dionysius the Elder c. 430–367 B.C.
Dionysius the Younger 397–343 B.C.
Aristoxenus of Tarentum fl. fourth century B.C.

CHAPTER 9

Socrates c. 470–399 B.C.
Plato 427–347 B.C.

CHAPTER 10

Aristotle 384–322 B.C.
Theophrastus 372–287 B.C.
Alexander the Great 356–323 B.C.
Heracleides Ponticus 387–312 B.C.
Dicaearchus of Messina fl. c. 320 B.C.
Euclid fl. c. 300 B.C.

CHAPTER 11

Cicero 106–43 B.C.
Numa ruled c. 715–673 B.C.
Ennius c. 239–c. 160 B.C.
Marcus Fulvius Nobilior 2nd century B.C.
Cato the Elder 234–149 B.C.
Pliny the Elder A.D. 23–79
Posidonius c. 135–51 B.C.
Sextus Empiricus fl. 3rd century A.D.
Eudorus of Alexandria fl. c. 25 BC
Nigidius Figulus fl. no later than 98–27 B.C.
Vitruvius fl. 1st century B.C.
Occelus of Lucania after Aristotle

CHAPTER 12

Eudorus of Alexandria fl. c. 25 B.C.
Sotion 1st century A.D.
Seneca c. 4 B.C.–A.D. 65
"Sextians" 1st century A.D.
Apollonius of Tyana 1st century A.D.

Alexander of Abonuteichos c. A.D. 110–170
Julia Domna died A.D. 217
Philostratus A.D. 170–c. 245
Philo of Alexandria 20 B.C.–A.D. 40
Ovid 43 B.C.–A.D. 17
Plutarch A.D. 45–125
Moderatus of Gades 1st century A.D.
Theon of Smyrna c. A.D. 70–130/140
Nicomachus fl. c. A.D. 100
Numenius of Apamea fl. late 2nd century A.D.
Ptolemy c. A.D. 100–c. 180

CHAPTER 13

Diogenes Laertius fl. A.D. 193–217
Porphyry c. A.D. 233–306
Iamblichus of Chalcis c. A.D. 260–330
Longinus A.D. 213–273
Plotinus A.D. 204–270
Macrobius A.D. 395–423
Boethius c. A.D. 470–524

CHAPTER 14

Hunayn 9th century
Brethren of Purity 10th century
Al-Hasan 10th century
Aurelian 9th century
John Scotus Eriugena c. 815–c. 877
Regino of Prüm died 915
Raymund of Toledo 1125–1152
King Roger of Sicily 1095–1154
Bernard of Chartres 12th century
Nicole d'Oresme 14th century
Nicholas of Cusa 1401–1464
Franchino Gaffurio 1451–1522

CHAPTER 15

Petrarch 1304–1374
Nicholas of Cusa 1401–1464

Leon Battista Alberti 1407–1472
Marsilio Ficino 1433–1499
Pico della Mirandola 1463–1494
Giorgio Anselmi 15th century
Nicolaus Copernicus 1473–1543
Andrea Palladio 1508–1580
Tycho Brahe 1546–1601

CHAPTER 16

Philipp Melanchthon 1497–1560
Tycho Brahe 1546–1601
Michael Mästlin 1550–1631
Johannes Kepler 1571–1630

CHAPTER 17

Vincenzo Galilei late 1520s–1591
Galileo Galilei 1564–1642
William Shakespeare c. 1564–1616
John Milton 1608–1674
John Dryden 1631–1700
Joseph Addison 1672–1719
René Descartes 1596–1650
Robert Hooke 1635–1703
Robert Boyle 1627–1691
Isaac Newton 1642–1727
Gottfried Leibniz 1646–1716
Carl Linnaeus 1707–1778
William Wordsworth 1770–1850
Pierre-Simon de LaPlace 1749–1827
Filippo Michele Buonarroti 1761–1837
Hans Christian Oersted 1777–1851
Michael Faraday 1791–1867
James Clerk Maxwell 1831–1879

CHAPTER 18

Bertrand Russell 1872–1970
Arthur Koestler 1905–1983

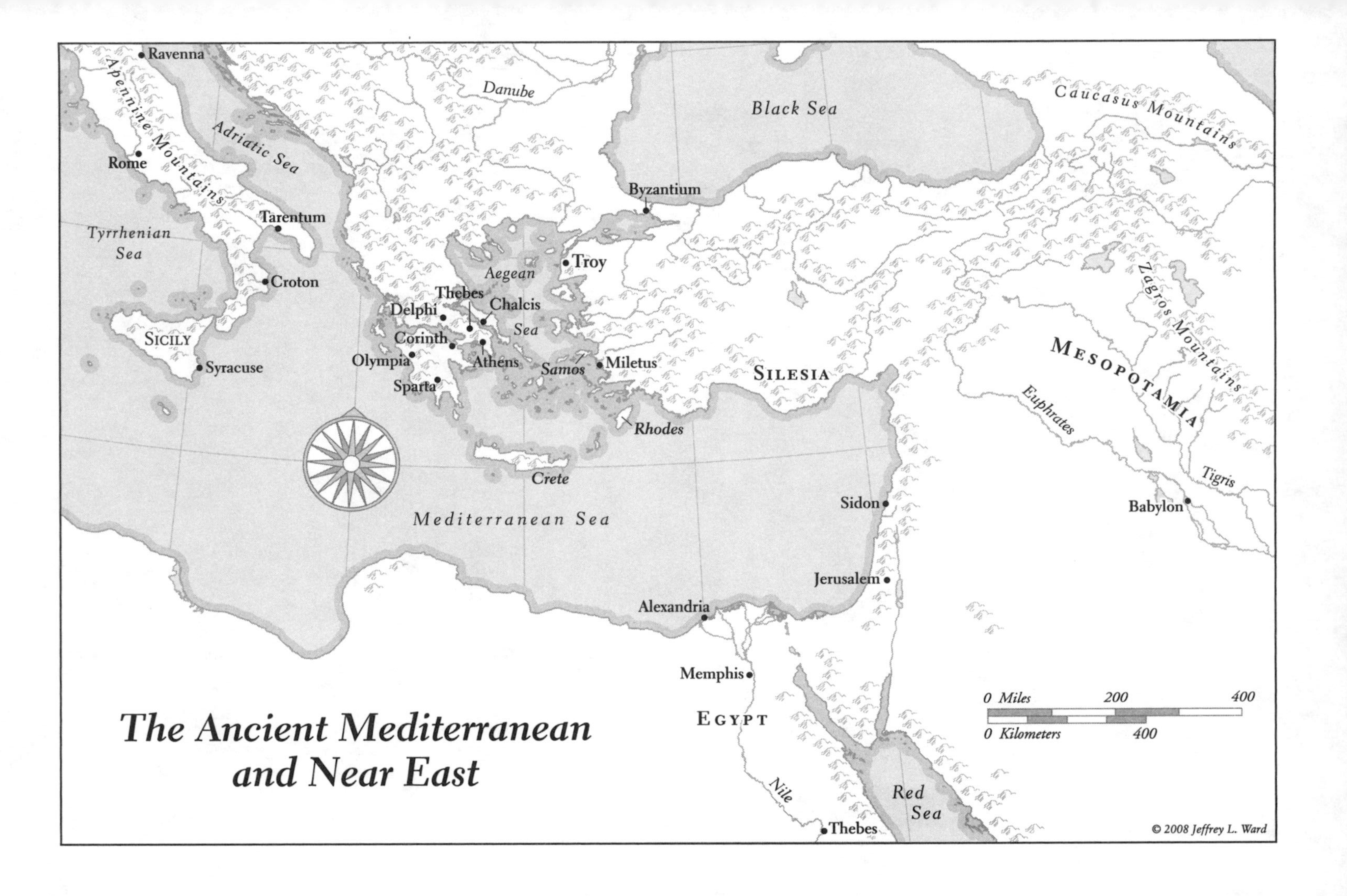
The Ancient Mediterranean and Near East
Ravenna
Apennine Mountains
Rome
Adriatic Sea
Danube
Black Sea
Caucasus Mountains
Byzantium
Tarentum
Tyrrhenian Sea
Croton
Troy
Aegean Sea
Thebes
Chalcis
Delphi
Corinth
Athens
Olympia
Sparta
Samos
Miletus
Sicily
Syracuse
Silesia
Rhodes
Crete
Mediterranean Sea
Zagros Mountains
Mesopotamia
Euphrates
Tigris
Babylon
Sidon
Jerusalem
Alexandria
Memphis
Egypt
Nile
Thebes
Red Sea
0 Miles 200 400
0 Kilometers 400
© 2008 Jeffrey L. Ward

PART I

Sixth Century B.C.

CHAPTER 0

"At the hinge of legend and history"

On the Aegean island of Samos, on the narrow arm of the harbor that juts farthest out to sea, there is a stark, skeletal structure. Immense shards of iron look as though they have fallen from the sky in the shape of a huge right triangle. One end of the diagonal has buried itself in the ground. Instead of a vertical line rising from the right angle, there is the statue of a man—lean, elongated, taller than life. He is reaching up with his right arm as though to conjure down the broken piece of iron that, if it were complete, would form the vertical of the triangle. Between his fingers and its lowest tip is a gap, such a gap as separates the finger of God from the finger of Adam in the ceiling of the Sistine Chapel. The triangle is not this man's creation. It is as old as the universe, as old as truth.

There is no argument but that this monument captures Western civilization's image of Pythagoras, a native son of this magical island. The triangle is his classic symbol . . . but, more authentically, he has become the icon of an unexplained but undeniable gift: the ability of human minds to connect with the bedrock rationality of the universe.

Behind all the veneration of Pythagoras and the undeniably great heritage attributed to him and his followers, behind the assumptions about his accomplishments, the uncritical early biographies, the legends, the debunkings, the forgeries, there was a real person. Who he was, actually—except for illusive wisps of information—is lost in the past.

Pythagoras and the devotees who surrounded him during his lifetime were obsessively secretive. As far as is known, they left no writings at all. There is no scroll, no text, no fragment, no firsthand account by any witness, no artifact for archaeologists to scrutinize, no tablet to decipher. If such ever existed, they no longer did by late antiquity. The earliest written evidence about Pythagoras himself that modern scholarship accepts as genuine consists of six short fragments of text from the century after his death, found not in their originals but in works of ancient authors who either saw the originals or were quoting from earlier secondary copies. The Pythagorean doctrine of reincarnation is the subject of three of these fragments, two of which also mention Pythagoras' courage, knowledge, and wisdom. Two others are scornful and derogatory. The sixth is a backhanded compliment in the middle of an unrelated story by the historian Herodotus, who termed Pythagoras "by no means the feeblest of the Greek sages." None name any discoveries, pinpoint any quotable wisdom or scientific contribution, or give biographical details. Though some treatises about Pythagoras tell you that his contemporaries seem not to have been aware of his existence, that was not the case, for all these fragments assume that Pythagoras was a famous man whose name readers would recognize. That, of course, has continued to be true for two thousand, five hundred years, in spite of the fact that as early as the time of Plato, in the fourth century B.C., Pythagoras and the Pythagoreans were already a mystery, and today they are often described as "an ancient cult about whom almost nothing is known."

Those six early fragments are not, however, the full extent of the available evidence about the Pythagoreans—those men and women who followed Pythagoras during his lifetime and who in later generations went on trying to live out his teachings. Philolaus, a not-so-secretive Pythagorean, wrote a book fifty to seventy-five years after Pythagoras' death, revealing that early Pythagoreans proposed that the Earth moves and is not the center of the cosmos. Plato knew Pythagoreans in the fourth century B.C., was strongly influenced by the idea of the role of numbers in nature and creation, and tried to incorporate what he thought of as a Pythagorean curriculum—the "quadrivium"—at his

Academy in Athens. Aristotle and his pupils wrote extensively about the Pythagoreans a few years later, relying on earlier material that still existed then but has since vanished, and on carefully chosen living spokesmen for the oral tradition, before a time when that became contaminated by forgeries. This present book will return frequently to the issues of evidence and how it was and is evaluated. It seems no other group has ever made such an effort to remain secret, or succeeded so well, as the Pythagoreans did—and yet become so celebrated and influential over such an astonishingly long period of time.

In an attempt to cut through the multilayered veil of twenty-five centuries that hangs between us and whatever happened on the ancient isle of Samos and in the harbor city of Croton, skeptical twentieth-century historians insisted on discarding all but the most concrete, "hard" historical evidence. Though certainly they were right to believe a corrective was needed, they arguably pruned too much, applying standards of their own time to an era for which it was inappropriate and even misleading to do so. The tiny "core of truth" left after discounting all folk wisdom, semi-historic tradition, legend or what might be only legend, and blatant forgeries and inventions can be stated in one paragraph:

Pythagoras of Samos left his native Aegean island in about 530 B.C. and settled in the Greek colonial city of Croton, on the southern coast of Italy. Though the date of his birth is not certain, he was probably by that time about forty years old and a widely experienced, charismatic individual. In Croton, he had a significant impact as a teacher and religious leader; he taught a doctrine of reincarnation, became an important figure in political life, made dangerous enemies, and eventually, in about 500 B.C., had to flee to another coastal city, Metapontum, where he died. During his thirty years in Croton, some of the men and women who gathered to sit at his feet began, with him, to ponder and investigate the world. While experimenting with lyres and considering why some combinations of string lengths produced beautiful sounds and others did not, Pythagoras, or others who were encouraged and inspired by him, discovered that the connections between lyre string lengths and human ears are not arbitrary or accidental. The ratios that underlie musical harmony make sense in a remarkably simple way. In a flash of extraordinary clarity, the Pythagoreans found that there is pattern and order hidden behind the apparent variety and confusion of nature, and that it is possible to understand it through numbers. Tradition has it that, literally and figuratively, they fell to their knees upon discovering that the universe is rational. "Figuratively," at least, is surely accurate, for the Pythagoreans embraced this discovery to the extent of allowing numbers to lead them, perhaps during Pythagoras' lifetime and certainly shortly after his death, to some extremely farsighted and also some off-the-wall, premature notions about the world and the cosmos.

One might assume that the above paragraph is a summary merely touching the highlights of what is known about events in sixth-century B.C. Croton, but it is, in fact, *all* that is known. Though you and I might wish to ask many more questions, the answers are irretrievably lost. No one can claim to tell how Pythagoras and his followers arrived at the religious and philosophical doctrines they espoused, or even precisely what these were . . . or in what specific ways Pythagoras and his followers influenced and changed the culture and civic structure of Croton and the surrounding area . . . or whether whatever caused Pythagoras and his followers to make such volatile enemies was something we would condemn or applaud today . . . or whether the great discovery in music of the power of numbers to reveal truth about the universe was made by Pythagoras himself. It may come as a particular surprise that there has been no mention of a Pythagorean triangle or a Pythagorean theorem in this "core of knowledge" about Pythagoras.

While historians in the twentieth century were clearing the deck, archaeologists were also playing a role in bringing down the legendary Pythagoras. They uncovered evidence that the "Pythagorean theorem" (or the "Pythagorean rule," for "theorem" implies a concept that was unrecognized this early) was known long before Pythagoras. Those revelations were not the end of the discussion, for with regard to such knowledge, there is more to be answered than the question of who had it "first." The way it passed—or may have passed—or failed to pass—from society to society and era to era is a complex, fascinating subject. Was it known and then lost? Or only partly lost? Were there separate discoveries? Equally significant is the way different societies and eras regarded such knowledge, what meaning they attached to it. Was it useful for surveying and building? Was it valued for the way it helped produce beautiful design? Was it considered holy? Was it something to be shared, or to be held in strictest secrecy, or taught only to a few? Was it intriguing in and of itself? Or did it imply something about—or raise questions about—the nature of all being? Did it buttress, or tear down, a trust in the power of numbers to uncover secret truth about the universe? Was there a "proof"? What constituted "proof" before the modern concept of "proof"? With questions like those, the origin of the "Pythagorean theorem" becomes an extremely interesting and complicated issue.

Numbers and mathematics had been in use for eons before Pythagoras was born, sometimes with more sophisticated understanding than his and his followers'. Their insight in the realm of music was extraordinary in a different way—different from the practical use of numbers or from an artist's appreciation for a beautiful geometric figure. Different even from the more abstract thinking of an early Babylonian teacher or student who found it an interesting exercise to do the math for a grain pile far larger than could ever be constructed. Imagine a carpenter looking at the hammer and chisel that he holds in his hands, that he has been taking for granted as a useful part of his daily work, and in an instant of dumbfounded recognition seeing that he holds the keys to unlock the doorway to vast hidden knowledge. That was what numbers became for the Pythagoreans and, through them, for the future. With this fresh appreciation—indeed, veneration—of the power of numbers, Pythagoras and his followers made one of the most profound and significant discoveries in the history of human thought. They stood at the sort of threshold that humanity has crossed only a few times. This particular door would not close again.

The brutally pared-down picture of Pythagoras and the events of his life offered by the twentieth century was no more satisfactory a representation than the one that overcredulous earlier centuries had accepted. All that could be said for it was that it was probably *not wrong*. But, for me, it has caused a dramatic refocusing of my attention onto the enormous, rich, multilayered, continuously reimagined story of "Pythagoras"—as seen separately from the life and person of the historical Pythagoras. That is the reason this book ends in the twenty-first century rather than in antiquity.

Amazingly it is the uncertainty about what really occurred and who Pythagoras really was and what he accomplished that has allowed something astounding to happen through the centuries. One truly powerful idea did come authentically from Pythagoras and his earliest followers—the recognition that numbers are a pathway from human ignorance to an understanding of the deepest mysteries of a universe that on some profound level makes perfect sense and is all of a piece. That vision has been a premier guide in the development of science and remains so today. However, the scarcity of sure knowledge about nearly everything else connected with Pythagoras and the Pythagoreans has encouraged generation after generation, beginning as early as Plato and still continuing in the twenty-first century, to reimagine him, to recreate him, to fashion their own variations on the theme of Pythagoras. As composers do in music, such figures as Plato, Aristotle, Ptolemy, Copernicus, Kepler, heroes of the French Revolution, Bertrand Russell, Einstein, and those who are now seeking extraterrestrial intelligence have taken a very slim theme indeed and composed intricate, sometimes whimsical, sometimes weird, often magnificent variations—a metaphor not inappropriate for a story that began with the strings of a lyre.

Two and a half millennia of writing and thinking and myth-making and composing variations about Pythagoras in one context after another, with one agenda after another, have of course multiplied the difficulties for a "biographer." Even more difficult to sort out than the outspoken detractors and obvious distortions and forgeries are those who, encountering Pythagorean or pseudo-Pythagorean thought, have joyfully recognized its links with their own thoughts and taken off from there, calling it *all* Pythagorean, even attributing their best ideas to Pythagoras himself—as Isaac Newton, of all people, did. Or calling *none* of it Pythagorean, but leaving the way open for others to say it was. Perhaps an author should abandon all hope of nonfiction and write a novel. To a certain extent, that is what two and a half millennia have written.

All of which might cause one to conclude that this book must be a postmodern parable. It would be difficult to find a better example of ideas, a life story, or a person being re-imagined time after time, century after century. Instead, I have come to see "Pythagoras" as a cubist painting, a Picasso or a Braque—either of whom would have insisted that there is more truth in their cubist paintings than in a photolike portrait. Life and history are impossible to fit together in a completely satisfying, coherent picture—and are continually reinvented in the eye of the beholder.

This book begins with something resembling a conventional "biography," indulging in calculated speculation, recounting legends and rumors, reporting intriguing and sometimes conflicting information, trying to discern what most likely happened—or might have happened—given the time and place and context. Much of the information comes through the research of three authors who wrote biographies of Pythagoras seven to eight hundred years after his death, in the third and early fourth centuries A.D., who in their time pieced together second-, third-, and fourth-hand accounts, legends and hearsay, oral tradition, what people believed or guessed, and other writers' references to lost works—ancient material that ranges from the reliable to the well-meaning and intelligent to the ridiculous. Pythagoras was already a cubist painting, but these three accounts more than any other sources have influenced what the world has thought it knew and still thinks it knows about him.

From the time of those biographies, the Pythagorean story wound its way into the Middle Ages and eventually into the modern world. It followed what is by no means a satisfying linear path. There are threads and trends, but more remarkable is the unavoidable impression that the idea of Pythagoras existed and still exists on an almost subliminal level. It shows up not only where you might expect it, underpinning the work of Copernicus, Kepler, Newton, and Stephen Hawking, but also in odd, unlikely places such as the architecture of Palladio and the philosophical interpretation of the French Revolution, and a grandfatherly figure in a novel by Louisa May Alcott. In spite of all the twentieth-century skepticism, impressive thinkers like Bertrand Russell, Arthur Koestler, and Jacob Bronowski regarded Pythagoras as a towering, foundational figure. Pythagorean principles have become imbedded in our worldview, and the original Pythagorean cracking of the code underpins the continuing development of science.

Lament the lost story of the life and person of Pythagoras, if you will, but join me in attempting to understand why and how it has birthed and nurtured such a rich tradition and wealth of interpretation, and in celebrating what is not a myth or a lie or even a legend . . . but one beautiful instance of realization about the truth of the universe.

CHAPTER 1

The Long-haired Samian

Sixth Century B.C.

N IN IMPERIAL ROME, THERE WAS a popular myth that the ancient sage Pythagoras had been the son of Apollo. The story was spread in the first century A.D. by Apollonius of Tyana, an itinerant wonder-worker who claimed he was the reincarnated Pythagoras and could speak with authority. The empress Julia Domna, wife of the emperor Septimus Severus, saw to it that Apollonius' tales were well publicized, in the hope of rivaling Jesus of Nazareth, whose followers believed that he was the son of the god of the Hebrews.

A century after Julia Domna (eight centuries after Pythagoras), the story of Pythagoras' divine patrimony came into the hands of the neo-Platonist philosopher and historian Iamblichus of Chalcis, who was writing a book titled *Pythagorean Life*.[1] Living in a superstitious age, he was not a particularly skeptical biographer when it came to the miraculous. He weighed carefully not whether he should believe "marvelous" tales, but which to believe, and he balked at the report that Pythagoras was descended from a god. It was "by no means to be admitted." Iamblichus did not, however, merely ignore myths that he could not accept as truth, nor should a historian have done so when sorting out the sixth century B.C.—this era that Jacob Bronowski called the "hinge of legend and history." Iamblichus liked to speculate about why a myth had arisen. Here is his version of Pythagoras' birth story, sanitized of what he saw as unduly supernatural details:

In the first third of the sixth century B.C., a merchant seaman named Mnesarchus embarked on a voyage, unaware that his wife was in the early stages of pregnancy. As most important merchants of his time who had the opportunity would have done, he included Delphi on his itinerary and enquired of the oracle—the Pythian Apollo—whether the remainder of his venture would be a success. The oracle replied that the next portion of the journey, to Syria, was going to be particularly productive. Then the oracle changed the subject: Mnesarchus' wife was already pregnant with a son who would be surpassingly beautiful and wise, and of "the greatest benefit to the human race in everything pertaining to human achievements." This was an astounding pronouncement, but Iamblichus insisted it was no indication that the son was not Mnesarchus' child. It was to honor the oracle, not to imply the patrimony of Apollo, that Mnesarchus changed his wife's name from Parthenis to Pythais and decided to name the boy Pythagoras. The voyage continued, and Pythais gave birth at Sidon in Phoenicia. Then the family returned to their home on the island of Samos. As the oracle had predicted, the mercantile venture had been a success and added substantially to their wealth. Mnesarchus erected a temple to the Pythian Apollo. No identifiable trace of it has survived, but Samos is sprinkled with the ruins of temples and shrines from that period that cannot now be attributed either to a particular god or donor.

The two other authors who lived during the time of the Roman Empire and wrote "lives" of Pythagoras in the third and early fourth centuries a.d.—Diogenes Laertius and Porphyry—were in agreement with Iamblichus that there was ample evidence Pythagoras' mother Pythais was descended from the earliest colonists on Samos.[2]* However, there is no other part of Pythagoras' life story, until the events surrounding his death, about which the discussion among them became so animated and contradictory as it did regarding his father Mnesarchus' origins. Iamblichus' research indicated that both parents traced their ancestry to the first colonists on Samos. Porphyry was in possession of a conflicting report from a third century b.c. historian named Neanthes—a stickler for juxtaposing conflicting pieces of information—that Mnesarchus

* The stories of the three biographers themselves are in Chapter 13.

was not Samian by birth. Neanthes had had it from one source that Mnesarchus was born in Tyre (in Syria) and from another that he was an Etruscan (Tyrrhenian) from Lemnos. The similarity of the names "Tyre" and "Tyrrhenian" had perhaps caused the confusion. Porphyry referred to an additional source, a book with an enticing title, On the Incredible Things Beyond Thule, that also mentioned Mnesarchus' Etruscan and Lemnos origins. Diogenes Laertius, the earliest of the three biographers, pointed out that the responsible ancient historian Aristoxenus of Tarentum—with excellent contacts, such as Dionysius the Younger of Syracuse and Pythagoreans in the fourth century b.c.—also had said Mnesarchus was a Tyrrhenian. All three biographers agreed that if Mnesarchus was not Samian by birth, he was naturalized on Samos. Diogenes Laertius also threw in that he had learned from one Hermippus, a native of Samos in the third century b.c., that Mnesarchus was a gem engraver.

The island of Samos, Pythagoras' childhood home, is the most precipitous and thickly forested of the Greek islands. Jacob Bronowski called it a "magical island. Other Greek islands will do as a setting for The Tempest, but for me this is Prospero's island, the shore where the scholar turned magician."[3] The boy Pythagoras would have been familiar with forest-clad mountain slopes, deep wooded gorges, and misty outlines of half-barren coastlines on a cobalt sea. For a family of the landholding class, life in the countryside, in this climate where flowers bloom most of the year and grape vines and olive groves proliferate, was pleasant, probably luxurious, even more so with goods Mnesarchus brought home from trips abroad. In poetry of which only fragments survive, Asius described the Samian aristocracy as wearing "snow-white tunics," "golden brooches," "cunningly worked bracelets," and wrote of their "tresses" that "waved in the wind in golden bands."[4]

In the port city and the precincts of Samos' temple of the goddess Hera were goods, treasures, and curiosities to carry a young man's imagination to the borders of the world. The temple had acquired a collection of valuable ornaments from Iran, Mesopotamia, Libya, Spain, and even farther away. Archaeologists have found no other Greek site so rich in foreign material, no ancient site anywhere with so wide a geographical spectrum of offerings. Not only Hera acquired treasures. Imported household and luxury items brought foreign textures, smells, and colors into Samian homes and no doubt fed the dreams and adventurous spirits of young men like Pythagoras and his brothers. Samos was in close touch with the much more ancient and mysterious culture of Mesopotamia.

What is known of Samos' history is a combination of folk memory, oral history, and archaeology. By legend, the first settlers were led by Ankaios, a hero son of Zeus who had sailed with Hercules and Orpheus on the voyage of the Argonauts in pursuit of the Golden Fleece. At the behest of the Pythian oracle at Delphi, Ankaios had decided to establish a colony and brought families from Arcadia, Thessaly, Athens, Epidaurus, and Chalcis. The oracle dictated the name of the future great city of the island, Samos. "*Sama*" implied great heights, and Samos has high mountains. Ancient stories traced Pythagoras' family's lineage to Ankaios himself.

Today, more than thirty centuries after Samos was pioneer territory, archaeologists are able to put dates to the stories. They agree that the ancient history of Samos was largely consistent with legend. Ionians from Epidauria arrived in the late second millennium B.C., and the Pythian oracle at Delphi was busy in operation then, though Apollo was not yet associated with it. The colonists who came, perhaps led by Ankaios, were part of large migrations from mainland Greece to the islands of the eastern Aegean and the shores of Asia Minor.

Archaeologists have also discovered that these Ionian settlers were not the first to set foot on Samos, which accords with another legend—that many of the Mycenaeans who besieged Troy and sent the great wooden horse into the doomed city settled on the Turkish coast and nearby islands. Excavations show that there were people living on Samos more than a thousand years before the Ionian settlers, and some were probably Mycenaean. Any who arrived after the Trojan War were actually relative latecomers.

Perhaps it helped smooth relations between that earlier population and the new Ionian colonists that the newcomers immediately recognized the prehistoric fertility "Mother Goddess" of Samos as the goddess they already knew and worshipped as Hera. So strong was the conviction that this was Hera, that a site sacred to the Mother Goddess, on the banks of the Samian river Imbrasos, was identified as Hera's birthplace. A wicker bush there was believed to have sheltered her birth. By the time Pythagoras was born, what for millennia had been a plain stone altar and a simple structure protecting a wooden effigy and a wicker bush had become one of the most magnificent temple complexes in the world. The great temple of Artemis at Ephesus, nearby across the Strait of Samos, did not quite succeed in copying its splendor.

Before the second millennium B.C. ended, another wave of settlers, this one led by a man named Prokles, from Pityous, disembarked on the beaches of Samos and seized control of the island. Prokles' people ruled for about four hundred years, until the eighth century B.C. Then the descendants of the earlier settlers turned the tables. These wealthy landowners called themselves Geomoroi, or "those who shared out the land." The period of their dominance was the "geometric" period, a term that applied not only on Samos but to a phase of history in the surrounding Greek areas as well.[5] The word "geometry" came from the way the Geomoroi "geometrically" divided up their land. Pythagoras' ancestors, at least on his mother's side, were among them.

The centuries of Geomoroi rule were an era of increasing prosperity for Samos, and also the time when the richest cultural interchange occurred between her and the peoples of Egypt and the Near East. Her location near the west coast of present-day Turkey placed Samos at the crossroads of the great sea-trading routes that linked the Black Sea with Egypt, and Italy and mainland Greece with the Orient. The mainland coast across the narrow Strait of Samos was the western terminus of overland trading routes that brought caravans bearing exotic goods from the East. Samos became a hub for ships that traveled all over the known world. Her sailors took larger, innovative new vessels, designed and constructed by Samian shipbuilders, beyond the Straits of Gibraltar, perhaps even to southern England. The semi-mythical Kolaios reputedly made that voyage and donated a tithe of his profits to Hera's temple. Samos controlled fertile areas across the strait on the mainland, ensuring an ample grain supply. By the sixth century B.C., when Pythagoras was born, she was founding colonies in Minoa, Thrace, and Cilicia. Samian expatriates were living in Egypt, bolstering trade relations with the pharaohs.

Though the island's prosperity continued to soar, the era of Geomoroi rule had ended by the time Pythagoras was born. In the late seventh century B.C., the aristocratic Geomoroi had succumbed to a tyrannical regime. The takeover reputedly occurred while most citizens were outside the city at the temple, enjoying a festival of the goddess.

Pythagoras was born in about 570 B.C., or perhaps a little earlier. Kolaios would have returned at about that time from his heroic voyage.

Though the Geomoroi had lost control of the island, Samos' climb toward her economic and cultural zenith continued. This was her golden age. For Pythagoras' mother's Geomoroi family, the ascent of the tyranny must have been a serious blow in terms of power and perhaps wealth. However, Mnesarchus was a merchant whose commercial situation would have improved rather than suffered in the upheaval. Theirs was surely a fortuitous marriage, with Parthenis bringing her family's ancient aristocratic heritage and lands, and Mnesarchus bringing a newer fortune earned in the thriving Samian mercantile empire.

Mnenarchus' profession makes it likely that Pythagoras did not spend his entire childhood and youth on Samos. According to the historian Neanthes (one of the most reliable sources used by the three biographers), he traveled to Tyre and Italy and elsewhere with his father. Also according to Neanthes, and others as well, he had two older brothers, Eunostus and Tyrrhenus, and perhaps a foster brother to share these adventures. If the story is correct that Pythagoras' father was not only a merchant but also a gem engraver, then his sons would have been trained in that craft. Iamblichus was sure that Pythagoras had the best possible schooling and studied with learned men on Samos and even in Syria, especially with "those who were experts in divinity." It is plausible that the family continued to have trading or personal connections with the area around Sidon, in Syria, where Iamblichus' biography said Pythagoras was born.

Describing Pythagoras as a youth, Iamblichus strayed into the overblown adulation that he would adopt in later chapters of his book, but a more realistic picture emerges of a young man gifted with a natural grace and manner of speech and behavior that made a good impression even on people much older than himself. Iamblichus wrote that he was serene, thoughtful, and without eccentricity. Statues in Samos' museums—*kouroi*, dating from that period—suggest that this was the ideal: a human youth, but hinting at something more centered, mysterious, and holy.

On Samos, Pythagoras was at the epicenter of the commercial world, but not at the epicenter of Greek science and natural philosophy. He was, however, only a narrow strait away from Miletus, where Thales, called "the first to introduce the study of nature to the Greeks," had his headquarters. About fifteen years before Pythagoras' birth, Thales observed and recorded an eclipse. That event has been taken to mark, or at least to symbolize, the beginning of Greek science and natural philosophy, and, because Thales' observation was an eclipse, it is possible to identify the date: May 28, 585 B.C.

Little is known about Thales except that he studied nature and astronomy and, unsatisfied with mythological explanations, pondered questions about how the world began and what was there before anything else. Plato, in his dialogue *Theaetetus*, used Thales as an example of a man too preoccupied with his studies:

> Thales, when he was star-gazing and looking upward, fell into a well and was rallied (so it is said) by a clever and pretty maidservant from Thrace, because he was eager to know what went on in the heaven but did not notice what was in front of him, nay, at his very feet.[6]

Thales did have a practical side. He was famous for coming up with simple, ingenious solutions to problems that stumped others. News probably reached Samos, if the story was true (and even if it was not), that when the army of King Croesus, of fabled wealth, was brought to a standstill for lack of a bridge over the river Halys, Thales had a channel dug upstream of their position that diverted the river to the other side of the army, so that without having moved a step they found they had crossed it.[7]

It might be said that Thales had a special affinity for water, be it in the river or the well, for he thought that water itself was the first principle from which all other things had sprung, and that the world itself floats on water "like a log or something else of that sort," as Aristotle later commented a bit dismissively. Pythagoras' biographer Diogenes Laertius wrote that Thales lived to be so old that he "could no longer see the stars from the earth." He was known as one of the "Seven Sages" of early Greek history, each of whom was connected with one great saying; Thales' was "Water is best." Would that all philosophers had been so concise.

Growing up on Samos, Pythagoras surely knew about Thales. Iamblichus thought that he made trips across the strait even in his early youth to sit at the feet of the elderly sage. Pythagoras acquired a nickname: "the long-haired Samian." Apollonius the wonder-worker provided Pythagoras' biographers with the information that Pythagoras also studied with the astronomer Anaximander, another scholar at Miletus. As was true of Thales, one date is fairly firmly associated with Anaximander: he was sixty-four years old when he died in 546. He would have been in his mid-twenties when Thales recorded the eclipse, and middle-aged to elderly by the time Pythagoras could have been his pupil.

Anaximander himself may have been a pupil of Thales, but their ideas were not alike. Anaximander used mathematics and geometry in attempts to chart the heavens and the Earth, and he drew one of the earliest maps of the world. To a young man eager to acquire cutting-edge knowledge, it would have been intriguing to learn that Anaximander rejected ideas that the Earth floated on anything or hung from anything or was supported from elsewhere in the heavens. The Earth, said Anaximander, remains motionless and in place because the universe is symmetrical and the Earth has no reason to move in one direction and not another. He introduced the notion of the "limitless" or "unlimited" as fundamental to all things. This idea surfaced again prominently when Pythagorean doctrine was written down by Philolaus in the next century.

For Anaximander, when the "unlimited" was "separated," the result was contrasts, such as male-female, even-odd, hot-cold. Contrasts were central to his creation scheme. Separation into opposites later became a major element in Pythagorean thinking. Most significantly, Anaximander believed that there was unity underlying all the contrasts, diversity, and multiplicity in the universe—an idea that would emerge much more strongly with the Pythagoreans. The parallels between Anaximander and the Pythagoreans might seem to indicate that Pythagoras must have studied with Anaximander, but Anaximander's ideas could have reached Pythagoras or Philolaus by other routes. The young Pythagoras may also have known Anaximander's pupil Anaximenes.

Iamblichus credited Thales with convincing Pythagoras to travel to Egypt. This kindly, modest teacher, wrote Iamblichus, apologized for his extreme old age and the "imbecility of his body" and urged his talented pupil to move on, claiming that his own wisdom was in part derived from the Egyptians and that Pythagoras was even better equipped than he had been to benefit from their teaching. Thales had either visited Egypt or knew it from the accounts of others, for he wrote a description of the Nile floods (water, again) and speculated that they were caused by winds blowing from the north in the summer, which prevented the waters of the river from flowing into the Mediterranean.[8] Porphyry thought that what Thales and Pythagoras had most to learn from the Egyptians was geometry: "The ancient Egyptians excelled in geometry, the Phoenicians in numbers and proportions, and the Chaldeans in astronomical theorems, divine rites, and worship of the gods." "It is said," Porphyry hedged, that Pythagoras learned from all of them.[9]

Recounting the tales and traditions about Pythagoras' associations with Thales, Anaximander, and possibly Anaximenes on the mainland coast near Samos, and the educational odyssey he was about to undertake, Porphyry and Iamblichus resorted often to those words "it is said," without revealing who said it. The stories were part of a long-standing semi-historical tradition. Unfortunately, in the centuries preceding Iamblichus, Porphyry, and Diogenes Laertius, this tradition had been embellished to the point of pollution by a spate of "pseudo-Pythagorean" literature. The three historians tried to circumvent this problem by using earlier sources, but they could not, or at least did not, completely disregard some information that was probably spurious.

The tradition that Pythagoras studied with Thales, Anaximander, and Anaximenes and even visited Egypt and Mesopotamia is not farfetched. Samos' position in the world geographically and economically, and what seems probable about Pythagoras' own economic circumstances and family, make these stories credible. He had reason to feel comfortable in the wider world because of his father's trading ventures and connections, was wealthy enough to travel and have the leisure to pursue an adventurous, eclectic self-education, and was probably insatiably curious. If Pythagoras did not make journeys like these, what could have prevented him?

Iamblichus wrote that Thales did not stop at telling Pythagoras he should go to Egypt. He warned him to be sparing of his time and careful about what he ate. Pythagoras confined himself to "such nutriment as was slender and easy of digestion" so that his sleep could be short, his "soul vigilant and pure," and his body in a state of "perfect and invariable health." Perhaps he did follow his old teacher's advice and succeed in maintaining this enviable conditioning, but according to Iamblichus, he did not immediately hasten to Egypt. He went by way of Sidon, probably his birthplace.

CHAPTER 2

"Entirely different from the institutions of the Greeks"

Sixth Century B.C.

YOUNG PYTHAGORAS' JOURNEY, as Iamblichus recounted it, was the ancient equivalent of a high-risk modern junior year abroad. He bedded down in a temple on the Mediterranean coast, at the foot of Mount Carmel, a mountain associated with the prophet Elijah and his God as well as with local pagan deities. There is a much-disputed claim by the historian Josephus that Pythagoras was influenced by Jewish teaching. He could have encountered it here, although many of the Jewish population were in exile in Babylon. Iamblichus wrote that he "conversed with prophets" and was initiated into the mysteries of Byblos and Tyre, not for the sake of superstition, but "from an anxiety that nothing might escape his observation which deserved to be learnt in the arcane or mysteries of the gods." For a man who himself lived in a superstitious age, Iamblichus was surprisingly eager to emphasize that Pythagoras was not influenced by the "superstition" of this area, though he made no such disclaimer about what Pythagoras might have picked up in Egypt or Mesopotamia. Iamblichus was writing at a time when many feared that Christianity, with roots in Jewish belief, would destroy Greek philosophy.

After a while, Pythagoras continued his journey to Egypt, and Iamblichus went into greater narrative detail than usual to relate an adventurous, delightful story. Fortuitously, or so it seemed at first, an Egyptian ship landed on the Phoenician coast near the temple where Pythagoras was living. The sailors were pleased to welcome him aboard, thinking they could sell such a comely young man at a good price. During the voyage, they changed their minds. There was something different about this modest youth from what one normally expected of a human being. The sailors reminded one another how he had appeared, descending the sacred Mount Carmel, how he had said nothing except to ask, "Are you bound for Egypt?" and then had come aboard and sat silently and out of their way for two nights and three days without taking food or drink, or sleeping—at least when any of them were watching. The voyage was, furthermore, going exceptionally well, with fair weather and favorable winds. The sailors delivered Pythagoras safely to the Egyptian coast and helped him off the ship (he was weak from fasting and lack of sleep), then built an altar in front of him and heaped it with fruit. When they left, he ravenously consumed the fruit. One may take this story as evidence of his godlike nature, or as suggesting that he was a canny young traveler, giving careful attention to self-preservation.

Iamblichus' sources indicated that in Egypt Pythagoras frequented temples, sat at the feet of priests and prophets, sought out men celebrated for their wisdom, and visited "any place in which he thought something more excellent might be found," "astronomizing and geometrizing." Isocrates, an older contemporary of Plato in the early fourth century B.C., eagerly latched on to the information that Pythagoras spent time in Egypt. Isocrates was intent on showing that the Greeks owed their learning to the Egyptians and had added very little. In his disparaging words, Pythagoras "went to Egypt, and having become their pupil was the first to introduce philosophy in general to Greece, and concerned himself more conspicuously than anyone else, with matters to do with sacrifices and temple purifications, thinking that even if this would gain him no advantage from the gods it would at least bring him high repute among men. And that is what happened." As in the tale of Pythagoras' sagacious handling of the Egyptian sailors, here is a hint that for all his reputed purity, he was not naive but perhaps even rather opportunistic.

Egypt at the time when Pythagoras could have been there was ruled by the pharaoh Amasis II (Ahmose II), later an acquaintance of Samos' tyrant Polykrates. It was unusual but not unprecedented for a Greek to visit Egypt. In the seventh century B.C., the pharaoh Psamtek I had hired Greek mercenaries, and in Pythagoras' day there were Greeks living in Naukratis in the Nile delta, for Amasis was eager to promote trade with the Greek cities and even made a donation toward a rebuilding project at Delphi. However, he restricted Greek merchants to the one city and did not allow them to move around the country as much as Pythagoras is supposed to have done.

Porphyry reported a different version of Pythagoras' Egyptian sojourn. His source was *On Illustrious Virtuous Men*, by Antiphon. By this account, Pythagoras set off with a letter of introduction from Polykrates to Amasis. This would place the journey too late, for Polykrates' reign began in 535, shortly before Pythagoras moved to Croton. Nevertheless, Porphyry's account is interesting: Pythagoras went first to the priests of Heliopolis, who sent him on to Memphis, saying the priests there were more ancient. These, in turn, on the same excuse, sent him to Diospolis (ancient Thebes), a journey of more than three hundred miles to the south. The priests of Diospolis had nowhere else to send him, but thought that if they made things difficult enough he would go away. They gave him "very hard precepts, entirely different from the institutions of the Greeks," which he doggedly performed, winning their admiration to the extent that they taught him their secret wisdom and permitted him to sacrifice to their gods, something not normally allowed a foreigner. Pythagoras would later adopt the practice of secretiveness with respect to his own teachings, as was not common in the Greek world.

If Pythagoras did go to Egypt, what could he have learned? In the temple complexes there were "Houses of Life" with many learned men copying manuscripts, large libraries, and sometimes schools. The ruling classes were literate, as we must suppose Pythagoras was, but he did not know the languages of Egypt. If the priests accepted him, as Porphyry believes they must have, then Pythagoras, though older than the schoolboys, would have had to start on an elementary level with a language, alphabet, and numbers that were foreign to him, before he could begin to understand priestly liturgy and wisdom. He would have studied the cursive hieratic script, perhaps copied out books of Egyptian literature, then advanced to hieroglyphs. He would have learned a decimal system with numbers the equivalent of 1, 10, 100, 1,000, and

10,000, but no symbol for zero. To multiply, an Egyptian added a number to itself the necessary number of times. To divide, he subtracted a number from itself until the remainder was too small to continue. Pi was unknown, but one could come close to calculating the area of a circle by measuring the diameter, subtracting 1/9, and squaring the result.

Such mathematical knowledge was for practical use: for construction or—when it came to the circle—for measuring such things as the capacity of a granary—but this was a culture whose worldview seamlessly included what was tangible physical fact and what was mythological or metaphorical, drawing no boundaries between practical and esoteric knowledge, or between everyday reality and the holy. The Egyptians' elaborate preparations for another world after death had a practical motive: to supply what one needed to get there and live there. Magic was a high category of knowledge, as were religious ritual, myth, and medicine. Pythagoras would have studied the Egyptian hierarchy of gods and goddesses and beliefs about the afterlife, but not a doctrine of reincarnation.[1] He also would not have learned vegetarianism, for the upper classes ate beef and other meat fairly often.

The Egyptians had long excelled in surveying. The near perfect squareness and north–south orientation of the Great Pyramid of Khufu at Giza is evidence of their astounding precision, and Pythagoras could not have missed seeing that pyramid if he traveled as Porphyry thought he did. It dated from about 2500 B.C., two thousand years before him. We cannot know with certainty that the Egyptians in the sixth century still had the technical genius of those distant predecessors, but surveying for land boundaries, city plans, and buildings was routine, and the older, magnificent structures that are still wonders of the world today were much fresher and much more impressive to someone who had not encountered human-made objects on this scale.

From the temple roofs, Pythagoras might have assisted with observations of the cycles of the moon and the movements of the stars and learned how these were related to the Egyptian twelve-month calendar and 365-day year. Egyptians thought their country was the center of the cosmos and that there were definite connections between the stars and events on Earth. For example, the star Sirius (Sopdet), invisible for several months, reappeared in mid-July as a morning star, signaling the onset of the yearly inundation of the Nile and the beginning of the new year.

The Great Pyramid of Khufu at Giza with the Sphinx in the foreground

Different temples had different specialties. If Pythagoras did not move on too quickly from Heliopolis (in Porphyry's scenario) he might have learned a creation theology that explained the diversity of nature arising from a single source, the god Atum, meaning "the All." Atum existed in a state of unrealized potential not far different from the "unlimited" in Anaximander's teaching and later in Pythagorean thinking. At Memphis, where, as Porphyry told it, Pythagoras spent a little time before being sent on, he could have learned a more remarkable theology of divine creativity that provided an agent through which an idea in the mind of the creator became a physical reality. In many early cultures, a spoken or written word was understood to have creative power. In creation as viewed in Genesis, God spoke, and it was so. The theology of the priests at Memphis divided that creative "word" into two different roles. A link was required, a divine intermediary between an idea in the mind of the creator and the actual physical creation. Memphis theology had arrived at a concept that would later be expressed in the opening of the Christian Gospel of John, where the Logos—Jesus, the second member of a trinity—bridges the creative gap between God and man: "through him [not "*by* him"] all things were created, without him nothing was created that has been created." Plato's "demiurge" bridged the same gap. The god who performed that role in the theology of Memphis, Ptah, operated in similar manner on the human level, enabling an idea in a human mind (a craftsman or artist) to become a real-world product. This role or force was "effectiveness" or "magic." Without it you had speech or an idea or something written on a page. With it you had creative power. Pythagoras and his followers would later assign that creative role to numbers, though, by some interpretations, Pythagoreans would understand numbers to be the idea in the mind of the creator, *and* the creation, *and* the link between the two.

At Thebes, where Porphyry thought Pythagoras finally spent a long period and was accepted by the priests into their most secret mysteries, Egyptian theology had a monotheism close to that expressed in the Christian concept of the Trinity, but with more "members." The god Amun (meaning "Hidden") was the greatest among the gods—"unknowable" and transcendent. The others were different manifestations of him.

Porphyry had Pythagoras returning to Samos from Thebes, but Iamblichus wrote an exciting addition to the story: Pythagoras was taken captive by "soldiers of Cambyses" and brought from Egypt to Babylon. If Iamblicus was right, Pythagoras arrived there during the reign of the Chaldean dynasty, which began in 625 B.C., in the century before Pythagoras' birth, and lasted until 539, well into his lifetime. During this period, Babylon enjoyed the second golden age in its long history—an age scholars call neo-Babylonian. However, Iamblichus' timing, as implied by the words "soldiers of Cambyses," is a problem. Cambyses I was a Persian prince in a royal line ruling in the southwestern part of present-day Iran. He was the father of Cyrus the Great, to whom Babylon would later fall, and whose empire would far exceed hers. Cambyses reigned from about 600 to 559 B.C. Pythagoras was probably only eleven years old in 559. There were frequent clashes between the Egyptians and the Babylonians, and Babylonian soldiers surely took some captives, but not until after 529 (when Pythagoras was already in southern Italy) did Cyrus the Great's son Cambyses II conquer Egypt.

Iamblichus estimated that Pythagoras lived in Babylon for about twelve years. Any adventurous young man would have envied him this opportunity, for Babylon was a splendid, exotic, cosmopolitan city at the height of her power and wealth, far older than Samos, and far more worldly and sophisticated than Egypt. A period of supreme success and prosperity a thousand years earlier—the era of the 1894–1595 B.C. "Dynasty of Babylon" and especially the reign of Hammurabi—had been one of the pinnacles of ancient civilization. In the millennium that had passed between that period and Pythagoras' lifetime, Mesopotamia had experienced wave after wave of migration, military clashes, and dynastic shifts, and one city after another had grappled for its moment in the Mesopotamian sun. Now it was again Babylon's turn. If Iamblichus' dates were near correct, Pythagoras' visit probably caught the wake of the reign of Nebuchadnezzar II, when Babylon was ruled by lesser, short-lived kings of the same dynasty. Nebuchadnezzar had died in 562 B.C., when Pythagoras was about eight years old.

Pythagoras would have arrived in Babylon either by caravan across the plain or by boat on the Euphrates.[2] Either way, the towering seven-level ziggurat was visible long before the city came into view. Though young in comparison with the Giza pyramid (and no match for it in height—the ziggurat was about 300 feet high, the pyramid 481 feet), the ziggurat nevertheless was an exceedingly ancient monument, a relic of Babylon's earlier golden age. Nebuchadnezzar had made sure that it was splendidly restored to connect his own reign with that former glory. The principal approach to the city from the north was an avenue sixty-six feet wide, built of giant limestone paving slabs covering a foundation of brick and asphalt. On either hand, sixty lions—fashioned of red, white, and yellow tile on the high walls—stalked the men and women on the road. At the city's Ishtar Gate, bulls and dragons took over from the lions. This entrance was one of eight massive, bronze-armored portals in a double-walled, moated fortification system that surrounded the city. Inside, the avenue continued and crossed the Euphrates on a bridge with supports high enough and far enough apart to allow the largest ships to pass. A temple complex housed the jewel-studded shrine of Marduk, god of the city, in a chamber lined with gold. Pythagoras and others who were not royalty or among the most elite of the priests would not have entered this chamber, but they would have known about it.

Beyond the temple precincts, the city spread on both sides of the Euphrates and included a royal palace with state rooms, private quarters, courtyards, and a harem for the queen and concubines brought from all parts of the empire. If they are not only legendary (the archaeological evidence is ambiguous but not entirely absent), the Hanging Gardens were part of this complex, and they, like the ziggurat, were a prominent landmark visible from a distance above the surrounding buildings—a terraced hill of earth, supported by massive vaults built so that their floors were waterproof and could support enough soil to plant large trees, watered from the nearby Euphrates by complicated irrigation machinery. Similar irrigation wizardry and a series of canals watered gardens and orchards in the newer part of the city and carried water to distant suburbs. The practical knowledge of mathematics and geometry that made possible these buildings and the surveying for the irrigation was evidence of how well the scribes of Babylon understood these subjects—or, at least, had understood them many centuries before, when the building techniques were developed. It is likely that the theory and deeper mathematical understanding underlying the techniques had been forgotten by the time of Pythagoras, though the techniques themselves had become routine and were still in use.

Because people who came to Babylon for whatever reason often chose to stay, her streets and passages were a cacophony of languages. There were Hurrians, Cassites, Hittites, Elamites, Jews, Egyptians, Aramaeans, Assyrians, Chaldeans, and all mixes thereof. Centuries of captives (including the Jews brought from Judea and Israel, who were there during Nebuchadnezzar's reign), conquerors, and visitors had lived in the city long enough and mixed sufficiently well to interbreed, until Babylon had become, in the words of the twentieth-century scholar H. W. F. Saggs, "a thoroughly mongrel city." Ancient tablets give evidence of an astounding variety of jobs, careers, and crafts, and a rich array of goods that arrived, some by caravan but mainly by way of the river. Women had authority over slaves or servants in their households, but probably wore veils in public.

Pythagoras, exploring these streets and passageways and listening to all the languages, would have seen house walls that glowed in bands of light and shade, an effect ingeniously produced by a "saw-toothed" treatment that made the surface reflect the brilliant desert sunlight in this variegated manner. He would have stayed in private houses oriented almost entirely toward interior courtyards, their entrances guarded by a porter and a confusing, indirect entryway to discourage unwanted visitors and peeping toms. Whether he lived in a house like that or in the temple precincts—for his success among the priests and scribes should not have been any less here than in Egypt—his diet was probably mostly vegetarian, not by choice but because, in a city fed from irrigated fields surrounded by desert wastelands, meat was a luxury item.

What could Pythagoras have learned in Babylon? He was familiar with her art and design, for Hera's temple on Samos included many examples. He would have sought out the scribes. Writing and calculating were their primary activities. Some were part of governmental and temple communities, some worked for the military, others served private citizens or taught. Many freelanced, offering their services in the marketplace for people needing letters written, legal documents drawn up, calculations made. Besides the scribes, only the rare Babylonian could read, write, or calculate. At the top of the profession were the highest-ranking priests at the temple of Marduk, who had to be able to read the texts for the rituals they used. These texts were often written in ideograms, making them inaccessible to those not trained in this particular type of text, and they often included a warning that only the initiated should even see them. Such secretiveness might have seemed prudent to Pythagoras, who instituted it later in Croton.

Much of the information that modern scholars have about knowledge in ancient Mesopotamia comes not from this neo-Babylonian period but from the first great era of Babylon a millennium earlier (1894–1595 B.C.). Tablets that were school texts then show that teachers and scholars knew the value of pi, could calculate square and cube roots, and understood what is now known as the "Pythagorean" theorem. The system of mathematics they used was already fully developed and being taught routinely to scribal students. But was the "Pythagorean theorem," which had made it into the textbooks in the second millennium B.C., still known in Babylon at the time of Pythagoras? Experts on ancient Mesopotamia think not; but, if it was, Pythagoras of course might have learned it from the scribes. If he carried away with him knowledge of their sexagesimal number system—based on sixes rather than on tens—nothing of that showed up in later stories about him or his followers.

Pythagoras would have encountered a sophisticated astronomy if he sought out Babylonians who studied the stars. "Early Greek science and natural philosophy" may have begun with Thales' observation of the eclipse on May 28, 585 B.C. but Mesopotamian scholars had long known how to predict eclipses. Again, evidence is lacking whether the learning that had been so impressive, and that is so well documented on tablets originating a thousand years earlier, was still in the grasp of Mesopotamian scribes and astronomers at the time of Nebuchadnezzar. The fact that Babylonians set up a sacred kettledrum during an eclipse and beat on it to drive off the demons that were obscuring the moon is no indication that the earlier sophistication had been lost. It is difficult to imagine even a modern society giving up such spectacle and fun just because of a scientific explanation! Later, in the Persian and Hellenistic eras, a highly mathematical Mesopotamian astronomy used observational data that had been collected for centuries in the temples.

Pythagoras did not learn the doctrine of reincarnation in Babylon. A Babylonian—barring unusual circumstances that left him flitting around as a baleful ghost—died, went to a dismal netherworld, and stayed there.

THOUGH IAMBLICHUS CANNOT have been correct that Pythagoras spent about thirty-four years in Egypt and Babylon (no acceptable chronology allows that much time), he was probably right that when Pythagoras returned to Samos only a few inhabitants of his home island remembered him. Nevertheless, wrote Iamblichus, he made an excellent impression with the learning he had accumulated and the tales he could tell, and was publicly requested to share this knowledge with his countrymen. That seemed an excellent idea to many Samians, until they realized what mental effort it required. Pythagoras' audiences dwindled, those who stayed were lazy, and soon no one was listening to him. Iamblichus believed that he did not take umbrage. He was still determined to give his fellow citizens a "taste of the sweetness of the mathematical disciplines" and concerned that his skills and learning would desert him as he aged. He adopted a fresh strategy: Rather than teach a multitude, he chose one promising disciple.

Iamblichus described Pythagoras' choice as a poverty-stricken but talented young athlete, whom Pythagoras discovered playing ball in the gymnasium "with great aptness and facility." They struck an agreement. Pythagoras would provide him with the necessities of life and the opportunity to continue his athletics, on condition that the young man would, in easy doses (at least by Pythagoras' standards) allow Pythagoras to educate him. At first the youth seemed motivated mostly by rewards of three eboli for learning figures on the abacus. As time passed, Pythagoras observed that his interest became keener, so much so that Pythagoras suspected it would continue even without the eboli—even if he had to "suffer the extremity of want." As a test, Pythagoras pretended to have had a catastrophic change of fortune, requiring the association to end. As Pythagoras had hoped, the youth declared that he could learn without rewards and would find a way to provide for both himself and Pythagoras. Iamblichus wrote that this young man, to honor his mentor, took the name "Pythagoras, son of Eratocles" and, alone among Pythagoras' acquaintances on Samos, eventually moved with him to southern Italy. Iamblichus did not indicate where he got this information except to mention that "there are said to be" three books by Pythagoras, son of Eratocles, titled *On Athletics*, in which he recommended eating meat instead of dry figs. If he took this recommendation from his teacher, then the advice ran counter to information from other sources that Pythagoras was a vegetarian and required the same of his students and followers.

A story about another pupil also conflicts with Pythagoras' reputation as a strict vegetarian. Eurymenes was also an athlete, but he was small. It was the custom to eat only moist cheese, dry figs, and wheat bread while in training for the Olympic games. Pythagoras instead advised Eurymenes to eat meat. He also taught him not to go into the games for the sake of victory but for the exercise of training and the benefit to his body. Diet and Pythagorean sports psychology worked wonders. Eurymenes, in Porphyry's words, "conquered at Olympia through his surpassing knowledge of Pythagoras' wisdom."*

According to Porphyry, these two athletes were not Pythagoras' only pupils during this period. Porphyry had read of another in *On the*

* There was an Olympic victor in the 588 B.C. games whose name was Pythagoras of Samos, as recorded in the highly reliable lists of Eratosthenes, the famous librarian at Alexandria who first measured the circumference of the earth. Eratosthenes conjectured that this man and Pythagoras the philosopher were one and the same. In order for that to be true, Pythagoras the philosopher would have had to have been born a few decades earlier than is usually supposed, in the late seventh century. For the Olympian Pythagoras to have been Eurymenes (who might have adopted the name of his teacher as the son of Eratocles did), Pythagoras the philosopher would have had to have been earlier still. What the appearance of the name in the Olympic victory lists probably does mean is that the name Pythagoras was current on Samos before Mnesarchus decided to name his son in honor of the Pythian oracle at Delphi.

Incredible Things Beyond Thule, the book he used for information about Pythagoras' father, insisting that its author had "treated Pythagoras' affairs so carefully that I think his account should not be omitted." Porphyry did not say it should necessarily be believed. On a trading journey, Pythagoras' father, Mnesarchus, discovered an infant under a poplar tree, lying on its back, looking unblinkingly at the sun and sipping dew falling from the tree through a reed pipe in its mouth. This struck Mnesarchus as divine activity, and he arranged for the child to be fostered by a friend and native of that country, later paid for his education, named him Astraeus, and reared him with his own sons. Pythagoras took this younger adopted brother as his pupil. Porphyry also mentioned a fourth pupil, Zalmoxis of Thrace, who "some said" also took the name Thales. Though not an Olympian, he must have had an impressive build, for barbarians mistook him for Hercules and worshipped him.

On the Incredible Things Beyond Thule listed the qualities Pythagoras looked for in those who came to study with him. Its author had learned (his source is not known) that Pythagoras did not agree to teach everyone who came, nor were his choices based only on intelligence or kinship. He observed a candidate's facial expressions, body language, and disposition. He looked for modesty, ability to keep silent being more important than readiness to speak. He observed whether the prospective pupil was moved by any immoderate desire or passion, how anger affected him, whether he was contentious or ambitious, inclined more to friendship or to discord. After a candidate passed those tests, Pythagoras took note of his ability to learn, memorize, and follow rapidly what was said. Of primary importance was how strongly a youth was motivated by temperance and love. Natural gentleness and "culture" were essential; ferocity, impudence, shamelessness, sloth, and licentiousness were distinct negatives. Pythagoras expelled pupils "as strangers and barbarians" if they failed to live up to his expectations.

IN 535 B.C., when the tyranny that had wrenched control from the Geomoroi had ruled for several decades, the most infamous of the tyrants, Polykrates, came to power in Samos. At first he ruled with two brothers, but he soon disposed of them. Samos continued to grow in power and wealth, but not in popularity among her neighbors, for Polykrates became a much hated and feared player in the politics of the eastern

Mediterranean. Depending on who described it, his fleet was either one of the most superb navies of the ancient world or a supremely successful band of pirates. Polykrates traveled in person to other countries to seal new agreements and forge connections with rulers like the pharaoh Amasis, but such agreements had little meaning, for he made and shattered alliances with ruthless abandon.

Under Polykrates, Samos reached the pinnacle of her fortunes, not only in terms of economic and rather ugly political prominence, but also in art, literature, and engineering feats. For a time it was the most powerful of all the Greek city-states. Pythagoras lived on Samos for only part of this period, but long enough to experience the excitement and intellectual stimulation that characterized Polykrates' otherwise deplorable reign. Polykrates was the patron of the poet Anacreon and engaged the engineer Eupalinos to construct a new harbor and a water tunnel that was one of the most astounding engineering achievements of the ancient world. It brought water from alpine springs through the mountain above the city of Samos, ending any shortage of water there no matter how dry the summer.* The fleet grew to a hundred ships, each manned by a thousand archers. In spite of Polykrates' widespread unpopularity and long absences, no one unseated him until finally, in 522—after Pythagoras had left Samos—a Persian governor of Sardis trumped Polykrates' treachery. He invited him for a state visit and, when he arrived, had him crucified.

It is reasonable to believe, with Iamblichus, that Pythagoras did not remain on Samos without interruption during the years before he finally moved to Croton in Italy, but visited oracles, spent time at Delphi, and went to Crete and Sparta to learn their laws, which were different from Samos'. Iamblichus first mentioned Pythagoras' taking an interest in public affairs at this time. Porphyry also believed that Pythagoras left Samos briefly to undergo an initiation ceremony on Crete: The supplicant seeking initiation to "the priests of Morgot, one of the Idaean Dactyls," was purified with a meteorite ("the meteoric thunderstone"), lying at dawn face down on the seaside and at night beside a river, crowned with a wreath of black lamb's wool. Then,

* Both tunnel and harbor still exist, though the harbor is now hidden below sea level, under later construction. You can visit the tunnel and walk a good distance into it. In an ingenious design, the walkway is separate and above the watercourse.

wrapped in black wool, he descended into the Idaean cave and remained for twenty-seven days. After that, he made a sacrifice to Zeus and was allowed to see the couch the priests made up every year for Zeus. Pythagoras, having gone through the initiation, inscribed an epigram on the tomb of Zeus, which began "Zan lies dead here, whom men call Zeus"—implying, it would seem, that he knew or had known this god on a more personal basis than other men did.

As time passed, Pythagoras' renown spread, learned men came from abroad to confer with him, potential students flocked to him. He also served Samos in an administrative capacity, as was expected of important scholars in this era in the Hellenic world. However, he often retreated to a cavern outside the city for discussions with a few close associates. Samians today identify a cave on the steep, wooded slopes of Mount Kerketeas, the island's highest mountain, as Pythagoras' cave. Nevertheless, as public responsibilities increased, it became impossible for Pythagoras to continue his studies. Furthermore, Porphyry observed, he saw "that Polykrates' government was becoming so violent that soon a free man would become a victim of his tyranny," and "that life in such a state was unsuitable for a philosopher." Involvement at a court like Polykrates' was dangerous for a man who spoke honestly. Also motivated by a failure of the Samians in all things relating to education, Pythagoras departed for southern Italy. He had heard, said Iamblichus, that in Italy "men well disposed towards learning were to be found in the greatest abundance."

CHAPTER 3

"Among them was a man of immense knowledge"

530–500 B.C.

NEAR THE BEGINNING OF THE seventh decade of the sixth century B.C., a vessel with Pythagoras on board sailed across the waters west of Tarentum toward the toe of the Italian boot and the port city of Croton. The date is the best established in Pythagoras' life. One of the most reliable of the earliest sources, Aristoxenus of Tarentum, gave it as 532/531 B.C.* Just short of a promontory where Croton's men and women worshipped at their own sanctuary of the goddess Hera, the ship came into harbor. The passengers disembarked at docks bustling with other voyagers, slaves, sailors, and craftsmen and laborers from the shipyards, for Croton was a major port and shipbuilding center in this region of the Mediterranean. Goods traded or transferred there came from up and down the coasts of the Italian peninsula, not only from the Greek cities but also from the Latin communities farther north and from numerous other regions of the Mediterranean.

There have been few archaeological excavations within the city of Croton proper. Modern Crotone covers the ancient Croton of Pythagoras, and frustrated archaeologists have to content themselves

* Iamblichus linked the date with the Olympic victory of Eryxidas of Chalcis, his own home city. Diogenes Laertius agreed that it had to have been between 532 and 528 B.C.

with sporadic work during the excavation of foundations for new buildings. Nevertheless, enough is known to allow for an idea of the arrangement of the city as Pythagoras found it.[1] Behind the harbor the ground rose steeply to a hill where Achaean settlers had first built their homes two centuries before his arrival. This hill had later become an acropolis until Crotonians began lavishing more attention on the temple of Hera on the promontory. Sixth-century B.C. Croton apparently included three large blocks of houses oriented perpendicular to the coastline with a divergence of 30 degrees between them, an impressively geometrical layout but not unusual in its time, as evidenced by the Geomoroi. Pythagoras walked in narrow streets precisely aligned and crossing at right angles with narrower lanes isolating individual houses. Crotonians had constructed these buildings of rough blocks of stone, sometimes unbaked bricks, roofed with tile, with large pieces of pottery and tiles protecting the wall footings. They lived in rooms clustered around courtyards, with almost no windows facing the streets and lanes. A man who had also experienced Babylon would have drawn the impression that the people of Croton were more trusting and friendly than those who lived in similar houses there, for entryways in Croton opened straight into the courtyard.

Pythagoras may not have been a complete stranger here, for Croton's harbor and shipyards were on the coastal sea route from Greece to the Strait of Messina, Sicily, and the Tyrrhenian Sea, and there were stories connecting his merchant seafarer father with the Tyrrhenian coast. The climate in Croton was magnificent and the region famous for being particularly healthful. The sea was not the opaque cobalt of Pythagoras' native waters, but a transparent, cheerful blue, and the coastline seemed infinitely long, for every rise in the terrain revealed curving bays and coves and headlands as far as the eye could see. Forests clothed the low hills and some of the headlands and the shores of coves near the city, and grew thickly in the mountains on the northern and western horizons. Trees were one of Croton's most valuable economic resources, as they were for Samos, providing timber for the shipyards.

Pythagoras surely knew that his new city had produced at least one amazing athlete and a fine medical man. Croton's Olympic successes made her the envy of the Greek world, and no young Greek, no matter how sequestered in intellectual pursuits, could have escaped knowing about this athletic preeminence. Every four years, the city's athletes voyaged east to Olympia in mainland Greece to compete, and from about two decades before Pythagoras' birth had enjoyed a continuous spate of triumphs. Milo of Croton won the wrestling competitions in six Olympic games, covering a span of at least twenty-four years—a long success streak for any athlete, ancient or modern—and at six Pythian Games, a similar competition at Delphi.* Everyone had heard how he had hoisted an ox onto his shoulders and carried it through the stadium at Olympia. In the field of medicine, Democedes of Croton had practiced in Athens and become physician to Samos' tyrant Polykrates. Such was Democedes' reputation and success that he would later be employed by the Persian Darius the Great. However, if Pythagoras had indeed heard—as Iamblichus reported—that in Croton men were "disposed to learning," that must have meant they were "ready to learn," for Croton was not yet renowned for scholarship or thought.

Croton's most important religious site, Hera Lacinia, was situated on a promontory at the end of a peninsula that jutted out into the sea near the town. When Pythagoras first arrived, major construction at the temple had only barely begun, if it had begun at all, but soon the buildings at Croton's temple of Hera would rival Samos' temple to the same goddess. The treasures would include one of the most beautiful items still surviving anywhere from the ancient world, a diadem of exquisitely worked golden flowers, now in a glass case in Croton's museum. Pythagoras may have seen it wreathing the head of the goddess' statue. Crotonian donors to the temple were wealthy, cosmopolitan citizens who venerated her, the mother of Zeus, as the protector of women and all aspects of female life, and as Mother Nature, who looked after animals and sea travelers.

Crotonians ruled their city in a manner that must have seemed blessedly old-fashioned to a man accustomed to living under Polykrates. The government was an oligarchy, as Samos' had been before the tyranny. They called themselves the Thousand, and all of them claimed descent from colonists who had come two centuries before Pythagoras from Achaea, on the Greek mainland. The population there had outgrown the arable land in narrow mountain valleys and, led by a man named Myskellos, had taken ship to the west to try their luck around

* Milo is also known as Milon. His name has come to symbolize extraordinary strength. He was the most famous wrestler in the ancient world.

Hera's golden diadem, dating from the sixth or fifth century B.C., from the temple of Hera Lacinia at Croton

the gulf between the toe and the heel of the Italian boot.[2] They were not "colonists" in the sense of remaining subservient and connected to a mother country. What was true for many Greek cities—though no definition fit all—was true for Croton and her neighbors: "hiving off," as happens with bees, was a better descriptive word than "colonization." Greeks of the independent maritime cities of southern Italy and Sicily had done well to leave their tight mainland valleys and were likely, in Pythagoras' time, to be as rich and cosmopolitan as those who lived in Athens. Archaeological finds show that Myskellos' settlers were not the first people to live at Croton, but had pushed the earlier inhabitants into the hinterlands and mountains.

Relations among the cities around the instep of the boot were often antagonistic, but Croton was apparently not walled or fortified. Perhaps the considerable distances between the cities made that unnecessary. Nevertheless, Crotonians visited the other communities. They might have hesitated to go to Sybaris, Croton's chief rival and enemy during Pythagoras' time, basking in "sybaritic" languor on a broad, fertile coastal plain about seventy miles to the north. However, stories placed Pythagoras on several occasions in Metapontum, another seventy miles north of Sybaris. Both Sybaris and Metapontum had been, like Croton, Achaean settlements, while Spartans had settled Tarentum, about thirty miles beyond Metapontum following the coastline, or 140 miles across the water from Croton. The people who lived in these cities may have clung to some identity as Achaean or Spartan, but the wider Greek world lumped them together as Italiotai, while neighbors to the northwest, in the Latin and Etruscan regions, called them Graeci. To the Greeks the region was Megale Hellas; to the Latins, Magna Graecia. In the end the Latin name would stick, because one of those Latin neighbors, about 350 miles northwest on the western side of the peninsula, was Rome, destined later to dominate the entire region and much of the western world and near east.

Those who lived in southern Italy at the time of Pythagoras had no premonition that some unusually ambitious construction projects in Rome—transforming a small, centuries-old community into a city designed on Etruscan lines, outgrowing one hilltop after another and expanding down the slopes into marshier territory, draining the swamps in the valley and paving it to make a forum—were only the first manifestations of a proclivity for building and conquering and expanding that would eventually make Magna Graecia seem a near suburb. Greek

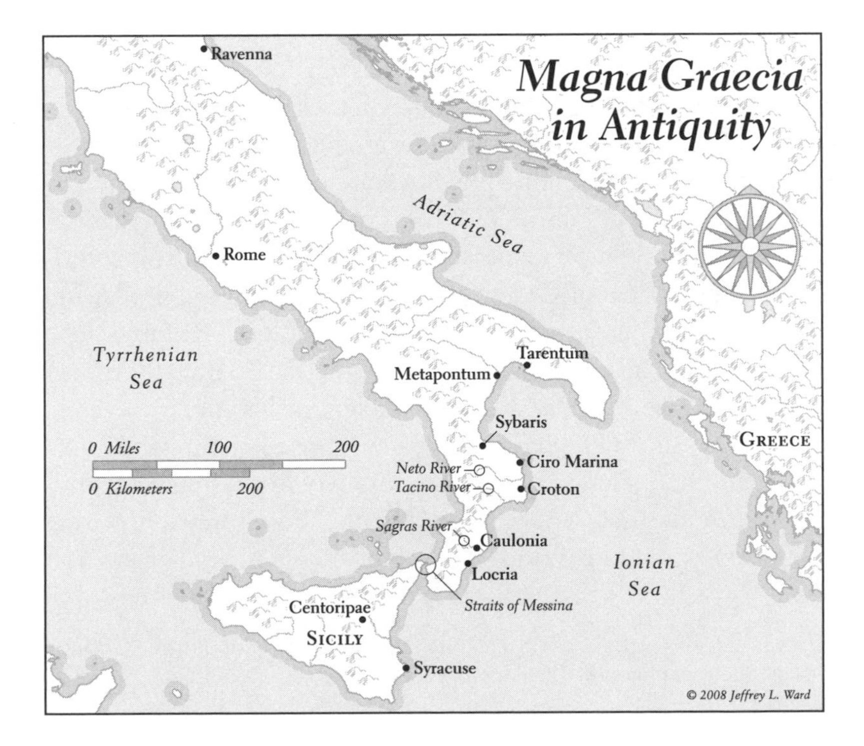

historians took no notice of Rome until she was in the process of completing her conquest of the Italian peninsula, 250 years after Pythagoras. Rome, for her part, was too busy with city planning, building, and wars during Pythagoras' lifetime to take much notice of what was happening in Magna Graecia. However, as Rome emerged as a world power, she would create for herself a tradition and history that traced her ancient ancestry to Greece's enemies at Troy, the Trojans, and made Pythagoras the teacher of one of her early kings, Numa. Pythagoras surely did not teach Numa, who died well before his birth, but well-educated Romans could not bring themselves to believe that their ancestors in Pythagoras' time knew nothing about this great sage. However, though Croton and her neighbors were trading actively and as equals, probably even superiors, with Rome and other Latin and Etruscan centers, Croton's more important friends and foes were closer to home on the southern coastline and in the wider Aegean and Mediterranean seafaring world to the east, south, and west. As the crow flies, and even by the more circuitous but safer coastal sea route, Croton was nearer to the Greek mainland than to Rome.*

PYTHAGORAS WAS ABOUT forty years old when he settled in Croton, where he would live for about thirty years. He rapidly gained respect and soon was gathering a loyal group of associates into a society that bore his name and treated him with reverence. "He said it himself" became a proverb among them—the last word on any subject. Those who joined him included ordinary citizens, noblemen, and women.

Iamblichus and Porphyry based their descriptions of Pythagoras' approach to the people of Croton on the writings of a pupil of Aristotle named Dicaearchus, one of the earliest sources available to any Pythagoras scholar. Originally from Messina in Sicily, a short voyage from Croton, Dicaearchus was at the height of his career in 320 B.C., about 180 years after Pythagoras died. When Iamblichus added details—and he included more than Porphyry or Diogenes Laertius—he gave no indication where he got them. The impression is that he could safely assume his readers knew—or thought they knew—a great

* In the twenty-first century, 2,600 years later, the people of former Magna Graecia still do not totally identify with the modern, centralized Italy. Old attitudes and identities die hard.

deal about Pythagoras. The name was the equivalent of modern figures who can be mentioned in the news or a sitcom, even in caricature, with no need to explain.

There was a possible lost source of information about Pythagoras' years in Croton that would add credibility to the details of the tradition, if one could be certain it existed. The most skeptical scholars disdain it, while others point out that it is improbable that it did not exist. Porphyry thought it did. Referring to a time after Pythagoras died and many of his associates had been killed, he wrote:

> The Pythagoreans now avoided human society, being lonely, saddened and dispersed. Fearing nevertheless that among men the name of philosophy would be entirely extinguished . . . each man made his own collection of written authorities and his own memories, leaving them wherever he happened to die, charging their wives, sons and daughters to preserve them within their families. This mandate of transmission within each family was obeyed for a long time.[3]

Such journals, as Porphyry implied, possibly gave the semi-historical tradition a better footing in fact than it would otherwise have had and were responsible for its being strong in details many of which are not the sort identifiable as the usual stuff of pure legend. Weighing against their existence is the fact that some pseudo-Pythagorean books later claimed to be such journals, and these forgeries may have been responsible for Porphyry's, and others', faith in the journals' reality. On the other hand, the existence of fictionalized journals does not necessarily mean there were no authentic ones, only that there was a strong rumor there had been, and that the claim to be a Pythagorean "memory book" could make a book a sure sell.

In Iamblichus' account, probably taken from Dicaearchus, Pythagoras began his approach to the Crotonians by conversing with some of the youth of the city whom he met in the gymnasium. There could hardly have been a surer way to endear himself to their elders than by advising young people to honor their parents, practice temperance, and cultivate a love of learning, but Pythagoras must have had amazing charisma, for such teaching seems unlikely to have aroused enthusiasm among the young.

Hearing of him from their sons, members of the Thousand invited Pythagoras into their assembly to share any thoughts that would be advantageous to Crotonians in general. Such an invitation was not unusual in a Greek city, especially when a man's pedigree in his native country was as unimpeachable as that of any of the local worthies. The Apostle Paul, soon after his arrival in Athens, was similarly invited to speak before the Areopagus, where Athenians and foreigners "spent their time talking about and listening to the latest ideas."[4] In a cosmopolitan city like Croton, high-ranking citizens were eager to meet a man recently arrived from an even more cosmopolitan area abroad.

Pythagoras complied with the request. Some of his advice (as reported by Iamblichus) was predictable, some unusual: Build a temple to the Muses to celebrate symphony, harmony, rhythm, and all things conducive to concord, he proposed. Symphony, harmony, and concord were going to be central to Pythagorean doctrine, and also to the neo-Pythagoreanism of Iamblichus' era. Consider yourselves the equals of those you govern, not their superiors, Pythagoras advised the rulers. Establish justice, with members of the government taking no offense when someone contradicts them. End procrastination. At home, make a deliberate effort to win the love of your children, for while other compacts are engraved on tablets and pillars, the marital compact is made incarnate in children. Never separate parents from their children—the greatest of evils. Avoid sexual relations with any other than a marital partner. If you seek an honor, seek it as a racer does, not by trying to injure competitors but merely by trying to achieve the victory for yourself. If you seek glory, strive to *become* what you wish to *seem* to be.

The simplicity and charm of this list—and its lack of pomposity—lend it an air of authenticity. These teachings may merely have been Iamblichus' late-Roman ideas put in Pythagoras' mouth, but it was the sort of advice that would have been remembered in an oral history or memory book and would have appeared, either from earlier Pythagorean sources or newly minted, in the teachings of the various groups that considered themselves Pythagorean in the centuries separating Pythagoras from Iamblichus. Iamblichus' account goes on to say that the elders were impressed. They built the temple and many sent their concubines packing. They asked Pythagoras to address the young men in a formal setting, and also to address the women of the city, whose inclusion was a strong theme in the Pythagorean tradition.

In Pythagoras' address to the young men, said Iamblichus, he repeated what he had taught those he met in the gymnasium, adding that they should not revile anyone or revenge themselves on anyone who reviled them, and that they should practice listening, as a way of learning to speak. Iamblichus interjected a personal opinion that because of these moral teachings to the youth, Pythagoras really did deserve to be called divine.

In Pythagoras' address to the women, wrote Iamblichus, he expressed high regard for female piety—particularly important in a city whose goddess was connected with all matters pertaining to women. He recommended equity and modesty and appropriate offerings rather than blood and dead animals or anything extravagant. Women should be cheerful in conversation and behave so that others could speak only good of them. A woman should know that it was all right to love her husband more than she loved her parents. She should not oppose her husband, but apparently it was acceptable to discuss matters with him and disagree, because Pythagoras said that if her husband gave way to her, she must not overinterpret that and think he had made himself subject to her. Again Iamblichus reported success that seems too good to be true: Marital faithfulness in Croton became proverbial. Women offered their costliest garments in the temple of Hera.

Though Iamblichus went into greater detail than Diogenes Laertius or Porphyry, the latter two were not silent when it came to what Pythagoras taught the Crotonians. Diogenes Laertius reported a teaching Iamblichus failed to mention: Some men have a "slavish disposition" and are "born hunters after glory," like men in a Great Game contending for prizes. Others are covetous, like those who come to the game for "purposes of traffic." Others are spectators. These are the seekers after the truth. Twenty-six centuries after Pythagoras (and about seventeen after Diogenes Laertius), Bertrand Russell would make much of this Pythagorean distinction. Diogenes Laertius also mentioned Pythagoras' advice not to pray for specific things, because you do not know what is good for you.

Iamblichus summed up Pythagoras' teaching in what he called the "epitome of Pythagoras's own opinions," which he would continue to stress in private and in public: one should by all means possible amputate disease from the body, ignorance from the soul, luxury from the belly, sedition from the city, discord from the household, and excess from all things whatsoever. Iamblichus also praised Pythagoras' teaching method—not to spout facts and precepts but to teach things (such as the power of remaining silent) that would prepare his listeners to learn the truth in other matters as well.

Porphyry described the splendid physical impression Pythagoras made: "His presence was that of a free man, tall, graceful in speech and in gesture." He was "endowed with all the advantages of nature and prosperously guided by fortune."*

Iamblichus numbered the followers who soon gathered around Pythagoras at six hundred. Members of the brotherhood were advised to regard nothing as "exclusively their own," wrote Diogenes Laertius. Friendship implied equality. They were to own all possessions in common and bring their goods to a common storehouse. Apparently, to judge from an incident later, in Syracuse, a good many Pythagoreans complied with this advice. Because of this "common sharing," Pythagoras's followers became known as Cenobites, from the Greek for "common life."

However, not all Pythagoreans had equal status within the community. The six hundred were Pythagoras' "students that philosophized," wrote Iamblichus, Porphry, and their source, Nicomachus. There was a much bigger group, called the Hearers, about two thousand men who along with their wives and children would gather in an auditorium "so great as to resemble a city" and built for the purpose of coming to learn laws and precepts from Pythagoras. It hardly seems a practical possibility that these people, presumably including many of Croton's most prosperous, influential citizens, all "stopped engaging in any occupation." However, according to the three biographers they did all live together for a while in peace, they held one another in high esteem, and they shared at least a portion of their possessions. Many, it seems, revered Pythagoras so greatly that they ranked him with the gods as a genial, beneficent divinity, but Iamblichus observed that, contra Nicomachus' account, they perhaps did not all think of Pythagoras quite as a god. In his treatise *On the Pythagoric Philosophy*, Aristotle wrote that the Pythagoreans made a distinction among

* Porphyry said he got this information from Dicaearchus.

"rational animals": There were gods, and men, and beings in between like Pythagoras.

WHEN PYTHAGORAS first arrived, Croton was at a low ebb of military prestige and clout. The communities of Magna Graecia were in a chronic state of conflict, internal and external, each attempting with varying success to dominate and enslave the next. The latest dismal chapter in this story had been Croton's embarrassing defeat by the army of the city of Locri at the Sagras river, a few miles to her south. Iamblichus called Croton "the noblest city in Italy," but in 530 B.C. she was licking her wounds from that disaster, while Sybaris was still a jewel in the crown of Greek colonial cities.

Croton nevertheless controlled considerable territory. Her normally acknowledged *chora* extended at least as far as what are now the river Neto to the north, in the direction of Sybaris, and the river Tacino to the south. The coastal lands between those two river mouths (with the city centered between) were hers, and away from the coast Croton's territory extended into the mountains, where the tributaries of the two rivers originate among precipitous slopes and deep, narrow valleys* reminiscent of the early colonists' homeland in Achaea. In the two centuries since Myskellos had brought those settlers, the coastal forests had begun to disappear, and the farmlands most vital to the life of Croton's people were large clayey plains to the south of the city, watered by numerous springs and two more rivers and divided into farmsteads that cultivated wheat and cereals. Other cleared areas to the north were suitable for livestock.

Inevitably a community expected a man like Pythagoras to assume a public role, and he and his associates soon did, either by advising the oligarchical leaders or as part of the oligarchy. They became influential, probably extremely so, not only in the city and its environs but in other communities of the region. Porphyry reported that Pythagoras was so extraordinarily persuasive that Simicus, the tyrant of Centoripa, "heard Pythagoras's discourse, abdicated his rule, and divided his property

* In some of the remoter villages of those mountains, the people in the twenty-first century still speak a form of Greek that linguists identify as neither modern Greek nor the Byzantine Greek that arrived with Byzantine Christian Greeks in late antiquity and the early Middle Ages, but as an ancient form of the language that is spoken almost nowhere else in the world.

between his sister and the citizens." Local lore still today agrees with the early historians that Pythagoras inspired a love of liberty in the cities of Magna Graecia and restored their individual independence, and that he and his followers were so successful in rooting out partisanship, discord, and sedition, and in establishing just laws, that the cities flourished in peace for several generations and became models for others before again falling into disputes and warfare. "Love of liberty" may be a later ideal attributed with hindsight to the Pythagoreans. Political thinking during Pythagoras' period in the Greek world saw good government not in terms of how much liberty was allowed but in terms of order and the well-being of the community.[5] Diogenes Laertius had information that Pythagoras gave the Crotonians a constitution, and that he and his followers were an "aristocracy" in the highest, literal sense of the word: "rule by the best."

In 510 B.C., twenty years after Pythagoras' arrival in Croton, Milo, of Olympic wrestling and ox-toting fame and by then a follower of Pythagoras, led Croton's army against her opulent neighbor Sybaris. Like a latter-day Thales, Milo reputedly exercised his own brand of military hydraulics, diverting the river Crathis to flood the enemy city, and the army of Pythagorean Croton razed Sybaris to the ground. Modern Sibari occupies a different site from Greek Sybaris. Because the more ancient Sybaris perished forever with the defeat by Croton, the archaeological site there, buried beneath a Roman town and part of the Appian Way, has yielded a treasure trove of artifacts. Among them are covered pots from the seventh century B.C. the size of modern sugar bowls, whose lids are decorated with what later would be called Pythagorean triangles. It was a super-wealthy, cultured—indeed, "sybaritic,"—city that Milo destroyed, but though archaeologists have done extensive work, the only trace visible to modern visitors is a water-filled hole beneath excavations of the Roman town.

With Sybaris gone, Croton's influence and power in the region reached a zenith, and historians credit Pythagoras and the teaching and training he initiated with bringing about this rise in Croton's fortunes. If Diogenes Laertius, Porphyry, and Iamblichus are to be believed—and modern scholarship does not say them nay—he was an ancient example, and arguably the most successful one in history, of Plato's "philosopher king."

Or was it all a sham? There is a darker version of the tradition that has Pythagoras and his followers ruling in an autocratic, repressive way. In this retelling, the war with Sybaris began when Croton, at Pythagoras' insistence, gave sanctuary to five hundred citizens of Sybaris who had been stripped of their property and banished. A social reform in Sybaris had justifiably confiscated the excessive wealth of these five hundred and distributed it to the poor, and Pythagoras' sympathy for the formerly rich exiles revealed him in an unfavorable light as a defender of an autocratic and repressive status quo. This story does not actually conflict with the reputed egalitarianism of the Pythagoreans, for there is no evidence that their egalitarianism applied to society in general outside the Pythagorean brotherhood. No one knows what reasons Pythagoras might have had for wishing to restore the status quo in Sybaris, or whether his reforms in Croton were motivated by personal demagoguery, a desire to strengthen the aristocratic class structure, or a wish to transform the communities to conform to higher moral standards. All the early biographers—and fervent revolutionaries of eighteenth- and nineteenth-century Europe—were sure it was the last.

Independent evidence speaks to Pythagoras's impact on the economics of Croton.[6] Numismatists credit him and his first followers with the introduction of a coinage with an incuse (hammered-in) design, the earliest coinage used in Croton and the area she ruled. These coins were both beautiful and difficult to create, and those familiar with the history of minting recognize the oddity and significance of their sudden appearance in this time and place, with apparently no gradual evolutionary process leading up to or explaining their emergence. The history of coinage does not normally work this way. Not that these were the first coins. There were earlier coins—for example, in Lydia, the region east of Miletus, before 700 B.C. But an innovation like the coins in Croton would seem to indicate a polymath—a "genius of the order of Leonardo da Vinci," in the words of the historian C. T. Seltman.[7] Given the area where the coins were used and the timing of their appearance, the inventor by default must have been Pythagoras, son of a prominent merchant with experience in a world-wide market, familiar (if his father was a gem engraver) with beautiful small design, and skilled with numbers. Aristoxenus, who had friends among the Pythagoreans of the fourth century B.C., wrote that Pythagoras introduced certain types of weights and measures but "diverted" the study of numbers from mere mercantile practice, implying that Pythagoras also understood the use of numbers in connection with such practice. It is difficult to believe that he had nothing to do with the invention and introduction of the remarkable Crotonian coinage.

Though Pythagoras undoubtedly made serious enemies, for many years that seemed not to hamper him or his supporters very much. Pythagorean leadership extended the area Croton dominated much further both while Pythagoras lived there and in the fifty years after his death or exile—as far as Caulonia in the south (almost to the doorstep of the old enemy, Locri) and to the sanctuary of Apollo Aleo at Ciro Marina in the north (well on the way to Sybaris). The acquisition of Ciro Marina was something to be celebrated, since already at this early date it was famous for its fine wine. To the west, Croton's influence extended almost to the Tyrrhenian Sea, to Terina. That was the best Croton would ever do. She was no Rome.

PORPHYRY, MORE THAN Iamblichus or Diogenes Laertius, stressed the silence of the Pythagoreans and recognized not only its value but also how disastrous it would prove for the Pythagorean tradition. It is frustrating to find that, though Porphyry mentioned Pythagoras winning over the Crotonian rulers and described the invitations to address the youth and women—and though it was Porphyry who identified Dicaearchus as the source of this information—he made no claim to be able to report with any certainty the details of what Pythagoras told his audiences. He attributed this lack of information to Pythagorean silence. Because all three biographers tended to err on the side of believing their sources too readily rather than too little, Porphyry's reluctance makes what he said on the matter of Pythagorean silence particularly credible. According to him, Pythagoras and those who followed him during his lifetime did not reveal their ideas, principles, or teachings, or the details of their discipline to others. They wrote nothing down, keeping "no ordinary silence." In great part because of this secrecy, much information about Pythagoras had come down through the centuries in scattered, fragmentary, hearsay form, consisting of what other people *thought* he and his associates taught and what their way of life was.

Porphyry was not alone in stressing Pythagorean silence. Diogenes Laertius made it clear that there were two kinds: On the one hand, "silence" meant keeping doctrine secret from outsiders; on the other, it meant maintaining personal silence in order to listen and learn—and that applied especially among followers in "training." For five years they were silent, listening to discourses. Only after that, if approved, were they allowed to meet Pythagoras himself and be admitted to his house. The advantage to be gained from remaining silent was an ancient theme that also appeared in the Wisdom chapters of the Hebrew Scriptures and was picked up by early Christian church fathers a few generations after Iamblichus.

Did the first type of silence extend to putting nothing in writing? Of the three third- and fourth-century biographers, Diogenes Laertius was the only one to insist that Pythagoras wrote down some of his doctrines, but the section of his biography titled "Works of Pythagoras" is confusing and unconvincing. He began on shaky ground with the words:

> Some say, mistakenly, that Pythagoras did not leave a single written work behind him. However, Heraclitus the natural scientist pretty well shouts it out when he says: "Pythagoras, son of Mnesarchus, practiced inquiry more than any other man, and selecting from these writings he made a wisdom of his own—a polymathy, a worthless artifice."

It would seem, contra Diogenes Laertius, that what Heraclitus "shouted out" was that Pythagoras could read and plagiarize, not that he wrote anything down. Diogenes Laertius was right, however, that Heraclitus' words were worth careful scrutiny, because his lifetime probably overlapped Pythagoras' and his comments about him are among the oldest that survive. Though in Heraclitus' own philosophy he often sounded like a Pythagorean, if he ever had anything good to say about Pythagoras there is no record of it. He had little better to say about anyone else. He was contemptuous of most of humankind, and in particular of polymaths, coming out with such disparaging remarks as "Much learning does not teach thought—or it would have taught Hesiod and Pythagoras, and again Xenophanes and Hecataeus." Be that as it may, there is no reason to take Heraclitus' diatribe as evidence that Pythagoras wrote a book.

Diogenes Laertius was not equally convinced about all claims for Pythagoras' authorship, but he believed that Pythagoras had written three books that still existed in his lifetime. If so, they then rapidly disappeared or were discredited, for Porphyry, only a few years later, wrote,

"He left no book." There was plenty of reason to be skeptical about the authorship of the books that Diogenes Laertius listed, considering the number of Pythagorean forgeries that had appeared during the Hellenistic and Roman eras. However, information that Pythagoras wrote poems under the name of Orpheus came from an earlier, more reliable source. Ion of Chios, a scholar, playwright, and biographer born shortly after Pythagoras died, tried to determine the true source of some poems that were widely supposed to have been written by Orpheus. He decided that the author was Pythagoras and that Pythagoras had attributed them to Orpheus.

CHAPTER 4

"My true race is of Heaven"

Sixth Century B.C.

A CHILDHOOD IN A PROSPEROUS agrarian family that was also involved in the mercantile world centered in Samos, with its temple of Hera, had placed Pythagoras at a crossroads of different beliefs about life after death. If there was an orthodox view of the afterlife and immortality in the ancient Greek world, it was that reflected in Homer's epics and later in the official cults of the cities and in much of the great literature. A human soul, or *psyche*, survived after death, but this survival was not an attractive one. For the Homeric heroes, the true "self" was the body, and the good life was closely tied with it. What good was survival in a form that could not enjoy feasting, combat, human love, sex, comradeship? Death was separation from these, leaving the soul in a weak, witless state—a shadow, a dream, smoke, a twittering bat. Only the gods had a better sort of immortality, but not in the sense that they survived death, for they never died. Furthermore, they jealously guarded their immortality. Woe betide any human who tried to overstep the limits and attain the immortality of the gods.

Alongside this mainstream, people who lived in the countryside, and some in the cities, too, clung to hundreds of small clusters of beliefs, so ancient that no one could trace their origins, that provided better answers to questions raised by an unfair world and suggested there would be future compensation for its injustice and suffering. One "mystery cult" had been centered in the town of Eleusis, and when Eleusis became part of Athens a few years before Pythagoras was born, the cult outgrew its local origins and spread across the Hellenic world. It required initiation into the mysteries of the earth mother Demeter and her daughter Persephone, an adoption into the family of the gods that carried with it a happier life in the next world. After initiation, normal everyday affairs could continue with no onerous new requirements.

The Orphic cult, by contrast, involved a complicated set of beliefs in which the soul was a mixture of the divine and the earthly. Developing the divine part and suppressing the earthly required a relentless pursuit of purity, including ceremonies of ritual cleansing and the avoidance of eating meat. This was the work of more than one lifetime. A soul was reborn again and again, with its conduct in one life determining its fate in the next. The ultimate goal was to become one with Bacchus, or "a Bacchus."

Orphism had roots before the historical era in the worship of Dionysus, another name for Bacchus, probably at first a fertility god and only much later connected with wine and drunkenness. He was a god of the Thracians, an agricultural people who lived north of mainland Greece in the area bounded by the Aegean Sea, the Black Sea, and the Danube River. The Greeks regarded them as primitive barbarians, and the fifth-century historian Herodotus described them as people who "led miserable lives and were rather stupid." When Dionysus/Bacchus worship reached Greece at about the beginning of the historical era, it was greeted with hostility, but its unorthodoxy and savagery gave it an irresistible fascination that was portrayed in Euripides' play *The Bacchae*. The cult exalted the status of women and, if the playwright is to be believed, married and unmarried women retreated into the mountains in large bands to dance in ecstasy and to tear apart wild animals and eat them raw. A tradition of strong, involved women may have come to the Pythagoreans through Orphism, but in a less bloodthirsty guise.

By the time of Pythagoras, Orphic communities were all over the Greek world, including southern Italy and Sicily. The primitive worship of Dionysus/Bacchus had evolved into something more ascetic, stimulating the mind instead of (or as well as) the body and psyche. Cult members attributed its reformation to Orpheus, whom frenzied Bacchic women had reputedly torn to pieces for his efforts. Orpheus was probably a real person clothed in legend. He seems to have been regarded first as a priestly figure, while his lyre and connection with music, and the status of a semi-mythical hero, came later. Some called him a god.[1]

If the stories about Pythagoras' youthful travels were genuine, he was familiar with religious traditions in Egypt and Mesopotamia, and perhaps (if Josephus was right) with the beliefs of the Hebrews near Mount Carmel or in Babylon. Regardless of the authenticity of the details, the impression that comes across, reinforced by the story of his initiation into the rites of the priests of Morgos on Crete, was of a man intent on exploring in depth and becoming personally involved in many religious ideas and beliefs.

In Croton, Pythagoras and his followers did not abandon the polytheism of the Homeric/Olympic tradition. Some thought Pythagoras was an incarnation of Apollo, and that god's association with moderation, intelligence, and order was in accord with Pythagorean ideals. As for other gods, the fact that the building boom at the temple of Hera occurred when Pythagoras' influence was strong in Croton is probably no coincidence. However, when Pythagoras chose what he would believe and teach with regard to immortality, he came down decisively with the Orphic cult, with the doctrine of transmigration of the soul or reincarnation. This was no secret. It was "very well known to everyone," wrote Porphyry.

An early fragment bears witness that Pythagoras believed a good man would be rewarded in the next life. The fragment is from Ion of Chios, the near contemporary of Pythagoras who attributed an Orphic poem to him, and who, though perhaps not a member of the Pythagorean community, adopted Pythagorean ideas:

> So he [a good human being], endowed with manliness and modesty, has for his soul a joyful life even in death, if indeed Pythagoras, wise in all things, truly knew and understood the minds of men.

Pythagoras went further than belief in reincarnation. He claimed he could remember his past lives. This, too, had roots in Orphism. An inscription on an Orphic document known as the Petelia tablet instructs a soul how to show itself worthy of joining the divine and worthy of "Memory," an Orphic reference to the special kind of memory that Pythagoras claimed to have.[2]

The earliest reference to Pythagoras' ability to remember his past lives is from the fifth century B.C. poet-philosopher Empedocles, who came from Acragas in Sicily and like Ion was born near the time Pythagoras died. He was often called Empedocles the Pythagorean, but much of his philosophy was different from Pythagorean teaching. On the doctrine of transmigration he was in enthusiastic agreement:

> *There was among them a man of immense knowledge*
> *who had obtained vast wealth of understanding,*
> *a master especially of every kind of wise deed [or*
> *"cunning act"].*
> *For when he reached out with all his mind*
> *he easily saw each and every thing*
> *in ten or twenty human lives.*

Iamblichus, without a murmur, accepted Pythagoras' ability to recall his past lives, but not all the details of how he acquired that ability and what he remembered. The memories began with Pythagoras' life as Aethalides, a son of the god Hermes—the sort of paternity Iamblichus found impossible to believe. However that may be, Hermes allowed Aethalides to choose a gift, anything short of the immortality of the gods. Aethalides asked to be able to remember everything that had happened to him in his former lives. So it came about that Pythagoras could recall not only his life as Aethalides but also as Euphorbus, as Hermotimus, and as Pyrrhus, a Delian fisherman, and much else besides. Euphorbus was a hero in the Trojan War who was immortalized in Homer's *Iliad*. Iamblichus and Porphyry both pictured Pythagoras singing the funeral verses Homer wrote for Euphorbus, accompanying himself "most elegantly" on a lyre:

> *The shining circlets of his golden hair*
> *Which even the Graces might be proud to wear,*
> *Instarred with gems and gold, bestrew the shore*
> *With dust dishonored, and deformed with gore.*
> . . . .
> *Thus young, thus beautiful Euphorbus lay,*
> *While the fierce Spartan tore his arms away.*[3]

Diogenes Laertius gave the full version of a tale that many thought constituted proof of Pythagoras' memories, but that Iamblichus rejected as being "too popular in nature" and Porphyry thought "too generally known" to require telling:* After Euphorbus died by the hand of King Menelaus in the Trojan War, his soul (either directly or after several other lifetimes) passed into Hermotimus. Hermotimus, in turn, was able to prove this had indeed happened. In some versions of the story it occurred at Branchidae in western Turkey; in others, at Argos on the Greek mainland; but, wherever it happened, Hermotimus entered a temple where a decaying shield was nailed up on the wall, little of it intact except an ivory boss. This relic had either been left by Menelaus as a tribute to Apollo or was simply among spoils of the Trojan War. At the sight of the rotten old shield, Hermotimus burst into tears. People standing near questioned him, and he muttered that he himself, as Euphorbus, had carried it at Troy. The bystanders thought he was insane, but he told them that they would find the name Euphorbus inscribed on the back. They unfastened the shield from the wall and discovered, in archaic lettering, that very name.[4] Hermotimus eventually died and became Pyrrhus, a fisherman of Delos, and, some time after Pyrrhus, Pythagoras. Nor was that the full extent of Pythagoras' memories. His soul had passed into many plants and animals, and he could recall his suffering in Hades, as well as the sufferings endured by the others there.

In the doctrine of transmigration as Pythagoras taught it, a soul was not irrevocably doomed to an eternal round of animal and vegetable existences. Escape was possible, as it was in Orphism. The possibility and method of this escape came to stand at the heart of the Pythagorean view of the world. There was a divine level of immortality from which each soul was a "torn off fragment," a mere "spark of the divine fire," held captive in a long train of dying bodies.[5] The goal of a wise human was to break free of bondage to this treadmill of earthly reincarnation and rejoin the sublime level.

By tradition, Pythagoras coined the term "philosopher," meaning "lover of wisdom," but it is probably more correct to say that he gave it

* Diogenes Laertius took the story from the writing of Diodorus, a scholar of the first century B.C. who in turn got the story from the writing of Plato's pupil Heracleides of Ponticus.

a new meaning. A philosopher did not merely love wisdom, he pursued it with all his might, because that was the way to regain the true, divine life of the soul. The historian Aristoxenus wrote of the Pythagoreans he knew: "Every distinction they lay down as to what should be done or not done aims at conformity with the divine. This is their starting-point; their whole life is ordered with a view to following God, and it is the governing principle of their philosophy."[6] All philosophy and inquiry—all use of the powers of reason and observation to gain an understanding of nature, human nature, the world, and the cosmos, including what would later be called "science"—was linked with, indeed *was*, the effort to purify the soul and escape the wheel of reincarnation. This connection, for the Pythagoreans, was the most exalted living-out of the doctrine of the "unity of all being."

Such a relentless pursuit had been recommended in much more ancient wisdom literature, including Proverbs in the Hebrew Scriptures ("Old Testament"). However, nowhere else did the search for the wisdom of God or the gods include so comprehensively the search for knowledge about the physical universe. As the scholar W. K. C. Guthrie put it:

> It is to this idea of assimilation to the divine as the legitimate and essential aim of human life that Pythagoras gave his allegiance, and he supported it with all the force of a philosophical and mathematical, as well as a religious, genius. In this lies the originality of Pythagoreanism.[7]

In a less reverent vein, Diogenes Laertius quoted the poet Xenophanes of Colophon, who lived most of his adult life in Sicily and Italy and was probably a contemporary of Pythagoras, though he survived him by many years. Xenophanes wrote satirical poems, and in these lines he made light of Pythagoras' belief in reincarnation:

> *And once when he passed a puppy that was being*
> *whipped*
> *they say he took pity on it and made this remark:*
> *"Stop, do not beat him; for it is the soul of a dear friend—*
> *I recognized it when I heard its voice."**

* Scholars regard this quotation as likely to be genuinely early, because it made light of Pythagorean belief, rather than extol it as would have happened later, in an overly adulatory period.

This verse is usually taken to mean that Pythagoras claimed to recognize the voice of a friend who had died and been reincarnated as the puppy, but for a Pythagorean it would have had a more profound meaning. A "dear friend" was any member of a vast kinship, embracing all of nature including animals and vegetables and the souls of humans. In no other Greek society was that kinship so celebrated as among the Pythagoreans, or so firmly believed to be not a melting pot but a beautifully ordered unity: in the words of W. K. C. Guthrie, "a *kosmos*—that untranslatable word which unites, as perhaps only the Greek spirit could, the notion of order, arrangement or structural perfection with that of beauty."[8] Some Pythagoreans extended the unity to time. Aristotle's pupil Eudemus wrote that "if we are to believe the Pythagoreans and hold that things the same in number recur—that you will be sitting here and I shall talk to you, holding this stick, and so on for everything else—then it is plausible that the same time too recurs."[9]

The belief that souls, at death, pass into other persons, animals, or plants might be expected to have had implications for what Pythagoreans did and did not eat, just as it did for the Orphic cult. However, the particulars of the Pythagorean diet have never been clear to anyone except Pythagoras and his immediate followers and have, since early times, been subject to much speculation, many opinions, and irreverent humor. Any abstention must have been for reasons other than the avoidance of eating another soul, for a human was just as likely to be reincarnated as a vegetable, and you had to eat *something*. Empedocles is supposed to have remarked that if you could choose your next life, a lion or a laurel bush would be good choices. Iamblichus thought Pythagoras ordained abstinence from animal flesh as "conducive to peaceableness." A man trained to abominate the slaughter of animals "will think it much more unlawful to kill a man or engage in war."

Aristotle felt sure that Pythagoras and his followers did eat the meat of animals except the womb and heart and sea urchins. Possibly they also avoided mullet, added Plutarch. Diogenes Laertius insisted that *red* mullet, blacktail, and the hearts of animals were forbidden but reported that Aristoxenus said Pythagoreans ate all other animals besides lambs, oxen used in agriculture, and rams. Porphyry, basing his conclusion on an early source from the fourth or early third century B.C., believed Pythagoras held a double standard: Someone not engaged in the lifelong Pythagorean pursuit of wisdom—an athlete or soldier, for instance (recall Pythagoras' advice to the young Olympians)—could eat meat. But for a member of his own school Pythagoras allowed only a ritual taste of meat being offered as a sacrifice to the gods. According to Porphyry, this abstinence was motivated by reverence for the unity and kinship of all life, and Pythagoras' preferred diet included honey; bread of millet; barley; and herbs, raw and boiled. Porphyry even provided recipes he said were favorites of Pythagoras:

> He made a mixture of poppy seed and sesame, the skin of a sea-onion, well washed until entirely drained of the outward juices, of the flowers of the daffodil, and the leaves of mallows, of paste of barley and chick peas, taking an equal weight of which, and chopping it small, with honey of Hymettus he made it into a mass. Against thirst he took the seed of cucumbers, and the best dried raisins, extracting the seeds, and coriander flowers, and the seeds of mallows, purslane, scraped cheese, wheat meal and cream, all of which he mixed up with wild honey.

Porphyry wrote that Pythagoras did not claim to have invented these recipes; they had been taught by Demeter to Hercules when he was sent into the Libyan desert.

Information about the diet of later Pythagoreans, though not necessarily the diet advised by Pythagoras himself more than a century before, comes from fourth century B.C. comic plays by Antiphanes, Alexis, and Aristophon.[10] Their portrayals may have been accurate or perhaps were only commonly accepted stereotypes, but these were all highly respected playwrights. Antiphanes, who was renowned for his parody and astute criticism of literature and philosophy, wrote that "some miserable Pythagorists were in the gully munching purslane and collecting the wretched stuff in sacks." In his play *The Sack*, he had a character who "like a Pythagorizer, eats no meat but takes and chews a blackened piece of cheap bread." In Alexis' *The Men from Tarentum* "'Pythagorisms' and fine arguments and close-chopped thoughts nourish them" while they eat daily only "one plain loaf each and a cup of water—a prison diet! Do all wise men live like that?" Apparently not, for another character replied that some Pythagoreans "dine every four days on a single cup of bran." Aristophon, in *The Pythagorist*, wrote:

> For drinking water [not wine], they are frogs; for enjoying thyme and vegetables, they are caterpillars; for not being washed, they are chamber-pots; for staying out of doors all winter, blackbirds; for standing in the heat and chattering at noon, cicadas; for never oiling themselves, dust-clouds; for walking about at dawn without any shoes, cranes; for not sleeping at all, bats.

Alexis, in *The Men from Tarentum*, offered a witticism that became so current it was probably eventually greeted with groans: "The Pythagorizers, as we hear, eat no fish nor anything else alive; and they're the only ones who don't drink wine." —"But Epicharides eats dogs, and he's a Pythagorean." —"Ah, but he kills them first and then they're no longer alive." Diogenes Laertius took up the same theme centuries later in a "jesting epigram" in his biography of Pythagoras:

You are not the only man who has abstained
From living food; for so have we;
And who, I'd like to know, did ever taste
Food while alive, most sage Pythagoras?
When meat is boiled, or roasted well and salted,
I do not think it well can be called living.
Which, without scruple therefore then we eat,
And call it no more living flesh, but meat.

THE BEST-KNOWN CONTROVERSY about Pythagoras' diet had to do with his attitude toward beans—not such a trivial question as it might seem, for this attitude may later have contributed to his death.

The poet Callimachus lived in the third century B.C. and, in addition to much splendid poetry, produced a critical and biographical catalog of the authors whose works were in the collection of the Alexandria Library. He was familiar with much literature that was no longer available to later scholars because it perished when the library burned. Callimachus agreed with an idea that he attributed to Pythagoras himself: that beans are "a painful food." Cicero wrote, citing Plato, that the Pythagoreans were forbidden to eat them because they cause flatulence and hence are not conducive to peace of mind and a good night's sleep. Other reports had it that flatulence was an indication beans contained air. Since it was widely held that the soul itself was air, this might have been interpreted to mean that when one ate a bean one was eating a soul. Diogenes Laertius said that avoiding beans made for gentle dreams, "free from agitation." He also reported several reasons given by Aristotle why the Pythagoreans did not eat beans, including that they were "used in elections in oligarchical governments." Plato's pupil Heracleides Ponticus connected the avoidance of beans with the discovery that a bean placed in a new tomb, buried in dung, and left for forty days took on the appearance of a human. One tale about Pythagoras' power to communicate with animals told of an ox that Pythagoras saw eating beans. When the herdsman mockingly refused to follow Pythagoras' advice to order the ox to abstain, Pythagoras whispered in its ear and the ox never again touched a bean. Pythagoras took it to live many years as the "sacred ox" at Hera's temple.

Aulus Gellius, whose second century A.D. writings preserve many fragments of otherwise lost works, vehemently disagreed with the idea that Pythagoras forbade the eating of beans. Aristoxenus, he pointed out, had insisted that Pythagoras ate plenty of them, in fact more than any other vegetable, because "they soothe and gently relieve the bowels." Aulus Gellius also believed he could explain the unfortunate misunderstanding: It stemmed from an overly naive interpretation of a poem by Empedocles that included the phrase: "Wretches, utter wretches, keep your hands from beans!" Gellius' scholarly approach revealed that the word "beans" here did not mean the vegetable; it meant "testicles." Pythagoreans used obscurely symbolic aphorisms that could only be deciphered by other Pythagoreans, and when Empedocles spoke of "beans," Gellius insisted, he intended them to symbolize the cause of human pregnancy and the impetus to human reproduction. The bean is, after all, a seed, with similar potential. Gellius, then, interpreted Empedocles' phrase to mean, "Avoid sexual indulgence!"[11]

Pythagoras apparently did not encourage celibacy, for several accounts had him urging his followers to beget children so as to leave servants of god to take their place in the next generation. But sex, it seems, stopped there. According to Diogenes Laertius, Pythagoras, "did not indulge in the pleasures of love" and advised others to have sex only "whenever you are willing to be weaker than yourself." In contrast to all the humor at the Pythagoreans' expense, Pythagoras himself came across, at least in Diogenes Laertius' biography, as humorless, though not necessarily joyless. He "abstained wholly from laughter, and from all such indulgences as jests and idle stories," advising others as well that "modesty and decorum consist in never yielding to laughter, without looking stern"—which, if true, indicates that Pythagoras would not have approved of Diogenes Laertius' "jesting epigrams." Porphyry's description showed Pythagoras not so much humorless as extremely even-tempered, not "elated by pleasure, nor dejected by grief, and no one ever saw him either rejoicing or mourning." Porphyry attributed this "constancy" to Pythagoras' careful diet.

According to the tradition, Pythagoras sired children. After introducing his paragraph concerning Pythagoras' family with the words "It is said," Porphyry recorded that Pythagoras' wife was Theano, from Crete, the daughter of Pythenax. Pythagoras and Theano had a daughter named Myia "who took precedence among the maidens in Croton and, when a wife, among married women," and also a son, Telauges, and perhaps a second son named Arignota. Iamblichus wrote that Pythagoras' "acknowledged successor," Aristaeus, married Pythagoras's widow Theano after Pythagoras died, "carried on the school," and educated Pythagoras' children. Among those children Iamblichus mentioned none of the names that Porphyry listed, but spoke only of another son named after Pythagoras' father, Mnesarchus, who, in turn, took over "the school" when Aristaeus became too old. Iamblichus confused matters still further by mentioning a "Theano" who was the wife of Brontinus of Croton and one of the "most illustrious Pythagorean women." Did Brontinus die and Pythagoras marry his widow, or was it the other way around? More likely there were two Theanos, mother and daughter. Diogenes Laertius recorded variously that Theano, the wife of Brontinus, was Pythagoras' pupil, and that Pythagoras' wife was probably Theano, daughter of Brontinus of Croton.

Theano's name was preserved on a list thought to have come through Aristoxenus of seventeen "most illustrious Pythagorean women" that also included Mya, the wife of the Olympic wrestler Milo. Women apparently played an active part in the Pythagorean "brotherhood." Diogenes Laertius said Theano had written books that still existed in his lifetime. Though these, sadly, were almost certainly some of the "pseudo-Pythagorean" books that appeared in antiquity, Diogenes Laertius felt confident enough of his source to quote Theano's outspoken advice: Asked how soon a woman becomes pure again after intercourse, she was supposed to have said, "The moment she leaves her own husband she is pure; but she is never pure at all, after she leaves anyone else." She advised that a woman going to her husband should "put off her modesty with her clothes"—which seems a great waste if these were indeed the words of Pythagoras' wife and Pythagoras really did entirely abstain from the pleasures of love!

PYTHAGOREAN LIFE IN Croton was, it appears, a good life—with the begetting of children who would be new Pythagoreans and could be schooled in a new, wondrous approach to the world and the universe . . . with properly chosen food, whatever it included, appearing on Pythagorean tables . . . with men and women engaging in fascinating studies that also improved their chances in the next life. Within the community, moreover, word got around of some occurrences that were difficult to explain and that indicated their leader was no ordinary man.

Unlike the ancient miracles in the Hebrew Scriptures and the Christian New Testament, the "wonders" attributed to Pythagoras were not associated with any teaching or divine revelation, nor were they examples of Pythagoras' helping or healing anyone. They were of a more random nature, chance glimpses of existence on a more divine level than that experienced by the men and women around him, a level on which the unity of all being—of all things, places, animals, and gods; of past, present, and future—could easily be seen. Aristotle told of reports that Pythagoras appeared on the same day at the same hour in Croton and in Metapontum, and that on one occasion, getting up from a seat in Olympia, he revealed his thigh, and everyone saw that it was made of gold. In Etruria (Tuscany), a poisonous snake bit him and he bit it back. The snake died; he did not. Several witnesses heard the river Casas greet him by name, and he correctly predicted that a white bear would be sighted in Caulonia. Once, after foretelling serious strife, he disappeared in Croton and appeared in Metapontum. "According to credible historians," wrote Iamblichus, and "ancient and trustworthy writers," wrote Porphyry, in each case without naming them, birds and beasts listened to Pythagoras and followed his advice—the same effect that Orpheus had on even the most savage animals.

Countering the miraculous reports were rumors that Pythagoras was a charlatan. Diogenes Laertius repeated a story from Hermippus, the third century B.C. native of Samos who had said Mnesarchus was a gem engraver. Pythagoras disappeared for a time into a set of subterranean rooms while his mother recorded everything that took place, marking the times and dates on tablets that she sent down to him. Eventually Pythagoras emerged, looking like a cadaver, and announced that he had arrived from Hades below. When he told the assembled people in detail all that had happened to them in his absence, they were awestruck, believed he was divine, wept and lamented, and "entrusted to him their wives," "who took upon themselves the name of 'Pythagorean women.'" That same story was told by Herodotus (who was skeptical about it himself) of a man who had been Pythagoras' slave when he lived on Samos, who was supposed to have used this strategy to create an aura of magical power among gullible people in Thrace. In an interesting turnabout, some scholars have suggested that the miraculous stories, as well as the rumors of charlatanism, were all inventions to discredit Pythagoras in an era when people scoffed at the "miraculous" in a way they no longer would in late antiquity.[12]

CHAPTER 5

"All things known have number"

Sixth Century B.C.

N THE PYTHAGOREAN DISCOVERY that "all things known have number—for without this, nothing could be thought of or known"—was made in music. It is well established, as so few things are about Pythagoras, that the first natural law ever formulated mathematically was the relationship between musical pitch and the length of a vibrating harp string, and that it was formulated by the earliest Pythagoreans. Ancient scholars such as Plato's pupil Xenocrates thought that Pythagoras himself, not his followers or associates, made the discovery.

Musicians had been tuning stringed instruments for centuries by the time of Pythagoras. Nearly everyone was aware that sometimes a lyre or harp made pleasing sounds, and sometimes it did not. Those with skill knew how to manufacture and tune an instrument so that the result would be pleasing. As with many other discoveries, everyday use and familiarity long preceded any deeper understanding.

What did "pleasing" mean? When the ancient Greeks thought of "harmony," were they thinking of it in the way later musicians and music lovers would? Lyres, as far as anyone is able to know at this distance in time, were not strummed like a modern guitar or bowed like a violin.

Whether notes were sung together at the same time is more difficult to say, but music historians think not. It was the combinations of intervals in melodies and scales—how notes sounded when they followed one another—that was either pleasing or unpleasant. However, anyone who has played an instrument on which strings are plucked or struck knows that unless a string is stopped to silence it, it keeps sounding. Though lyre strings may not have been strummed together in a chord, more than one pitch and often several pitches were heard at the same time, the more so if there was an echo. Even when notes are played in succession and "stopped," human ears and brains have a pitch memory that causes them to recognize harmony or dissonance. In truth, the ancient Greeks, including Pythagoras, heard harmony both ways, between pitches sounding at the same time and between pitches sounding in succession.

The instrument Pythagoras played was probably the seven-stringed lyre. He tuned it with four of the seven strings at fixed intervals. There were no options about what these intervals would be. The lowest- and highest-sounding of the fixed-interval strings were tuned to sound an octave apart. The middle string on the lyre (the fourth of the seven strings) was tuned to sound a fourth above the lowest string, and the one next higher was tuned to sound a fifth above the lowest string.* The intervals of the octave, fourth, and fifth were considered concordant, or harmonious. A Greek musician could adjust the other three strings on the seven-stringed lyre (the second, third, and sixth string), depending on the type of scale desired.

Pressing a string exactly halfway between the two ends produces a tone one octave higher than the open, unpressed string plays. The ratio of those string lengths is 2 to 1, and they always produce an octave. But the octave is not something a musician creates by pressing the string. Plucking an open string without pressing it at all causes it to vibrate as a whole, sounding the "ground note," but various parts of the string are also vibrating independently to produce "overtones." Even without the string being pressed at the halfway point to play an octave, the octave is present in the sound coming from the open string. Pressing the string releases tones at

* Think of having the lowest string tuned to C on the piano, the fourth string tuned to F above the C, the next to G a whole step above that, and then the top string tuned at C an octave above the lower C.

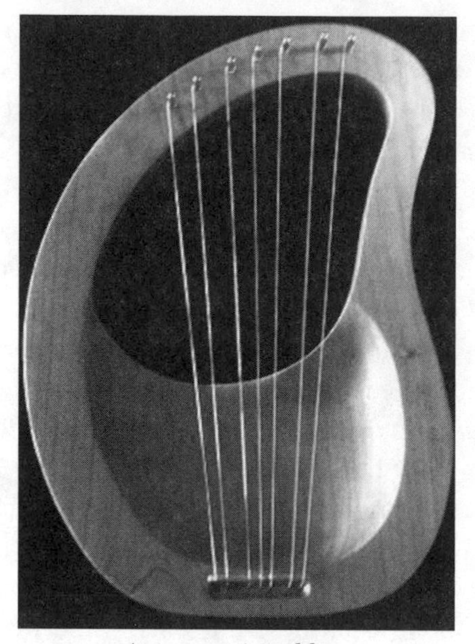

A seven-stringed lyre

the octave, fifth, fourth, and so on—depending on where you press it—that were always there in the ground note but more difficult to hear.*

Tradition credits Pythagoras with inventing the *kanon*, an instrument with one string, and using it to experiment with sound. He would have found that the notes that sounded harmonious with the ground note were produced by dividing the string into equal parts. Dividing it into two equal parts produced a note an octave higher than the open string. Pressed so as to divide it into three equal parts, the string played a note a fifth above that octave; in four equal parts, it played a note a fourth above that. The series goes on to a major third, then a minor third, then smaller and smaller intervals, but there is no indication the Pythagoreans took the process any further than the interval of the fourth.†

* Musical instruments and human voices, because of intricate differences in the way their structures resonate and amplify sound, emphasize or "bring out" certain overtones more than others, and that is what causes the great variety of sounds they make. That is how a trumpet ends up sounding like a trumpet while a clarinet sounds like a clarinet.

† On the piano, equivalent notes might be, for example, middle C (ground note); c (octave above that ground note); g (fifth above that octave); c (fourth above that g). For a demonstration using the piano: Press down gently on the c above middle C without

Looking beyond the task of getting good, practical results from a musical instrument to ask more penetrating questions about what was going on, and whether it could have wider implications, required an unusual turn of mind. Though with hindsight a shift of focus from useful knowledge to recognizing deeper principles can look simple, it is not a trivial change. A lyre sounded pleasant used one way and not another way . . . but *why*? Often, in writings about the Pythagoreans, a clause added to that question has them asking whether there was any meaningful pattern? . . . any orderly structure? but they were not necessarily looking for pattern or order yet, for no precedent would have led them to expect it. Nevertheless, they were about to discover it.

When Pythagoras and his associates saw that certain ratios of string lengths always produced the octave, fifth, and fourth, it dawned on them that there was a hidden pattern behind the beauty they heard in music—a pattern that they were able to understand, but that they had not created or invented and could not change. Surely this pattern must not be an isolated instance. Similar mathematical and geometrical regularities must lie concealed behind all the everyday confusion and complexity of nature. There was order to the universe, and this order was made of numbers. This was the great Pythagorean insight, and it was different from all previous conceptions of nature and the universe. Though the Pythagoreans hardly knew what to do with the treasure they had found—and modern mathematicians and scientists are still learning—it has guided human thinking ever since. Pythagoras and his followers had also discovered that there apparently was a powerful link between human sense perceptions and the numbers that pervaded and governed everything. Nature followed a fundamental, rational, beautiful logic, and human beings were tuned in to it, not only on an intellectual level (they could discover and understand it) but also on the level of the senses (they could hear it in music).

There are other mathematical relationships hidden beneath the experience of music that neither Pythagoras nor others of his era had any way of discovering. The ratios he found represent the rate at which a

---

allowing it to sound (removing the damper from the strings). Strike middle C (the ground note) and you will clearly hear the octave. Press carefully on the g above that octave. Strike middle C and you will hear that fifth above the octave. A piano is not tuned to the Pythagorean system, but it is close enough for you to hear these overtones.

string vibrates, but there was no way he could have studied the vibrations. However, after the initial discovery using a *kanon* or a lyre, Pythagoras and/or his early associates may well have begun listening for octaves, fourths, and fifths in other sounds and attempted to discover what could, and what could not, produce the intervals. Perhaps it is the memory of some of their experiments that lies behind several puzzling early stories in which Pythagoras made the discovery of the relationship in ways that he could not possibly, in fact, have made it.

According to one tale Pythagoras was passing a blacksmith's shop and noticed that the intervals between the pitches the hammers made as they struck were a fourth, a fifth, and an octave. That part of the story is possible, but the next part is not: The only differences between the hammers were their weights, and Pythagoras found that those weights were related in the ratios 2:1, 3:2, and 4:3, presupposing that the vibration and sound of hammers are directly proportional to their weight, which is not the case. Pythagoras then took weights equaling those of the hammers and hung them from strings of equal length. He plucked the taut strings and heard the same intervals—another supposed discovery based on false premises, for the account incorrectly assumes that the frequency of vibration of a string is proportional to the number of units of weight hanging from it. However, it is easy to imagine Pythagoras, or his followers, or both, performing such experiments and considering, with more understanding and skill than those who later ignorantly repeated the tales, what could be learned from the successes and failures. The manner in which these stories came down in history as the way Pythagoras *made* the discovery could be an example of how knowledge is sometimes preserved while the manner of its discovery, and true understanding of it, are lost. Such a loss would be explained if, as some have supposed, the more sophisticated knowledge of Pythagoras was largely forgotten with the breakup of Pythagorean communities after his death.

Aristoxenus told a story having to do with another harmonic ratio experiment that involved Hippasus of Metapontum, and this experiment has particular significance because it is one of the reasons scholars are willing to attribute the discovery of the musical ratios to Pythagoras and his immediate associates. Hippasus, himself a contemporary of Pythagoras, made four bronze disks, all equal in diameter but of different thicknesses. The thickness of one "was 4/3 that of the second, 3/2 that of the third, and 2/1 that of the fourth." Hippasus suspended the disks to swing freely. Then he struck them, and the disks produced consonant intervals. This experiment is correct in terms of the physical principles involved, for the vibration frequency of a free-swinging disk is directly proportional to its thickness. Whoever designed and executed this experiment understood the basic harmonic ratios, or learned to understand them from doing the experiment, and the way the story was told suggests that the musical ratios were already known and Hippasus made the four disks to demonstrate them. According to Aristoxenus, the musician Glaucus of Rhegium, one of Croton's neighboring cities, played on the disks of Hippasus, and the experiment became a musical instrument.

To Walter Burkert, a meticulous twentieth-century scholar, the blacksmith tales make "a certain kind of sense." In ancient lore, the Idaean Dactyls were wizards and the inventors of music and blacksmithing. According to Porphyry, Pythagoras underwent the initiation set by the priests of Morgos, one of the Idaean Dactyls. A Pythagorean aphorism stated that the sound of bronze when struck was the voice of a daimon—another connection between blacksmithing and music or magical sound. "The claim that Pythagoras discovered the basic law of acoustics in a smithy," writes Burkert, may have been "a rationalization—physically false—of the tradition that Pythagoras knew the secret of magical music which had been discovered by the mythical blacksmiths."[1]

WHEN THE PYTHAGOREANS, with their discovery of the mathematical ratios underlying musical harmony, caught a glimpse of the deep, mysterious patterned structure of nature, the conviction became overwhelming that in numbers lay power, even possibly the power that had created the universe. Numbers were the key to vast knowledge—the sort of knowledge that would raise one's soul to a higher level of immortality, where it would rejoin the divine.

However revolutionary, one of the most significant insights in the history of knowledge had to be worked out, at the start, in the context of an ancient community, ancient superstitions, ancient religious perceptions, without any of the tools or assumptions of later mathematics, geometry, or science, without any scientific precedent or a "scientific method." How *would* one begin? The Pythagoreans turned to the world itself and followed up on the suspicion that there was something special about the numbers 1, 2, 3, and 4 that appeared in the musical ratios.

Those numbers were popping up in another line of investigation they were pursuing.

They had at their fingertips a simple but productive way of working with numbers. Maybe at first it was a game, setting out pebbles in pleasing arrangements. Most of the information about "pebble figures" and the connections with the cosmos and music that the Pythagoreans found in them comes from Aristotle. He knew about Pythagorean ideas of "triangular numbers," the "perfect" number 10, and the *tetractus*.

The dots that still appear on dice and dominoes are a vestige of an ancient way of representing natural numbers, the positive integers with which everyone normally counts. Dots and strokes stood for numbers in Linear B, the script the Mycenaeans used for the economic management of their palaces a thousand years before Pythagoras, and also in cuneiform, an even older script. Pebble figures were a related way of visualizing arithmetic and numbers, but they seem to have been unique to the Pythagoreans.

By tradition, Pythagoras himself first recognized links between the pebble arrangements and the numbers he and his colleagues had discovered in the ratios of musical harmony. Two of the most basic arrangements worked as follows: Begin with one pebble, then place three, then five, then seven, etc.—all odd numbers—in carpenter's angles or "gnomons," to form a square arrangement.*

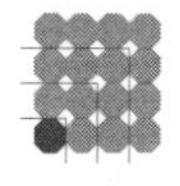

Or, begin with two pebbles and then set out four, then six, then eight, etc.—all even numbers—and the result is a rectangle.

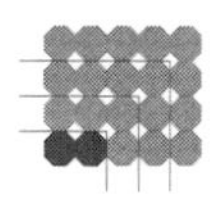

That is easier to understand visually than verbally, one reason to use pebbles.

Pythagoras and his associates were alert for hidden connections. The pebble figures of the square and rectangle dictated a division of the world of numbers into two categories, odd and even, and this struck

* A gnomon is an instrument for measuring right angles, like the device used by carpenters called a "carpenter's square."

them as significant. It was a link with what they were thinking of as the two basic principles of the universe, "limiting" and "limitless." "Odd" they associated with "limiting"; "even" with "limitless."

Another way of manipulating the pebbles was to cut a triangle from either the square or the rectangular figure.

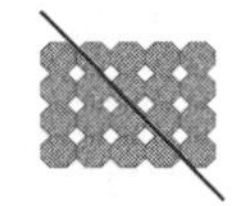

In the line of pebbles that then forms the diagonal or hypotenuse of the triangle, the pebbles are not the same distances from one another as they are in the other two sides, nor are they touching one another. Having all the pebbles in all three sides of a triangle at equal distances from their immediate neighbors, or all touching one another, requires a new figure: Set down one pebble, then two, then three, then four, with all the pebbles touching their neighbors. The result is a triangle in which all three sides have the same length, an equilateral triangle. Notice that the four

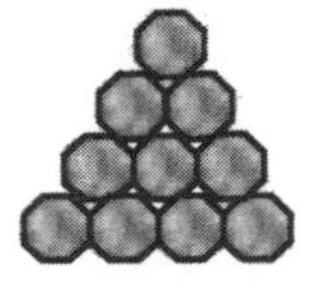

numbers in this triangle are the same as the numbers in the basic musical ratios, 1, 2, 3, and 4, and the ratios themselves are all here: Beginning at a corner, 2:1 (second line as compared with first), then 3:2, then 4:3. The numbers in these ratios add up to 10. The Pythagoreans decided 10 was the perfect number. They also concluded that there was something extraordinary about this equilateral triangle, which they called the *tetractus*, meaning "fourness." The *tetractus* was, in a nutshell, the musical-numerical order of the cosmos, so significant that when a Pythagorean took an oath, he or she swore "by him who gave to our soul the *tetractus*."

Most scholars think it was after Pythagoras' death that the Pythagoreans found they could construct a tetrahedron (or pyramid)—a four-sided solid—out of four equilateral triangles, and they probably knew this by the time Philolaus wrote the first Pythagorean book in the second half of the fifth century.* The word *tetractus*, however, was in use

* Not all pyramids have only four sides. The Great Pyramid that Pythagoras may have seen in Egypt is not a pyramid of this sort. It has five sides: a square base and four triangular sides.

during Pythagoras' lifetime. It hints that there was more "fourness" to the idea than the fact that 4 was the largest number in the ratios. The tetrahedron or pyramid is a solid in which each face is a *tetractus*, but which also uses the number 4 in other manners—4 faces, 4 points.

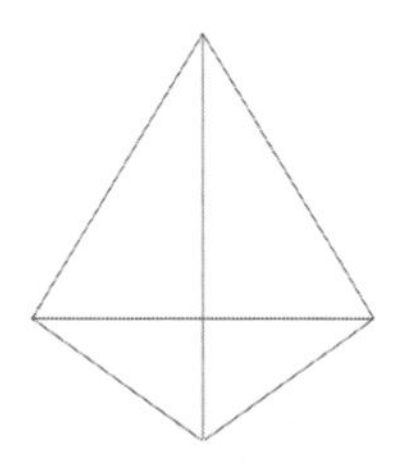

When Aristotle, in the fourth century B.C., was researching the Pythagoreans, he found a list of connections they made between numbers and abstract concepts. He apparently could not discover what they connected with the numbers 6 and 8.

1 Mind
2 Opinion
3 The number of the whole
4 Justice
5 Marriage
6 ?
7 Right time, due season, or opportunity
8 ?
9 Justice
10 Perfect

It is not difficult to understand how Mind might be 1 and Opinion 2. Justice appears twice because of an association with squareness. The Greeks did not think of 1 as a number. "Number" meant plurality, more than 1. So, for them, the smallest number that is the square of any whole number was 4.* The first number that is the square of an odd number is 9, and that, too, they associated with justice. The idea that

* In some later ancient mathematics, whose roots can be traced to the "Pythagorean" tradition and which by some scholars' interpretation existed separately and in parallel with the Euclidean tradition, the number 2 also had no status as a "number." It was not considered even or odd or prime. Like "1," it was not a number at all, but the "first principle of number."

"square" means an evened score—with all need for retaliation at an end—still shows up in the colloquial phrase "That makes us square." Marriage (5) was the sum of the first odd and even numbers (2 and 3). The link between 7 and "right time" or "due season" reflected wider Greek thought. Life happened in multiples of 7. A child could be born after 7 months in the womb, cut teeth 7 months later, reach puberty at 14, and (if a boy) grow a beard at 21.

The Pythagoreans followed one line of thought that seems particularly odd today, accustomed as most of us are to thinking of squares and cubes of numbers but not of other geometric shapes possibly connected with them in a similar manner. The "square" of 4 was 16, but the "triangle" of 4 was 10, the perfect number. Both ideas were equally picturable with

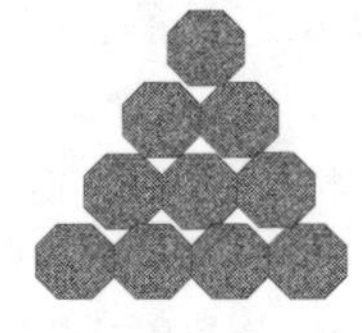

pebbles. Stacking the pebbles so as to discover that the "cube" of 4 was 64, you might just as easily pile them up another way so that the "pyramid" of 4 was 20. Montessori teaching exploits the delight of playing games like this with little objects like pebbles—in the case of Montessori, beads.

Having come to the conclusion not only that numbers, but the specific numbers 1, 2, 3, and 4 and the ratios between them were the primordial organizing principle of the universe, Pythagorean thinking moved in other directions, some of which seem strange and primitive, but it is not surprising that they overestimated the simplicity of the rationality they had glimpsed and were too expectant of immediate applications and results. They were not unlike the earliest followers of Jesus, coming away from what was for them a transforming experience and trying to apply it to the everyday world, thinking all would be resolved soon. The Pythagoreans had discovered a new road to "truth." Great thinkers thought about truth and proposed answers. Only a shaman—and many regarded Pythagoras as what we today would call a shaman—was sure he had the answer. In fact, Pythagoras and his followers *did*, but they traveled their new road weighted down with ancient baggage. Still in the age of oracles, divination, and mystic utterances, with its preconceptions about the universe and nature, their naive conception of the world carried over into a naive conception of the power of numbers.

THE HALCYON DAYS in Croton lasted thirty years. Iamblichus' biography included long lists of names, which he probably got from Aristoxenus, of Pythagoras' first followers, who sat at his feet, heard his teaching, argued points and worked out problems with him, played with the pebbles, and experimented with the *kanon* and with hanging disks. Was the young physician "Alcmaeon" really one of them? Was there actually a "Brontinus" who was husband and/or father of Theano? Were "Leo" and "Bathyllus" real people? And what of the "Pythagorean women," about whom nothing is known but their names on these lists? Frustratingly, there is no specific surviving information about how the new coinage affected the economy or, except the story of Milo's defeat of Sibaris, about Pythagorean leadership in Croton and the surrounding territory, what offices the Pythagoreans held, or exactly in what capacity they wielded their power—only that they did wield it and that the results were by most accounts beneficial to the region. What is clear is that in about 500 B.C., three decades after Pythagoras arrived in Croton, hostility among the populace and perhaps a coup within the ranks of his followers brought it all to an end. The information is confused and contradictory, with common themes being others' suspicion that Pythagoras and his followers were either becoming too powerful politically or aspiring to too much power—and, oddly, an unusual respect for beans.

According to Diogenes Laertius, Pythagoras was visiting with friends in Milo's home when someone deliberately set fire to the house. The arsonists were either Crotonians who feared that Pythagoras might "aspire to the tyranny" or envious, disgruntled people who thought they should have been included in this gathering but had not been deemed "worthy of admission." Pythagoras escaped but was captured and killed when he avoided crossing a bean field and took a longer way around. He must have decided, Diogenes Laertius said, that death was preferable to trampling on beans or speaking with his pursuers. About forty of his companions died as well.

Diogenes Laertius was interested in conflicting accounts, so he also reported a story he got from Hermippus, portraying Pythagoras and his "usual companions" in a militaristic light. They had joined the Agrigentine army to fight the army of Syracuse. The Syracusans put them to flight and captured and killed Pythagoras as he was making a detour around a bean field. Being less squeamish about trampling on beans did not help his companions. About thirty-five were caught and burned at the stake in Tarentum, accused of trying to set up a rival government in opposition to the prevailing magistrates.

Diogenes Laertius showed he had a rather macabre sense of humor by casting part of this story into verse in another of his "jesting epigrams."

Alas! alas! why did Pythagoras hold
Beans in such wondrous honor? Why, besides,
Did he thus die among his choice companions?
Here was a field of beans; and so the sage,
Died in the common road of Agrigentum,
Rather than trample down his favorite beans.

Two other endings to the story came through Diogenes Laertius from the trustworthy Dicaearchus and Heracleides Ponticus; in these, Pythagoras escaped his pursuers but died soon thereafter in Metapontum of self-imposed starvation.*

Porphyry gave a more detailed description of what supposedly happened, based on Aristoxenus, naming names and filling in the gaps in the other stories, and Iamblichus had some of the same specifics. According to this fuller account, the huge success of Pythagoras and his associates, and particularly their role in the administration and reform of the cities, aroused envy, most notably and ominously from one Cylon. He was a wealthy community leader of impeccable breeding, but also of a "severe, violent and tyrannical disposition," and he controlled a large group of loyal supporters. He had a high opinion of himself, "esteemed himself worthy of whatever was best," and assumed he would be welcomed to the Pythagorean fellowship. When Cylon approached Pythagoras, "extolled himself," and tried to converse, Pythagoras peremptorily "sent him about his business." Pythagoras, Porphyry pointed out, "was accustomed to read in the nature and

* Heracleides Ponticus is not to be confused with the earlier Heraclitus who so severely criticized Pythagoras. Heracleides Ponticus lived in the fourth century B.C. and was a pupil of Plato.

manners of human bodies the disposition of the man." Cylon did not take the rebuff gracefully. He assembled his cronies and instigated a conspiracy against Pythagoras and his followers. According to Iamblichus it took some time for Cylon to bring his plans to fruition because of the Pythagoreans' power and the trust placed in them by the citizens of the various cities. Accounts more sympathetic to Cylon had him as the leader of a group that opposed the oppressive ultra-conservatism of the Pythagoreans.

Iamblichus suggested that Hippasus—who invented the demonstration of the musical ratios using the disks of different thicknesses—may have played a subversive role. Before Cylon's attack, Hippasus was, according to Iamblichus, one of a faction, among the insiders, who disagreed with Pythagoras and the more orthodox members of the school. He urged Pythagoreans who were playing prominent roles in governing the cities to adopt more democratic policies. He may have attempted to stir up popular feeling against Pythagoras' leadership, playing into Cylon's hands.

W. K. C. Guthrie described, with good understanding of human nature, the complicated political situation that probably contributed to the death or exile of Pythagoras:

> This combination of forces seems to have been due on the one hand to popular discontent with the concentration of power in the hands of a few, coupled with the ordinary man's dislike of what he considers mumbo-jumbo, and on the other to the native aristocracy's suspicion of the Pythagorean coteries, whose assumption of superiority and esoteric knowledge must at times have been hard to bear.[2]

Porphyry took the longest and most dramatic version of the story from Dicaearchus. Pythagoras was with his friends in Milo's house when Cylon's men set it afire. Pythagoras' most devoted followers threw themselves into the flames to make a bridge with their bodies for the elderly sage to cross and escape. He and a remnant of survivors then tried to reach the city. Fleeing along the road, the others were gradually picked off by their pursuers, but Pythagoras, protected by them as much as possible during their flight, managed eventually to make his way to the harbor of Caulonia and from there to Locri. The Locrians refused him sanctuary. Perhaps they sensed that the days of Pythagorean preeminence had come to an end and feared retribution from Cylon if they sheltered him. Or perhaps they feared Pythagoras himself, for, as the story goes, their message to him as they turned him away was that they admired his wisdom but liked their present condition and way of life and did not wish to change.* In any case, the story has it that they sent some old men to intercept him before he could reach their gates and tell him that the Locrians would give him food and supplies but he must "go to some other place." Pythagoras sailed to Tarentum, then back to Croton. The Crotonians also sent him away. Everywhere, as Porphyry reported Dicaearchus' words, "mobs arose against him, of which even now the inhabitants make mention, calling them the Pythagorean riots." In Dicaearchus' account, Pythagoras eventually found asylum in the temple of the Muses in Metapontum, where he starved himself to death, grieving for the friends who had perished trying to save him.

The people of Metapontum prefer another ending. According to the tradition in that city, after Pythagoras arrived as a refugee from Croton, he settled down and established a school. After his death, his house and school were incorporated into a temple of Hera. Fifteen columns and sections of pavement from that temple still remain today in Metapontum, called the Palatine Tables, because knights (paladins) in the Middle Ages assembled there before setting off on the Crusades. In the first century B.C., when Cicero visited Metapontum, people could still identify the house where they believed Pythagoras had lived. Cicero wrote about how moved he was when he visited it.

Porphyry lamented that most of what Pythagoras taught died with him and his closest followers. "With them also died their knowledge," he wrote, "which till then they had kept secret except for a few obscure things which were commonly repeated by those who did not understand them." Iamblichus wrote that the cities hardly mourned Pythagoras at all or took much notice of what had happened, though in truth they had lost "those men most qualified to govern." "Then science died in the breasts of its possessors, having by them been preserved as something mystic and incommunicable."

* Part of their "present condition" was an economy that was more primitive than Croton's. They used no coinage, and would not until more than a century later. See W. K. C. Guthrie (2003), p. 178 n.

CHAPTER 6

"The famous figure of Pythagoras"

Sixth Century B.C.

IN THE FIRST OR EARLY SECOND century A.D., Plutarch, the author of the famous *Parallel Lives*, and his team of researchers tried to find the earliest reference connecting Pythagoras with the "Pythagorean theorem."* They came upon a story in the writing of a man named Apollodorus, who probably lived in the century of Plato and Aristotle, that told of Pythagoras sacrificing an ox to celebrate the discovery of "the famous figure of Pythagoras."† Plutarch concluded that this "famous figure" must have been the Pythagorean triangle. Unfortunately Apollodorus was no more specific than those words "the famous figure of Pythagoras"—which probably indicates that it was so famous he had no need to be.

A modern author could also write "the famous figure of Pythagoras" and be as certain as Apollodorus apparently was that no reader would think of anything but the "Pythagorean triangle." Even nonmathemati-

* "Theorem" has implications, in modern terminology, that do not apply to the earliest knowledge of this rule. With that in mind, this book will nevertheless continue to use "theorem" to avoid seeming to mean something different from what everyone calls the Pythagorean theorem.

† There were more than one Apollodorus, but this one was probably Apollodorus of Cyzicus, who lived in the fourth century B.C.

cians can often recall the "Pythagorean theorem" from memory: the square on the hypotenuse of a right triangle is equal to the sum of the squares on the other two sides. For millennia, anyone who had reason to know anything about this theorem thought Pythagoras had discovered it.

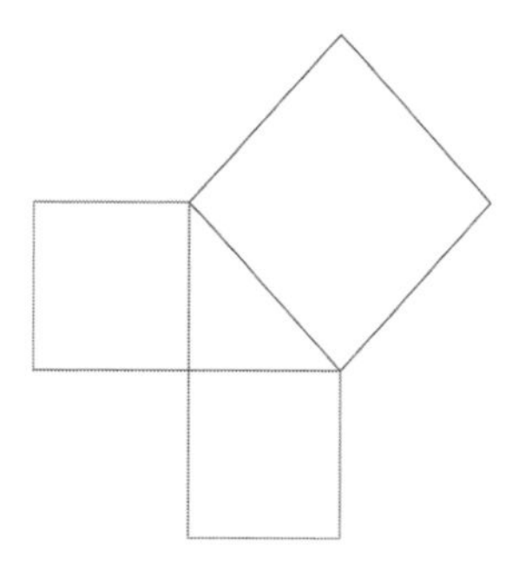

For many who learned the formula in school and always thought of it only in terms of squaring numbers, rather than involving actual square shapes, it came as an almost chilling revelation when Jacob Bronowski in his television series *The Ascent of Man* attached a square to each side of a right triangle and showed what the equation really means. The space enclosed in the square "on the hypotenuse" is exactly the same amount of space as is enclosed in the other two squares combined. The whole matter suddenly took on a decidedly Pythagorean aura. Clearly this was something that might indeed have been discovered and is true in a way that does not require a trained mathematician or even a mathematical mind to recognize. In fact, using numbers is only one of several ways of discovering it and proving it is true.

Bronowski pointed out that right angles are part of the most primitive, primordial experience of the world:

> There are two experiences on which our visual world is based: that gravity is vertical, and that the horizon stands at right angles to it. And it is that conjunction, those cross-wires in the visual field, which fixes the nature of the right angle.[1]

Bronowski did not mean that experiencing the world in this way necessarily leads immediately, or ever, to the discovery of the Pythagorean theorem. Indeed, all over the ancient world, long before Pythagoras, right angles were used in building and surveying, and right triangles appeared decoratively.* Without drawing tools, a draftsman can produce right triangles, and a skilled draftsman can produce right triangles that no human eye can see are not absolutely precise—this without knowledge of the Pythagorean theorem. Just as tuning a harp is an "ear thing"—and was, long before anyone understood the ratios of musical harmony—the use of right triangles in design was an "eye thing." Such judgments of harmony, figures, and lines are intuitive for human beings, and the mathematical relationships that lie hidden in nature and the structure of the universe often manifest themselves in the everyday world in useful ways long before anyone thinks of looking for explanations or deep relationships.

Yet at certain times and places in history and prehistory—for reasons about which it is only possible to speculate—circumstances have been right to call forth a longing to look beyond the surface. Among the Pythagoreans there was a strong and unusual motivation. Investigation like this was the road by which one could escape the tedious round of reincarnations and rejoin the divine level of existence. One cannot summarily dismiss the tradition that they discovered the theorem, though, contrary to popular belief for centuries, they were definitely not the first to do so.

No one knows how or when the "Pythagorean theorem" was first discovered, but it happened long before Pythagoras. Archaeologists have found the theorem on tablets in Mesopotamia dating from the first half of the second millennium B.C., a thousand years before his lifetime. It was already so well known then that it was being taught in scribal schools. In other regions, evidence of early knowledge of the theorem is less conclusive but still interesting. Egyptian builders knew how to create square corners with an astounding and mysterious degree of precision, perhaps by using a technique that earned them the nickname "rope pullers" among their Greek contemporaries. There is a hint about what that meant, perhaps, from circa 1400 B.C. in a wall painting in a tomb at Thebes, where Porphyry's story had Pythagoras spending most of his time while in Egypt. The painting shows men measuring a

* The claim has never been that Pythagoras discovered the right angle or right triangle, but that he discovered the relationship between the three sides of a right triangle—what we call the Pythagorean theorem.

field with what looks like a rope with knots or marks at regular intervals.[2] Possibly they were using the rope to create right angles, taking a length of rope 12 yards long, making it into a loop, and marking it off with three notches or knots so as to divide it into lengths 3, 4, and 5 yards long. Three, four, and five are a "triple" of whole-number unit measurements that create a right triangle, and holding the loop at the three marks and pulling it tight would have given them one. The knots or marks in the Thebes wall decoration are not clearly spaced at those intervals, but that could be because the artist was no surveyor.

1400 B.C. wall painting at Thebes depicting men measuring a field

The Egyptians left no instructions about "rope puller" techniques, and knowledge of the 3–4–5 triplet is no clear indication that they knew the theorem that made deeper sense of it. They had another method of getting right angles that involved no ropes at all. The *groma* was a wooden cross suspended from above so that it pivoted at the center. A plumb bob was hung from the end of each of the four arms; a surveyor or builder sighted along each pair of plumb bob cords in turn, then turned the entire device ninety degrees and repeated the sighting, and finally adjusted one of the cords to make up half of the difference. The result was a precise right angle.

In India, right triangles appeared in the designs on Hindu sacrificial altars dating from as early as 1000 B.C.[3] A collection of Hindu manuals called the Sulba-Sûtras ("Rules of the Cord"), dating from between 500 and 200 B.C., told how to construct these altars and how to enlarge them while retaining the same proportions. In times of trouble, enlarging the altar was a way of seeking surer protection from the god or gods, and getting the right response depended on keeping the exact proportions. Builders attached cords to pegs set in the ground, as bricklayers do today, hence "rules of the cord." The Pythagorean theorem does not appear in the manuals, but the writers seem to have been aware of it. Knowledge originating in Greece in the sixth century B.C. could possibly have reached India, for instance with Alexander the Great's armies in about 327 B.C. It is not too far-fetched to speculate that it did. The Cynic philosopher Onesicritus traveled with Alexander, and in his records he mentioned being questioned by an Indian wise man about Greek learning and doctrine. One of the matters they discussed was the Pythagorean avoidance of eating meat.[4] By the time Onesicritus had that discussion, the Pythagorean theorem was well known in the Greek world and almost certainly known to him. However, there is more to the Indian case. Though the written manuals date from after Pythagoras' lifetime, records exist of similar altars, and of their proportional enlargement, from several centuries earlier. No instruction manuals survive from that time and it is plausible that the writers of the later manuals were applying new understanding to an ancient art. One only need witness the astounding facility with which the humblest, most isolated, illiterate Indian woman today is able to create highly elaborate symmetrical geometric designs with painted powders on her doorstep, referring to a small pattern held down by a stone nearby, to question whether an understanding of mathematical geometry was necessary to create an intricate design and enlarge it while retaining the original proportions.

In Mesopotamia, however, the evidence is irrefutable that the theorem was known and understood in the early second millennium B.C.[5] We have not the vaguest hint about who discovered it or how, or how useful it was. School lessons on tablets measured gates and grain piles, and one grain pile was so amazingly large that the lesson problem was clearly set out only as an exercise, not with a real pile in mind—though probably with the goal of equipping pupils to put the same number skills to work in real-life, practical situations.[6]

The twentieth-century discoveries about the theorem's Mesopotamian origins began in 1916 when Ernst Weidner studied a Mesopotamian school tablet labeled VAT 6598, dating from the Old Babylonian period in the early second millennium B.C. The two final problems that he could read on the tablet, part of which was missing, required calculating the diagonal of a rectangle and showed methods for doing that. These did not include the Pythagorean theorem, but Weidner assessed the accuracy of the methods and compared them with the theorem, alerting

The text tablet labeled Plimpton 322

archaeologists and mathematicians to the possibility that it was known more than a thousand years before Pythagoras.

In 1945, a text that archaeologists have labeled Plimpton 322 came to light, listing fifteen pairs of what would later be known as Pythagorean triples—three whole numbers that, when used as the measurements of the sides of a triangle, produce a right triangle.[7] The smallest Pythagorean triples are 3–4–5 and 5–12–13.* The list took the ancient scribes into large numbers. While the Plimpton 322 list was not airtight evidence that its makers knew the Pythagorean theorem, it was further evidence of the possibility.

In the 1950s, the Iraqi Department of Antiquities excavated a site known as Tell Harmal near the location of ancient Babylon†—a town called Shaduppum that had been an administrative complex under kings ruling just before the great lawgiver Hammurabi, during the First

* You can think of 3–4–5 as 3 inches, 4 inches, and 5 inches, though it could just as well be centimeters, miles, parsecs or any other unit of measurement.

† Unfortunately, most of Babylon of the early second millennium B.C. cannot now be excavated because it is well below the water table.

Babylonian Dynasty (1894–1595 B.C.).* Modern Baghdad has sprawled out so far that the area where Tell Harmal is situated is now one of its suburbs, but during the First Babylonian Dynasty, Shaduppum was a heavily fortified independent community. The Iraqi archaeologists uncovered massive walls buttressed with towers, a temple with life-sized terra-cotta lions at its entrance, captured in mid-roar, and, across the street from the temple, buildings that had been the primary administrative center and had included a school for scribes. The cuneiform documents buried among its rubble were not only administrative texts, letters, and a law code, but also long lists of geographical, zoological, and botanical terms, and mathematical material. Many of these tablets were, like Weidner's VAT 6598, school texts, used and copied by pupils with differing degrees of skill and sloppiness at the scribal school. They amounted to a cross section of Babylonian knowledge at its height, four thousand years ago. One tablet revealed that the scribes of that era understood right triangles, square roots, and cube roots, and were using them in a manner that implied familiarity with the Pythagorean theorem.[8]

In the 1980s, Christopher Walker of the British Museum made an extraordinary find, not at an archaeological dig but in the museum's vast, disorganized collection of tablet fragments. A piece labeled BM96957 turned out to be a "direct join" to the tablet Weidner had written about in 1916. The two pieces together present three problems and three methods of solving them. The third method, found only on Walker's BM96957, is the Pythagorean theorem. (See the box for a near translation of a part of the text.)

Was the mathematical knowledge that scribal students were mastering in the first half of the second millennium B.C. still available in Babylon in the sixth century B.C., in the neo-Babylonian era, when Iamblichus thought Pythagoras visited Babylon? We tend to assume that knowledge once discovered stays discovered, but much can happen to knowledge in a thousand years, particularly in as politically unstable a region as this. For example, sophisticated building techniques used routinely by the Romans were unknown to even the most brilliant architects and builders in the Middle Ages and early Renaissance, and were being discovered as though for the first time as late as the fifteenth

* The tablet is in the Iraq Museum in Baghdad, listed in the register as 55357.

The Babylonians used the sexigesimal place value system, not the decimal—that is, their number system was based on sixes, not tens. (The modern system of counting hours, minutes, and seconds is derived from it.) In the drawing and text below, the portions in brackets are a conjectural reconstruction by Eleanor Robson, based on the contents of the rest of the tablet. The italicized numbers in brackets give the equivalents in the decimal system. The drawing is not to scale, nor was it on the tablet. The length of the diagonal is an irrational number. It is 41 plus an infinite string of numbers after the decimal point. The author of the tablet satisfied himself with an imprecise measurement of the diagonal. The measurement is of a rectangular gate, lying on its side, so that "height" refers to the longest side.[9]

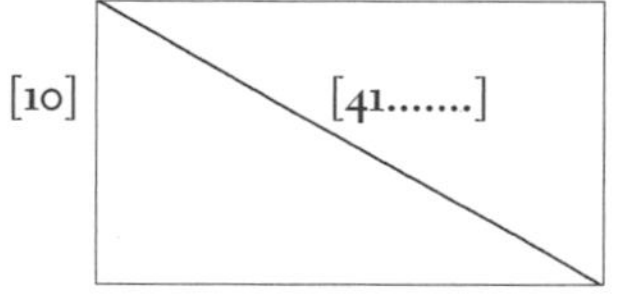

[What is the height? You:] square [41. . . . . . , the diagonal]. 28 20 (*1700*) is the squared number. Square [10, the breadth]. You will see 1 40 (*100*)

[Take] 1 40 from 28 20 (*1700 minus 100*) [26 40 (*1600*) is the remainder.]

What is the square root? The square root is 40.

This solution definitely used the theorem we now call Pythagorean. In modern terminology: The breadth of the gate is 10, which squared is 100. The height of the gate is 40, which squared is 1600. The length of the diagonal of the gate is a number close to 41. The square of that number is 1700. 1600 + 100 = 1700

century.[10] Knowledge of a thirty-geared, hand-operated mechanical computer known as the Antikythera Mechanism used by the Hellenistic Greeks in 150–100 B.C., and the technological understanding necessary to manufacture and use it, were likewise lost, and a thousand years passed before anyone even thought of the possibility of such an invention again.[11]* The tablets at Shaduppum disappeared in the rubble before 1600 B.C. and were, in the time of Pythagoras, lying right where archaeologists would find them in the twentieth century A.D.

There are very few school mathematics tablets dating from 1600–1350 B.C., and another evidential gap 1100–800 B.C. The historian and Assyriologist Eleanor Robson, who has given these issues more thought than perhaps any other modern scholar, listed several possible explanations, but concluded that "the collapse of the Old Babylonian state in 1600 B.C.E. entailed a massive rupture of all sorts of scribal culture. Much of Sumerian literature was lost from the stream of tradition, it seems, and most of Old Babylonian mathematics too."[12]†

Although Robson believes that the later Babylonians were probably ignorant of the achievements of Old Babylonian mathematics, it is likely that useful fallout from that lost knowledge, such as a triple that was handy for finding right angles, would have remained in use in Mesopotamia and elsewhere for centuries, without those who utilized it remembering the hidden relationship among the numbers.[13] And even if Pythagoras never visited Babylon, Greece was no wasteland when it came to building and surveying: Eupalos' astounding water tunnel on Samos was built in Pythagoras' century, as were many magnificent Greek temples. Though Pythagoras and his followers were not the first to know the theorem, their discovery might have been an independent discovery, or linked only by some surviving vestige of the more ancient, lost knowledge.

Pythagoreans in possession of the triple 3–4–5, wherever they learned it, and recognizing its usefulness, would not have let matters

* This mechanism, used probably in preparing calendars for planting, harvesting, and religious observances, was discovered in the wreck of a Roman ship that sank off the island of Antikythera in about 65 B.C. It was more technically complex than any known instrument for at least a millennium afterward.

† Political and social upheaval may have created disruptions. Or the fault may lie with modern scholarship, for few sites have been dug from these periods. They do not attract many scholars, partly because the documents are terribly difficult to decipher. Furthermore, as the very complicated cuneiform script gave way to alphabetic Aramaic, documents tended to be written on perishable and recyclable materials. The old Sumerian, Akkadian, and the cuneiform script were used for fewer purposes, mathematics apparently not being one of them, and even where cuneiform was used, it was often on wax-covered ivory or wooden writing boards that were erased for reuse or have not survived.

Think of land: Pythagoras had, after all, grown up in a Geomoroi family on Samos, and the Geomoroi got their name from the way they laid out their land. Take 9 plots of land, add 16 more, and you have 25 plots, as you can see if you draw them.

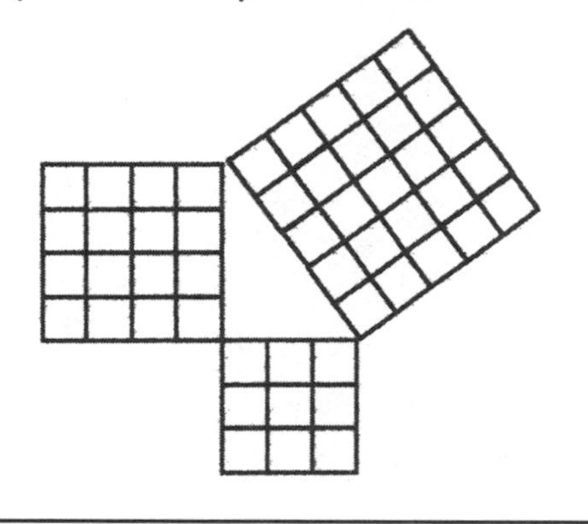

rest there. And if they set their minds to looking for a meaningful connection among the three numbers, it would arguably not have taken long to find the theorem. They could not have done it with pebbles, which they used as counters rather than as units to measure distances, but the same visualization that made pebbles interesting would soon have arrived at something like the diagram above, where squaring the numbers in the triple reveals the hidden relationship.

If the Pythagoreans found this relationship, having already discovered the harmonic ratios, they must have felt as though lightning had struck twice, for here was another stunning example of the hidden numerical rationality of the universe. Believing so strongly in a unity of all things, they would have been quick to jump to the correct conclusion that this same pattern of hidden connections had to apply to all right triangles—perhaps even to the incorrect conclusion that it had to be true of all triangles.

The second part of the tradition that has Pythagoras and his early followers discovering the theorem was that afterward a sword of Damocles hung over their heads. The universe had a cruel surprise in store for them, "incommensurability." Most right triangles have no whole-number triplet like 3–4–5. For example, according to the theorem, a right triangle with sides measuring 3 inches and 3 inches, each of which squared is 9, must have a third side—a hypotenuse—the length of which squared is 18.

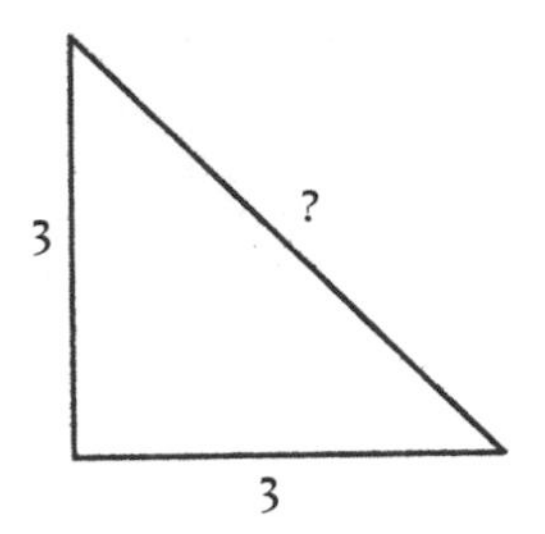

However, it is no simple matter to find the square root of 18 and the length of that hypotenuse, for the square root of 18 does not exist among whole numbers or fractions. An isosceles triangle like this one was a nightmare for a community of scholars who believed in a rational universe based beautifully and neatly on numbers. They could see that it did indeed exist and was a right triangle. It was not something hypothetical hanging fuzzily out in conceptual space. It was the triangle they got when they drew a diagonal from corner to corner of a square. But no subdivision of the length of the sides (neither inch, nor centimeter, nor mile, nor any fraction thereof) divided evenly into the length of the diagonal. More generally, though it might seem that for any two lengths you might try to measure there would be some unit that would divide into both of them and come out even with no remainder—maybe not the inch or centimeter or any length that has a name, but some unit, however small—the fact is that this is not the way reality works in this universe. Nor does the problem of incommensurability exist only with isosceles triangles. It was also true of the "gate" measured on the Babylonian tablet. The Babylonians knew about the triples and also apparently accepted that the units measuring the diagonal most of the time did not come out even.

The early Pythagoreans may well have discovered the problem, but it is far less likely that they found the solution—irrational numbers—or that they would have liked it if they had.* Irrational numbers are not neat or beautiful like whole numbers. An irrational number has an

* A rational number is a whole number or a fraction that is made by dividing any whole number by another whole number: ½, 4/5, 2/7, etc. An irrational number is a number that cannot be expressed as a fraction, that is, as a ratio of two whole numbers. The square root of 2 was probably found by Pythagoreans, working from their theory of odd and even numbers, possibly as early as about 450 B.C., and surely by 420, fifty to eighty years after Pythagoras' death. Plato knew of the square roots of numbers up to 17.

infinitely long string of digits to the right of the decimal point with no regularly repeating digit or group of digits.

The suggestion that the only information Pythagoras learned elsewhere was a vestige of the theorem—the triple 3–4–5—has in its favor that it solves a problem with the sequence of the Pythagoreans' discoveries. In order to have had a devastating crisis of faith resulting from the discovery of incommensurability, they had to have had the faith first, not the crisis, and indeed the tradition is strong that this was the order in which the discoveries were made. However, it is difficult to imagine anyone discovering the theorem from scratch (without the triple) while not simultaneously discovering the problem. Another possible sequence would have the Pythagoreans discovering incommensurability first, as they struggled with right triangles, and only later realizing that a few right triangles were, in fact, not incommensurable; but that is not the traditional sequence. A similar issue undermines the possibility that Pythagoras learned the whole theorem elsewhere: He would also have learned about incommensurability, so that it could have come as no surprise later.

Eleanor Robson is convinced, from evidence in the mathematics itself, that Old Babylonian mathematics was not the arithmetical precursor to early Greek mathematics; but a triple used in construction and surveying—the origin of which no one remembered—hardly represented the bulk of Old Babylonian mathematics.[14] It could have waited on the shelf while the Pythagoreans investigated harp string lengths and discovered musical ratios. The study of those was a new kind of thinking about numbers, what Aristoxenus meant when he wrote: "The numbers were withdrawn from the use of merchants and honored for themselves." A love of this sort of thinking could have led the Pythagoreans, after their musical discovery, to consider the 3–4–5 triple more carefully. Granted, this particular triangle was, by Pythagorean standards, not very interesting. Five does not show up in the basic ratios of music; 3, 4, and 5 do not add up to 10 or make the *tetractus*. No one is going to swear by this triangle! But the hidden connection . . . *that* was another reason to fall to one's knees, and perhaps to have a huge crisis of faith when you began looking at other right triangles.

Some Pythagoreans reputedly found other beauty in the triplet: They designated 5 as "marriage." The 5-unit side of this right triangle connected the 3-unit side and the 4-unit side. Thus, 5, or "marriage," connected 3 (which is odd) and 4 (even). "Odd" was male and "limiting";

"even" was female and "limitless." So this triangle was a manifestation of the harmony reconciling limiting and limitless. In the modern world, we associate such weak links with a different sort of mind from that which would come up with the Pythagorean theorem. In the ancient world, whose people were taking the first tentative steps toward understanding nature and the cosmos and the human condition, that distinction is invalid.

If Pythagoras discovered or knew the rule, did he prove it? Most historians of mathematics believe that the concept of "proof" as later understood was unknown before the Alexandrian Euclid introduced it in his *Elements*, around 300 B.C. The decision that something would "be true for every right triangle," for someone living as early as Pythagoras, would probably have been made on grounds other than a Euclidian proof. It would either have been an unsupported assumption or a guess, or a decision made in a scientific rather than a mathematical way—by testing it, as many times and with as many different examples as possible. The idea that mathematical statements should apply generally, though taken for granted today and implied in the Babylonian work, was not usually part of the ancient mind-set before Pythagoras. It is considered to be one of the great contributions of early Greek mathematics and probably a contribution of Pythagoras and his followers. With them, it might have been only an assumption based on their belief in the unity of all being, not something they could demonstrate at all or even thought it necessary to demonstrate.

There are, nevertheless, simple proofs of the theorem that some would like to attribute to the Pythagoreans, and one argument that they used such proofs is that these same thought sequences are good ways to discover the theorem, even if you had no concept of "proofs" after the fact. There is a geometry lesson in Plato's *Meno* that some think is traceable to Pythagoras. The clue is that Plato used it to demonstrate the "recollection" of what one learned before birth, an idea related to the Pythagorean doctrine of memory of past lives. The triangle in the *Meno* proof is the troublesome isosceles triangle, but the proof sidesteps the problem of incommensurability by using no numbers. It is admittedly difficult to imagine Pythagoreans being satisfied with any "truth" that used no numbers. It would have seemed the universe was thumbing its nose at them, with this triangle that provided such a clear and unequivocal demonstration of their rule, and that contained incommensurability.

The discussion of Plato's proof fits better in the context of a later chapter. Bronowski, in the book from his television series, showed another numberless proof that he believed Pythagoras may have used. Bronowski's clever proof is in the Appendix.

The right triangle was not the only pitfall in Pythagorean thinking where incommensurability lurked, but it was the most obvious. The scholarly argument about whether the Pythagoreans discovered it, and whether that caused a crisis of faith in the rationality of the universe, rambles on until it resembles that string of digits after the decimal place in an irrational number. However, in truth, an intelligent person thinking along Pythagorean lines and dealing with right triangles could hardly have missed discovering incommensurability. But only someone who reverenced numbers and the rationality of the universe would have been deeply troubled. Some have thought that Pythagoras and his followers reacted by retreating to a geometry without numbers—that what had early been an "arithmetized geometry" was reformulated in a nonarithmetical way, and this carried over into Euclid. In spite of the passage in Plato's *Meno*, and the suggestion that it reflected Pythagoras' proof of his theorem, nothing could seem more blatantly un-Pythagorean than a retreat from numbers![15]

Porphyry would have been pleased to learn that the earlier Mesopotamians knew about right triangles, the triples, and the theorem. His choice of possibilities would almost certainly have been that the theorem was known earlier but that Pythagoras' was an independent discovery, for he believed that several ancient peoples—he named the Indians, Egyptians, and Hebrews—possessed primeval, universal wisdom (*prisca sapientia* was the later term), and Pythagoras was the first to possess it in the Greek world.[16] The theorem is so intrinsic to nature, so beautifully simple, that it would be odd if no earlier triangle user in prehistory or antiquity got curious and figured it out.

What about a more startling suggestion: that Pythagoras had nothing whatsoever to do with the discovery? Could it be that it was later credited to him only because such legends tend to become associated with famous people? Over two thousand five hundred years, numerous achievements that were not remotely Pythagorean have been carelessly credited to Pythagoras. "Pythagorean" or "of Pythagoras" have become descriptive words connoting something clever that shows mathematical insight, with an overlay of wisdom, fairness, or morality.

A "Pythagorean cup," sold on Samos, punishes the immoderate drinker who fills it above a marked line, by allowing the entire cup of wine to drain out the bottom. Modern citizens of Samos are surprised—or at least pretend to be—that anyone would doubt this was an invention of Pythagoras. A "Pythagorean" formula predicts which baseball teams in America are likely to win. No one is insulted by doubts about that one.

A WORSE POSSIBILITY for Pythagoras' image is that he took the theorem from the Babylonians and claimed it as his own. According to Heraclitus, he "practiced inquiry more than any other man, and selecting from these writings he made a wisdom of his own—much learning, mere fraudulence."* It would certainly not have surprised Heraclitus if Pythagoras had stolen the Pythagorean theorem lock, stock, and barrel from the Babylonians. However, the fragments in which Heraclitus dismissed him as an imposter also placed Pythagoras high in the echelon of thinkers. Two of Heraclitus' other targets, Xenophanes and Hecataeus, were renowned polymaths. "Inquiry" meant not study in general but Milesian science. Most scholars think that Heraclitus had no basis for his attacks. He had an aversion to polymaths, and he was simply an ornery and contentious man being ornery and contentious. On another occasion he commented that "Homer should be turned out and whipped!"

If the Pythagoreans did come up with the theorem independently, the question remains whether credit should go to Pythagoras and his contemporaries or to later generations of Pythagoreans. Intemperate Heraclitus would not have been pleased to know that evidence coming from his own work places the appearance of the Pythagorean mathematical achievements in Pythagoras' lifetime: Heraclitus followed up on Pythagorean ideas about the soul and immortality and continued to develop the idea of harmony. For him, the lyre and the bow—Apollo's musical instrument and weapon—symbolized the order of nature. The bow was "strife," the lyre *harmonia*. The significance of the bow

* Though Heraclitus seems forthright and outspoken in the fragments about Pythagoras, he was known to be no easy read. His contemporaries dubbed him Heraclitus the Obscure and Heraclitus the Riddler. A story circulated in the time of Diogenes Laertius that when Socrates received a copy of a book by Heraclitus, he commented: "What I understand is splendid; and so too, I'm sure, is what I don't understand—but it would take a Delian diver to get to the bottom of it."

("strife") was original with Heraclitus, but the role of the lyre and *harmonia* were developments from Pythagorean thought, which suggests that the idea of connections between numerical proportions, musical consonances, and the Pythagorean numerical arrangement of the cosmos dated from the time of Pythagoras himself. Heraclitus was only one generation younger than Pythagoras.

In the first century B.C., the theorem seems to have been widely attributed to Pythagoras. A case in point: The great Roman architect Marcus Vitruvius Pollio, better known as Vitruvius, knew it well, attributed it without question to Pythagoras, and, in Book 9 of his ten-volume *De architectura*, mentioned the sacrifice to celebrate it. Apparently Vitruvius could write about Pythagoras as the discoverer of the theorem and assume that no one would gainsay him. He knew other methods of forming a right triangle, but found Pythagoras' much the easiest:

> Pythagoras demonstrated the method of forming a right triangle without the aid of the instruments of artificers: and that which they scarcely, even with great trouble, exactly obtain, may be performed by his rules with great facility.
>
> Let three rods be procured, one three feet, one four feet, and the other five feet long; and let them be so joined as to touch each other at their extremities; they will then form a triangle, one of whose angles will be a right angle. For if, on the length of each of the rods, squares be described, that whose length is three feet will have an area of nine feet; that of four, of sixteen feet; and that of five, of twenty-five feet: so that the number of feet contained in the two areas of the square of three and four feet added together, are equal to those contained in the square, whose side is five feet.[17]

WHERE, THEN, DOES this discussion end? In spite of the certainty that Vitruvius and his contemporaries shared, the most skeptical modern scholars think Pythagoras had nothing to do with the theorem at all. Others do not close the door to the possibilities that Pythagoras and/or his early followers may have made the discovery independently, unaware of previous knowledge of the theorem, or that they learned it elsewhere but were the first to introduce it to the Greeks.

My own conclusion is that there is no good reason to decide that

Pythagoras and the Pythagoreans had nothing to do with the theorem, and several meaningful hints that they did, including the fact that Plato chose to assume that right triangles were the basic building blocks of the universe when he wrote his *Timaeus*, the dialogue most influenced by Pythagorean thinking.* If earlier knowledge of the theorem had indeed been lost, then someone had rediscovered it at about the time of Pythagoras. Of all those who were aware of right angles and triangles and used them in practical and artistic ways, the Pythagoreans were unique in their approach to the world, apparently having the motivation and leisure to give top priority to ideas and study. Their intellectual elitism kept them focused beyond the nitty-gritty of "what works" on the artisan level, and their musical discovery led them to think beyond number problem solving for its own sake—causing them to turn their eyes beneath the surface and view nature in an iconic way. For the Pythagoreans (as for no others among their contemporaries), the theorem would have represented an example of the wondrous underlying number structure of the universe, reinforcing their view of nature and numbers and the unity of all being, as well as the conviction that their inquiry was worthwhile, and that their secretive elitism was something to be treasured and maintained. Has any other ruling class—and the Pythagoreans seem also to have been busy ruling—had that same set of priorities? Regarding the possibility that they began with the triple, I like the fact that their having it would not imply a continuum with the Old Babylonian mathematical tradition—a continuum that scholars like Robson have convincingly argued did not exist. And in this scenario, the Babylonian evidence, instead of pulling Pythagoras off the pedestal, actually suggests a way that he and his followers could have rediscovered the theorem in the time and place that tradition has always said they did without being disillusioned too soon by the discovery of incommensurability.

Bronowski credited Pythagoras with discovering the link between the geometry of the right triangle and the truth of primordial human experience. He echoed Plato's reverence for right triangles as the basics of creation when he wrote, "What Pythagoras established is a fundamental characterization of the space in which we move. It was the first time that was *translated* into numbers. And the exact fit of the numbers

* See Chapter 9.

describes the exact laws that bind the universe." Bronowski for that reason thought it not extravagant to call the "theorem of Pythagoras" "the most important single theorem in the whole of mathematics."[18]

But what of the ox? Did Pythagoras sacrifice it, or perhaps forty of them (some stories say), in thanksgiving for the discovery of the theorem? That Apollodorus referred to this "famous" story does not necessarily mean he believed it. Many dismiss the tale as impossible on the grounds that Pythagoras, who ate no meat, would not have sacrificed an ox. However, there is plenty of evidence that he had no objection to the slaughter of animals for ritual purposes. If vegetarianism is a clue, it may point in a different direction: Later Pythagoreans were more ready than early ones to believe that Pythagoras was a strict vegetarian. Burkert thought the existence of the sacrifice story "ought rather to be considered an indication of antiquity," weighing in on behalf of the argument that Pythagoras or his earliest followers made the discovery that spawned the tale. A later generation would not have made up this story about their hero.

PART II

Fifth Century B.C.–Seventh Century A.D.

CHAPTER 7

A Book by Philolaus the Pythagorean

Fifth Century B.C.

After Pythagoras' death, the demise of the Pythagorean brotherhood in southern Italy did not take place overnight or in a few short years. Many Pythagoreans survived the violence that ushered in the fifth century B.C., and the Pythagorean drama, minus its lead character, continued in the colonial cities. Only one Pythagorean is known for certain by name from that period: Hippasus of Metapontum. He was a scholar and perhaps a brilliant one, apparently part of the Pythagorean inner circle, who worked in music theory, mathematics, and natural philosophy and considered fire a first principle. One report credits Hippasus with constructing the dodecahedron, the twelve-sided solid—he "first drew the sphere constructed out of twelve pentagons." He may have taught the cantankerous Heraclitus. However, after Pythagoras' death, Hippasus fell from grace, and he is chiefly remembered as an ill-fated, perhaps ill-intentioned figure.

When the Pythagoreans discovered that mathematical relationships underlie nature, they did not announce this to the world. Secrecy was their custom. Hippasus, however, had to have been privy to the discovery, because he performed the successful experiment with bronze disks. Some stories connected him with the discovery of incommensurability, or even made him the unlucky discoverer. Accounts differed about how he erred, but somehow, while all good things were attributed to Pythagoras, all bad things seemed to get hung on Hippasus. His transgression was revealing a secret of geometry, or discovering incommensurability, or effrontery to the gods by making a discovery in geometry (that could have been the dodecahedron), or taking credit for a discovery instead of attributing it to Pythagoras.

The most nuanced and authentic-sounding material about Hippasus comes from Aristotle and Aristoxenus, and it links Hippasus with a fault line that developed in the Pythagorean brotherhood between two factions calling themselves the *acusmatici* and the *mathematici*.[1] The antagonism, which may have had its roots in a two-level hierarchy initiated by Pythagoras, the better to organize his brotherhood according to interests and abilities, split the community into opposing camps.

The *acusmatici* were devoted to rote learning. Their philosophy (quoting Iamblichus) "consisted of unproven and unargued aphorisms, and they attempted to preserve the things Pythagoras said as though they were divine doctrines." Scholars surmise that these aphorisms were relics of the most elementary, easily remembered part of Pythagoras' teaching. Some were folk maxims with added interpretations, with knowledge of the interpretations possibly serving as passwords or signifying rank in the community. There were three kinds of aphorisms (this according to Iamblichus): Some asked what something was: "What are the Isles of the Blessed? The sun and the moon." A second kind indicated superlatives: "What is most wise? Number." "What is most truly said? That men are wretched." A third concerned minutiae about "what one must do or not do." Many sounded pointless to anyone unaware of the secret interpretations. "Do not turn aside into a temple" meant "Do not treat God as a digression." "Do not help anyone put down a burden; rather, help him take it up" meant "Do not encourage idleness." "Do not break a loaf of bread" because "it is disadvantageous with regard to the judgment in Hades." Iamblichus threw up his hands at that and called it "far-fetched." *Acusmatici* evidently understood the connection. They claimed the title "Pythagorean" for themselves exclusively.

The *mathematici*, on the other hand, were willing to admit the *acusmatici* under the banner of the brotherhood, but they preserved and extended a different kind of Pythagorean knowledge. Though not always agreeing among themselves, they shared a conviction that the *acusmatici*'s refusal to allow knowledge to develop further was contrary to the spirit in which Pythagoreanism had been practiced when Pythagoras was alive.

According to Aristotle and Aristoxenus, Hippasus was one of the *mathematici*, or one of those who would be labeled *mathematici* when the groups became fully polarized. The opposing camp—those who would be known as *acusmatici*—frowned on his work as new and subversive. The *mathematici* might have been expected to defend Hippasus, but they were engaged in delicate maneuvers, insisting they were not introducing new doctrines, merely working on explication of the doctrines of Pythagoras. They disassociated themselves from Hippasus, to no avail since the *acusmatici* continued to accuse them of following him rather than Pythagoras. Hippasus was caught in the crossfire. His punishment, from the gods or the Pythagoreans, depending on which story to believe, was drowning at sea, expulsion from the community, and/or the construction of a tomb to him as though he were dead.

Hippasus' story provides a clue for dating some of the Pythagorean discoveries. Modern historians are skeptical about the claim that Hippasus taught Heraclitus, but they think the fact that many believed he did dates Hippasus reliably. He was supposed to have taught Heraclitus, not the other way around, and since Heraclitus' lifetime overlapped Pythagoras', Hippasus must have been an even earlier contemporary of Pythagoras. Furthermore, Hippasus' disk experiment had to have happened after the discovery of the ratios and before Hippasus' disgrace, which occurred (dated by the split in the brotherhood) shortly after Pythagoras' death. This chronology makes it impossible for the discovery of the musical ratios to have been made later, in the next generation. They were an authentically early Pythagorean discovery.

Hippasus' disgrace and the *mathematici/acusmatici* conflict are not the only evidence that some Pythagoreans who survived the turn-of-the-century upheavals stayed in Magna Graecia. Remnants of the brotherhood persisted throughout the region. Iamblichus had information that Pythagoras' "successor" was Aristaeus, who married his widow Theano, "carried on the school," and educated Pythagoras' children. A son of Pythagoras named Mnesarchus reputedly took over the school when Aristaeus became too old. If folk memory in Metapontum had it right, Pythagoras survived for a while in exile and established a school there.

Some Pythagoreans continued to hold, or rapidly regained, positions of political importance and possibly extended their influence over an even wider area than before, but as these leaders became influential again in the government of the cities, they courted disaster by ruling more and more autocratically. A revolution unseated them in mid-century, about 454 B.C. The second century B.C. historian Polybius repeated a description he found in earlier accounts: "The Pythagorean meeting places were burned down and general constitutional unrest ensued—a not unlikely event, given that the leading men in each state had been thus unexpectedly killed. The Greek cities in these regions were filled with bloodshed and revolution and turmoil of every kind." The result this time was a Pythagorean diaspora—to Thebes, to Phlius (near Corinth), to Syracuse, and elsewhere. The curtain fell for the second and final time on the Pythagorean golden age in Magna Graecia. The original community that Pythagoras had taught no longer existed.

In a larger context, the story had only begun. From about this time, there were two discernible contrasting strands of thought in the ancient Mediterranean world: "Ionian," from mainland Greece and that area of the Mediterranean; and "Pythagorean" or "Italian," stemming from southern Italy. Through members of Pythagorean refugee communities and their intellectual descendants—and men like Plato who were drawn to their ideas—the remnants of the thinking of an obscure ancient group became a powerful worldview. By late antiquity, no one could claim to be a serious thinker and ignore the "Pythagorean" or "Italian" school.

Meanwhile, the *acusmatici/mathematici* split nevertheless continued to infect the scattered brotherhood, and the disagreement about who reflected the spirit and work of the first Pythagoreans still causes difficulty for anyone trying to discern the truth about that earliest era. Most educated people through the centuries would insist that the *mathematici* were the true Pythagoreans, preserving and extending the great Pythagorean mathematical legacy. The reason for this certainty is that it was the *mathematici* tradition that Plato handed down to the future. He made the choice for Western civilization.

Aristotle, a generation later than Plato, was well acquainted with both Pythagorean varieties and described an *acusmatici* legacy that in addition to the aphorisms included the miraculous legends, the doctrine of reincarnation and Pythagoras' memory of his past lives. The *mathematici* legacy accepted most of that, too, but emphasized the different approach to the world and the soul through numbers, mathematics, and music. The *mathematici* had preserved historical information: that Pythagoras came to Croton during the reign of Polycrates on Samos and had a powerful influence on the leaders of his new home city. Aristotle never traced a heritage of knowledge and mathematics to Pythagoras himself by naming names in succeeding generations, but he had no quarrel with the *mathematici*'s claim that this unbroken heritage existed.[2] Plato attributed a sophisticated version of the Pythagorean *mathematici* number theory not just to Pythagoreans but to Pythagoras himself.

The second half of the fifth century B.C. (450 to 400) is still much alive in the cultural memory of the modern world. Greek tragedy had blossomed with Aeschylus and was continuing with the plays of Sophocles and Euripides, raising issues that need no modern context to make them relevant today. Aristophanes was scandalizing his delighted audiences, satirizing public affairs and leaders in brilliant, flagrantly indecent comedies. Though these would soon be dubbed "Old Comedy" as newer forms and subjects become fashionable, in the twenty-first century his *The Frogs* became a Broadway musical. The physician Hippocrates was working and writing, and medical school graduates more than two millennia later repeat the oath attributed to him. Athens, in mainland Greece, was picking up the pieces after a long conflict with the Persians and enjoying an interval of peace, growing rich from silver mines and tribute from other members of the Delian League, former allies in the Persian Wars. Unaware how short this respite would be before they became involved in the Peloponnesian Wars, Athenians restored their city, which the Persians had burned, and erected the Parthenon. In this half-century, Plato was born and grew to manhood, and Philolaus the Pythagorean, nearly fifty years Plato's senior, wrote the first Pythagorean book—or at least the first that was destined to survive.

Philolaus was one of the refugees who left Croton or Tarentum at mid-century. He settled in about 454 in Thebes, a powerful old city northwest of Athens whose ancient origins made her a favorite setting for Greek dramas. She had once been the seat of the real King Oedipus. Politically, Thebes' only consistent policy was hatred of Athens. She had sided against Athens in the Persian Wars and then collaborated with Sparta against her, an alliance that would last until nearly the end of the Peloponnesian Wars at the close of the century. Thebes and Sparta would finally part ways when Thebes suggested the defeated Athenians be totally annihilated and Sparta disagreed.

It would appear that Thebes was not a particularly serene location for a fledgling brotherhood to pursue peaceful studies, but Philolaus founded a new exile Pythagorean community there. He had either died or moved elsewhere by the end of the century—information that comes indirectly through Plato, who in his dialogue *Phaedo* had a character named Cebes comment, "I heard Philolaus say, when he was living in our city. . ." Cebes' city was Thebes, and this conversation was supposed to be taking place the day Socrates died in 399 B.C. If Philolaus was still alive then somewhere else, he was seventy-five, but that reference to him was the last that has survived.

Some time between 450 and 399 B.C., probably in Thebes, Philolaus set down an extensive written record of Pythagorean thought, something no Pythagorean had done before as far as anyone has been able to discover. The only traces of it today are fragments, mostly references in the writing of scholars during the first century A.D., long removed from his time.* In the nineteenth century there was controversy about whether Philolaus wrote a book and whether the fragments are genuine, but in 1893 a papyrus came to light with excerpts from a medical history by Menon, a pupil of Aristotle in the fourth century B.C., referring to a book by Philolaus that already existed then. Since that discovery, scholars have analyzed the Philolaus fragments in the context of the fifth century B.C., Philolaus' century, and they largely agree about which are authentic.[3]

Though it hardly seems fair to Philolaus, anyone looking for specifics about Pythagoras and what he taught is frustrated by the fact that Philolaus was a splendid thinker in his own right. He was writing his own book, not recording the discoveries or words of another man, and included his own thinking as well as what had evolved in the Pythagorean *mathematici* communities since Pythagoras' death. Nevertheless, Philolaus definitely considered himself a Pythagorean, and, given the time frame, much of the science and doctrine in his book must have been a direct reflection of Pythagoras and his earliest

* For historians, one use of the word "fragment" is for a quotation or reference in the writing of another author who had access to material that has since disappeared.

An anachronistic, late fifteenth-century A.D. drawing, from a music theory book by Gaffurio, reveals how scholars of that era envisioned Pythagoras (and Philolaus, who was not actually Pythagoras's contemporary) studying the ratios of musical harmony.

followers. Philolaus was almost a direct link, for Pythagoras had died or disappeared from public view in 500 B.C., only twenty-five years before Philolaus' birth. Philolaus' teachers and acquaintances as he grew up in Croton or Tarentum must have been almost exclusively Pythagoreans, and some of the older of them would have known Pythagoras.

Unfortunately, Philolaus treated all of his material as a unified body of knowledge, making no distinctions between earlier and later, between the time Pythagoras was alive and the time of Philolaus' writing, or between himself and others. He was not being careless. For a Pythagorean there was unity to truth, and unity to the search for it. The path to knowledge about the universe and the path to reunion with the divine were one and the same path. Truth about nature, and divine truth, were one and the same truth. In such a context, even if Pythagoras himself had not made a particular discovery, one could assume it had been implicit in his teachings. Furthermore, there was a form of ancient one-upmanship that Pythagoreans like Philolaus shared with their contemporaries. It was demeaning to an idea or discovery to call it new or original. Knowledge became more credible the older it was and the more it could be attributed to a great figure. Philolaus would have been loath to identify any source other than Pythagoras, even if it was himself.

Nevertheless, Philolaus was not without an agenda of his own. One of the clues that place his writing in the late fifth century B.C. was that he was trying to present Pythagorean ideas in a way that responded to a stalemate arising from "Eleatic" teaching.

The philosopher Parmenides was from Elea (hence "Eleatic"), a Greek colony north of Croton on Italy's west coast. According to Plato he was born in 515 B.C., but Greek chronicles say about 540. In either case, he was a younger contemporary of Pythagoras, but remarkably, in spite of the overlap of their lifetimes, the close proximity of Elea to Croton, and a passage in Plutarch that says Parmenides "organized his own country by the best laws," only one early source gave Parmenides even the remotest link with Pythagoras or the Pythagoreans. The link was indirect, in Diogenes Laertius' third-century-A.D. biography of Pythagoras:

> [Parmenides] was also associated (as Sotion said) with Ameinias, son of Diochaites, the Pythagorean, a poor man but of good character. It was rather Ameinias that he followed: when Ameinias died he set up a shrine for him (Parmenides came from a famous, wealthy family); and he was led to calm by Ameinias and not by Zeophanes.

It would seem that if Parmenides "followed" Ameinias, was "led to calm" by him, and thought so highly of him as to set up a shrine, then Parmenides' own thinking would show traces of Pythagorean ideas. Lured by this clue, scholars have repeatedly attempted to find elements of Pythagoreanism in Parmenides' writing, with no success.

In a paradoxical twist, history celebrates Parmenides for insights that he did not claim were correct; for example, that the light of the Moon "always gazing at the rays of the Sun" is reflected light.[4] He laid out such ideas in Part 2 of a beautiful, enigmatic poem, after he had warned in Part 1—a guide to the Way of Truth—that what he was going to present in Part 2 was "deceitful." He was not claiming to present "facts" or even opinions, only what human opinion on these matters might plausibly be at best.

He argued that those setting out on a voyage of "inquiry" probably mistakenly believed they had a choice between two subjects, things that existed and things that did not exist. But nonexistent things were unthinkable and unsayable, and inquiry into them was "a trail of utter ignorance." As for what existed, certain things had to be true about it: It had always to have existed, and it had to be indestructible. Otherwise there would be a chance it might at some time not exist, which was unthinkable and unsayable. It had also to be continuous in space and time (no gaps), unchanging and unmoving, and finite. Human senses told one otherwise, admitted Parmenides, but they could not be trusted. So much for any possibility of learning about the world by observing and experiencing it!

Melissus, another "Eleatic" philosopher, was an admiral from Samos, though his and Pythagoras' lives there did not overlap. As Aristotle told the story, in 441 B.C. Athens declared war on Samos. The Samians defeated Pericles himself in a sea battle, but Pericles survived and hostilities continued. When a stalemate dragged on, Pericles, bored and underestimating the Samians, led some of his ships away on an expedition. Melissus, commanding Samos' fleet, took this opportunity to attack, "despising the small number of their ships and the inexperience of their commanders."[5] This time the Athenian fleet suffered a devastating defeat. Samos destroyed many enemy ships, captured war supplies, and gained control of the eastern Mediterranean.

Melissus also found the time to write a prose version of Parmenides' *Way of Truth*, introducing new arguments to support Parmenides but disagreeing with him on key points. Melissus argued that whatever existed had to be infinitely extended in all directions, not be finite as Parmenides thought. For that reason, no more than one thing could be in existence. Melissus believed even more strongly than Parmenides that sense perception was an illusion, that reality was completely different from the way it appeared.

Zeno, like Parmenides, was from Elea. He came up with forty different arguments to support Melissus' assertion that only one thing could exist, produced four arguments to show that motion was impossible, and carried similar issues—including the concept of infinity—to even greater extremes, some have said to the point of intellectual nihilism. Zeno is believed to have been the author of a book called *Against the Philosophers*, which almost certainly meant "Against the Pythagoreans." His criticisms may have influenced a change in their way of thinking that showed up in the work of Philolaus, about whether a point in geometry has any dimensions.

The Eleatics' penchant for a strictly abstract, logical approach and their distrust of sense perceptions was, in turn, a reaction against thinkers like Thales, Anaximander, and Anaximenes, all of whom may have taught Pythagoras. Observational evidence had seemed no illusion to them, and they suggested that substances such as water (Thales) and air (Anaximenes) were fundamental reality. Anaximander had been more abstract, but he implied a "many" that did not gel with the concept that only one thing exists. These ideas set the stage for Philolaus. He chose not to take on the Eleatics directly, but continued to value the possibility of studying nature using the tools of the five senses.

Philolaus dealt with old questions: How did everything (the cosmos, or "world order") begin? What basics—"first principles" (*archai*)—had to be in place, or had to be true, in order for anything else to happen? He answered in the first sentence of his book: "Nature in the cosmos was fitted together harmoniously from unlimited things and limiting things, both the cosmos as a whole and all things within it." The key words were "harmoniously," "unlimited," and "limiting."* The ideas of the "unlimited" and the "limiting" were older than the Eleatics or the Pythagoreans, the first known mention having come from Anaximander. "Harmoniously" was uniquely Pythagorean.

If Pythagoras did study with Anaximander, he learned that for him the first principle was something more abstract than Thales' water. It was the "limitless" (or the "unlimited")—characterless, indefinite, unbounded by time or space. The primordial description in Genesis of the earth "without form and void" is close to the same idea. So is the late-twentieth-century scientific concept that describes the universe (or "pre-universe") as a state of wobbling quantum nothingness from which anything (or nothing) could have emerged. The "limitless" was a situation with no differentiation, no choices made, no orders given, no laws laid down that would allow or compel some things to happen but not others. It was the "limiting" that was responsible for differentiation. Anaximander did not, however, think of the limitless only as a situation that preexisted the world, that came first chronologically and ended when the heavens and the world emerged. The limitless was a fundamental background to eternally continuing cycles of destruction and generation. He associated the limitless with *time*.

* Ancient authors (and later translators) also called "unlimited" "limitless." They called its opposite "limiting," "limit," or "limited."

If Philolaus can be taken as an example, Pythagoreans also believed that the fundamental principles limitless and limiting were both needed in order for anything else to exist, which raised a problem. The two principles were discordant and opposed to one another; how could they work together to produce anything? There had to be another first principle. The limitless and the limiting "must necessarily be locked together by a *harmony* if they are to be held together in a world." Harmony had also to be a "first principle," one of the *archai*, maybe the most fundamental of all.

The word *harmonia* was not coined by Pythagoras or Philolaus. As early as Homer's *Odyssey*, it meant joining or fitting together. In carpentry it meant a wooden nail or peg. In music it referred to the stringing of a lyre with strings of different tension. The Pythagoreans gave it new importance. In the ratios of music, they felt they had found an actual link between *harmonia* on the everyday level and the *harmonia* that helped create the universe and that bound it together. They had come to think of the ratios of musical harmony as exemplifying the primordial organizing principle of the universe.

The musical interval of the octave was the "first consonance," which Philolaus identified by the name *harmonia*. The "second consonance" was the interval of a fifth; the next was the interval of a fourth. Add the four numbers in these ratios (1, 2, 3, 4) and the result is 10, the perfect number.*

The numbers 1, 2, 3, and 4 had additional significance for Philolaus. They underlay the progression from point to line to surface to solid:

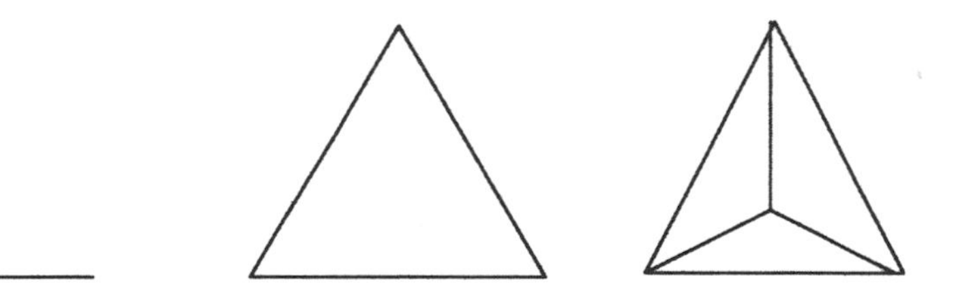

A point (on the left) is 1; a line is 2 (defined by two points, one at each end); a surface is 3 (defined by three points, one at each corner); a solid is 4 (defined by four points, one at each corner).

* Note that 10 is *not* a perfect number as the term is defined in modern mathematics. We will get to those later.

Later, Speusippus, Plato's nephew and successor as head of his Academy, explained what he understood Philolaus to have meant: "The point is the first principle leading to magnitude, the line the second, the surface the third, the solid the fourth." This sounded more complicated, but it allowed the progression to apply to other shapes and solids besides the triangle and pyramid; for example, a square, with four corners, led to a cube, with eight.

According to Philolaus, the Pythagoreans took the number 10 and ran with it. Aristotle later commented that they "construct the whole heavens out of numbers." Ten being the perfect number, there had to be ten major heavenly bodies, though no one could see ten in the sky. Also, said Philolaus, there had to be fire both at the center of the universe and at the highest point, surrounding everything, at the outermost circumference or "uppermost level." This was partly observable, for the stars were fires out on the periphery, but what about the center? Here, according to Philolaus, the Pythagoreans made a leap that set them far ahead of their contemporaries. The Earth could not be the center of the cosmos, nor, for that matter, could the Sun. The center had to be a "central fire," a fiery "hearth of the universe" around which the Earth, the Moon, the Sun, the five planets, and the stars revolve. As the scholar Aëtius—probably of the second century A.D.—described it, "Unlike other philosophers, who say that the Earth is at rest, Philolaus the Pythagorean said that it revolves about the fire in an inclined circle like the sun and moon." In the centuries when no one—with the singular, brilliant exception of Aristarchus of Samos, who proposed a sun-centered cosmos in the third century B.C.—was willing to consider a moving Earth that was not the center of the cosmos, scholar after scholar tried to show, or simply assumed, that the Pythagoreans could not really have meant this.

Earth, Moon, Sun, five planets, and the "outer fire" (stars) added up to nine things to "dance around the center." Since there had to be ten, the Pythagoreans decided there was a "counter-earth," closer to the central fire than the Earth. The central fire and counter-earth were never visible from the Earth, because in their revolutions Earth and counter-earth were always "opposite" one another. Aristotle commented, not too approvingly:

> Any agreements that they found between number and harmony on the one hand, and on the other the changes and divisions of

> the universe and the whole order of nature, these they collected and applied; and if something was missing, they insisted on making their system coherent. For instance they regarded the decad as something perfect, and as embracing the whole nature of number, whence they assert that the moving heavenly bodies are also ten; and since there are only nine to be seen, they invent the counter-earth as a tenth.[6]

The central fire and counter-earth were certainly consistent with the Eleatic view that human sense perceptions were not trustworthy for finding out what was true about the universe, for neither could be perceived with any of the five senses.

The Pythagoreans could have picked up the idea that the Moon shone by reflected light from Parmenides, or perhaps from Anaxagoras, but according to Philolaus the light it reflected was not the Sun. Instead, both Moon and Sun caught the light and heat of the central fire and the outer fire. The Sun, like glass, filtered these through to the Earth. Living beings inhabited the Moon and probably the counter-earth, though because of their positioning the inhabitants of Earth and counter-earth never saw one another. The Moon was home to "living creatures and plants that are bigger and fairer than ours. Indeed the animals on it are fifteen times as powerful and do not excrete, and the day is correspondingly long." This must have been calculated from the fact that the lunar "day" lasts fifteen Earth days. It would be consistent with Earth and counter-earth orbiting the central fire for Philolaus to have thought the Earth was a sphere. Though this does not appear in any of the fragments, Aristotle and another later author, Alexander Polyhistor, wrote that Pythagoreans in the late fifth century (that would have included Philolaus) and early fourth century B.C. believed that the Earth was spherical.

Night and day on Earth, wrote Philolaus, were produced by the Earth's and Sun's positions relative to one another, and the apparent rotation of the planets and Sun were in part the result of movement of the Earth. The Pythagoreans were not only the first to realize that what we see, from Earth, as the heavenly motions is a combination of movement in opposite directions; they were also far ahead of their contemporaries in recognizing that the movement of the Earth itself contributes to the picture.

Philolaus linked all of this to the origin of the cosmos, when *harmonia* reconciled the limitless and the limiting. The discovery of the ratios of musical harmony had provided a brilliant metaphor for the interaction of the limited and the unlimited. The whole range of musical pitches, stretching infinitely in opposite directions, higher and lower, and including an infinite number of possible pitches "between" the tones usually heard in music, represented the unlimited. When this infinite continuum of possible pitches was sorted out (limited) according to one series of ratios and not another, the result was order and beauty. The infinite possibilities still existed, higher, lower, and between the notes, but the "unlimited" was thus disciplined and brought into harmony within an order, a kosmos.[7]

The "first thing to be harmonized," wrote Philolaus, was the central fire. The central fire was the number 1; the outer fire, the number 2. The ratio 2:1 represented the musical octave, so an octave separated the two extremes of the cosmos. Some Pythagoreans went so far as to suggest that the periodic motions of the nine orbiting bodies around the central fire were related to the musical ratios, and their revolutions produced the "music of the spheres," but that idea did not appear in Philolaus' book, at least in the fragments that survived.

Philolaus' cosmic arrangement was odd and imaginative. In spite of Aristotle's disparagement, it must be admitted that the Pythagoreans were clearly capable of independent, outside-the-box thinking. This was not storytelling or myth-making, but drawing conclusions by deciding "This must be so, on the basis of what we already know about the cosmos and the numerical rules by which things work"—a giant step in the direction of what has become the time-honored way of developing scientific theories.

Philolaus made clear that Pythagoras believed in and taught reincarnation (transmigration of the soul) and that the soul was immortal, tied up with a divine, universal soul to which it might someday return. The way Philolaus applied the idea of *harmonia* to the soul showed up in Plato's dialogues and in Aristotle, who wrote, "There seems to be in us a sort of affinity to musical modes and rhythms, which makes some philosophers say that the soul is a *harmonia*, others that it possesses *harmonia*."[8]

Philolaus evidently played an important role in forming Plato's impression of Pythagoras and Pythagorean teaching, though judging from the dialogues they never met in person. Plato knew Philolaus' work through members of surviving Pythagorean communities, and from Socrates. In Plato's dialogue *Phaedo*, his characters are supposed to have learned of the soul's *harmonia* from Philolaus. Simmias, who has listened to him in Thebes, says:

> And in point of fact I fancy that you yourself are well aware, Socrates, that we mostly hold a view of this sort about the soul: we regard the body as held together in a state of tension by the hot, the cold, the dry and the moist, and so forth, and the soul as a blending or *harmonia* of these in the right and due proportion.[9]

Simmias must have been listening to others besides Philolaus, because his interpretation sounds more like that of the medical scholar Alcmaeon of Croton. Alcmaeon's lifetime may have overlapped Pythagoras', and he was probably a Pythagorean himself. If not, he was close to them and clearly reflected Pythagorean thinking about opposites when he wrote, "What preserves health is an equilibrium of the powers . . . health is a balanced mixture of opposites."[10]

Because a body could go so out of synchrony in sickness and death as to lose any suggestion of harmony, Plato's Simmias worried that his soul could not be immortal. Echecrates, another Pythagorean character in the dialogue, also comments uneasily,

> This teaching that the soul is a kind of *harmonia* has had, and still has, a strong hold on me, and when you mentioned it I was reminded that I too had believed it. Now, it is as if I were starting at the beginning again. I terribly need another argument that can persuade me that the souls of the dead do not die with them.[11]

The argument Echecrates and Simmias needed to hear—the best Pythagorean shoring up of their faltering faith—had appeared earlier in the same dialogue when Socrates expressed surprise that Simmias and his friend Cebes, both described as students of Philolaus, were ignorant of Philolaus' teaching about suicide. Socrates admits that this is part of a "secret doctrine": We are put into the world by the gods, who take care of us, and we must not leave the world until the moment they have chosen, even though death, when finally permitted, is like getting out of prison. A Philolaus fragment in the writing of the early Christian scholar Clement of Alexandria echoes the idea: "This Pythagorean Philolaus says: 'The ancient theological writers and prophets also bear witness that the soul is yoked to the body as a punishment, and buried in it as in a tomb.'"[12] Unfortunately, souls had a tendency to become too fond of bodily existence. Plato attempted to put this issue in its proper perspective in one of the most Pythagorean statements of his *Phaedo*:

> [The soul that is not completely purified] has always associated with the body and tended it, filled with its lusts and so seduced by its passions and pleasures as to think that nothing is real except what is bodily, what can be touched and seen and eaten and made to serve sexual enjoyment.[13]

Though Plato's characters Simmias, Echecrates, and Cebes had misgivings, Philolaus, Plato, Socrates, the Pythagoreans before them—including Alcmaeon—and Pythagoras himself all clearly believed that souls were immortal. Bodily health was harmony; sickness and death a breakup of that harmony, but this physical harmony was not the *ultimate* harmony. There was a universal harmony to which every Pythagorean aspired to escape from the tedious round of earthly reincarnations. The soul was set in the body by means of numbers and an immortal *harmonia*, and its quest for the divine level was dependent on number. Plato's thinking took off from there.

Most of the ancient world regarded natural phenomena as beyond human understanding or explanation, subject to the whims of capricious deities and best dealt with in imaginative stories. Philolaus referred to the central fire as the "home of Zeus," perhaps to make his contemporaries feel comfortable with the notion. But what we learn from him is that the first Pythagoreans, led by a man who was, by some descriptions, more shaman than scientist or mathematician, were trying a new way of securing a foothold on the climb to understanding nature and the universe, through numbers. The earliest pre-Socratic philosophers—Thales, Anaximander, Anaximenes—for all their yearning to get at the roots of things did not connect or confirm their

Bust of Plato

philosophical ideas with numbers or mathematics. The Mesopotamians of the First Babylonian Dynasty had found numbers useful and enjoyed using impressive mathematics in exercises that had no practical applications, but apparently did not think that numbers and mathematics were a way to reach a profounder, all-encompassing truth. Philolaus wrote that "nature itself admits of divine and not of human knowledge," but he was convinced that number relationships underlay the origin of the universe and the soul's relationship with the divine, making it possible for humans to figure such things out. This insight was a fresh departure, a sea change of enormous proportions, and Pythagoreans such as Philolaus regarded the relationship of rational humans to a rational universe with awe. The kinship was reflected in a doctrine of the unity of all being. A fragment in *Against the Mathematicians*, by the skeptic philosopher and physician Sextus Empiricus (second–third century A.D.) states: "The Pythagoreans say that reason is the criterion of truth— not reason in general, but mathematical reason, as Philolaus said, which, inasmuch as it considers the nature of the universe, has a certain affinity to it (for like is naturally apprehended by like)."

That "certain affinity"—the fact that human mathematical reasoning does match up with what is really happening in nature—was not something that the Pythagoreans, or Philolaus, or anyone since them could or can explain. It was enough to know that numbers were tied in a fundamental way to the origin and nature of the cosmos.

CHAPTER 8

Plato's Search for Pythagoras

Fourth Century B.C.

IN ABOUT THE YEAR 389 B.C., Plato left his home in Athens and boarded a ship setting sail westward into the Ionian Sea. His destination was Tarentum, one of the old colonial cities of southern Italy, in the coastland known to him as Megale Hellas. He was going in search of Pythagoras.[1]

In the 110 years since his death, Pythagoras had become the stuff of legend. Some believed he had been the wisest man who ever lived, almost a god. There were stories that a wealth of precious knowledge had perished with him and his followers in upheavals that had destroyed their communities in 500 and 454 B.C. Though no one alive was old enough to have known Pythagoras, Plato had heard that in Megale Hellas there were still men calling themselves Pythagoreans. So, in his thirty-eighth year, he sailed to the shores where Pythagoras at about that same age had preceded him and walked and taught and died. The stones of the promontories, the pleasant coastlines, the very dust of the roads, ought to remember him.

Plato's investigation began in Tarentum, on a small peninsula at the western extreme of the instep of the Italian boot, the first port of call for ships crossing from Greece.* The only story connecting Pythagoras with that city was that he had convinced a bull there not to eat beans, but Tarentum had been far enough from Croton for refugees from the fifth-century attacks to have settled, felt reasonably safe, and started their own exile Pythagorean community. It had survived, and Plato knew that its most prominent member now was Archytas of Tarentum—"Archytas the Pythagorean."

In Archytas, Plato found a man who embodied Pythagorean ideals both in his lifestyle and his studies. Archytas was an outstanding scholar and mathematician working in the Pythagorean *mathematici* tradition, and also an able civic leader. Meeting him must have confirmed for Plato that the years of Pythagorean rule in Megale Hellas had been an era of peace and stability, strengthening his conviction that men who knew philosophy and mathematics made splendid rulers. Plato and Archytas were within a year of each other in age. The visit in 389 was the first of several during which Plato conversed with him and his Pythagorean friends, absorbing knowledge and information that only a handful of men in the world could have given him. Megale Hellas would continue to draw Plato, not only because of Archytas.

At the time of Plato's first visit, the southern Italian cities were living under the encroaching shadow of a formidable enemy—Dionysius, tyrant of Syracuse, close across the water in Sicily. "Tyrant" did not necessarily have negative connotations then. The term meant a ruler whose claim to power was not hereditary, and, indeed, Dionysius had begun in the lowly position of clerk in a city office. However, he also fit the later, ugly definition. Tactics that made him hugely successful shocked even his contemporaries. Dionysius reigned for nearly forty years, preserving Syracuse's independence during repeated invasions while most of the rest of Sicily fell to the Carthaginians from North Africa. Syracuse became one of the most powerful cities in the world,

* Tarentum was the only colony established by Sparta, and Plato greatly admired the Spartan system of government. However, the people who had colonized Tarentum in 706 B.C. had come there under unusual circumstances and might not have shared Plato's enthusiasm for Sparta. They were sons of officially arranged marriages uniting Spartan women with men who were not previously citizens. The purpose was to increase the number of male citizens who could fight in the Messenian wars. When the husbands were no longer needed as warriors, the marriages were nullified and the offspring forced to leave Sparta.

her fleet for a time the strongest in the Mediterranean. It was certain that if Dionysius chose to move against his Italian neighbors, no one could stop him. Plato had come to an unstable, dangerous region, but instead of heading directly back to safer Athens, he decided to experience at first hand the court of a powerful, gifted ruler. Here was no theoretical governance. It was the real thing.

Dionysius' capital was, or was in the process of becoming, a splendid, well-fortified city, built strategically on an island separated from the mainland of Sicily by a narrow swath of water. There was a Pythagorean community in Syracuse, begun like the one in Tarentum by fifth-century Pythagoreans who in this case had fled west across the Gulf of Messina, but Plato was more interested in the court of Dionysius. He was becoming increasingly intrigued with public affairs, and he seems to have enjoyed—perhaps too well for his own good—rubbing shoulders with powerful courtiers among whom he felt more than able to hold his own. On this first visit, Plato met one of the most influential men in Syracuse, the tyrant's brother-in-law Dion. Plato was impressed with Dion . . . and Dion with Plato.

Not long after Plato's visit, Dionysius' invading forces wreaked devastation on the south Italian cities, and the entire region fell to Syracuse. In terms of the map, the football had kicked the boot. Meanwhile, back in Athens, Plato went on to establish his Academy, adopting a "Pythagorean curriculum" that he had learned from Archytas: a "quadrivium" of arithmetic, geometry, astronomy, and music. The inclusion of music was an exceptionally Pythagorean touch.

The ruthless Dionysius died in 367, survived by his son, Dionysius the Younger. Unfortunately for Syracuse—though perhaps to the relief of many in the region—the son was a much less able leader than the father. Plato's acquaintance Dion, the new ruler's uncle, was dubious about his nephew's ability to keep Syracuse as dominant as the old tyrant had left it. For whatever well-meaning or devious reasons (history records the events but not the motivation) Dion decided to improve his nephew by seeing to his belated education. The father had been an innately brilliant leader with literary pretensions (though his writing was widely judged to be embarrassingly bad), but the son needed assistance if he was to rule effectively and continue to frustrate the Carthaginians' desire to complete their takeover of Sicily. Dion recalled his conversations with Plato twenty years earlier and some of Plato's dialogues that he had read since then, in which Plato had been developing the idea that men like Pythagoras and Archytas—philosophers for whom the "quadrivium" was bread and butter—should be the political rulers. To fill such shoes and be a "philosopher king," as Plato coined the term, Dionysius the Younger needed training only Plato could provide. Dion decided to try to convince Plato, by then sixty-one and famous in Athens and far beyond, to return to Syracuse and tutor him.

In spite of what must have been a yearning to foster a philosopher king in a world power like Syracuse, Plato was not initially keen about Dion's proposal, thinking it would be a risky undertaking and unlikely to succeed. Archytas convinced Plato to change his mind. Partly tempted by the opportunity for more conversations with Archytas, Plato sailed for Syracuse. For a while, he was on sufficiently good terms with Dionysius the Younger to do some networking on Archytas' behalf. A friendly relationship between Dionysius and Archytas was advantageous for the city of Tarentum. However, Dionysius did not study with Plato long. Before the year 366 ended, he banished Dion; Plato, suspecting that his own best interests did not lie in this court, prudently took his leave.

Yet five or six years later, in 361–360 B.C., Plato was back, invited by the tyrant himself. Dionysius sent an emissary named Archedemus, a friend of Archytas, on a special ship to summon Plato. The banished Dion also had a clandestine hand in his return. He asked Plato to engineer a reconciliation between him and Dionysius.

Plato arrived and Dionysius' lessons resumed, but any hope of transforming Dionysius into a philosopher king was, again, short-lived. It cannot have helped that Plato was at court partly at the behest of the banished Dion. Plato was soon not only out of favor but in danger for his life. He got word to Archytas, and that resourceful man, using the influence he retained with Dionysius, sent an ambassador with a ship from Tarentum and persuaded the tyrant to release Plato. Afterward Archytas was not only known as "Archytas of Tarentum" or "the Pythagorean" but also as "Archytas who saved Plato's life."

Dion captured Syracuse three years later and was assassinated three years after that at the behest of another Syracusan acquaintance of Plato. Dionysius regained control for a short period, but he seems never to have had much talent or inclination for ruling, and it may have come as a relief to him in 344 when the Corinthian general Timoleon compelled him to surrender and retire to Corinth. There he became a language teacher. Perhaps Plato's efforts had not been entirely wasted and a former tyrant was well qualified to teach.

In Corinth, Dionysius met Aristotle's pupil Aristoxenus, who was collecting information about Pythagoras and the Pythagoreans. Aristoxenus would turn out to be one of the earliest and most valuable sources, for Tarentum was his birthplace and he said his father knew Archytas. From Dionysius, who had been rather useless at nearly everything else, Aristoxenus was able to glean firsthand knowledge about Pythagoreans in the fourth century in Syracuse, not far from the area where the society had originated.

As Dionysius told the story to Aristoxenus, some of his courtiers in Syracuse had spoken disparagingly of the local Pythagoreans as arrogant, pious fakes whose rumored moral strength and superiority would evaporate in a crisis. Other courtiers disagreed, and the two sides contrived a way to settle the dispute. Would one Pythagorean be willing to stake his life on the dependability and faithfulness of another? Would the other's faithfulness and dependability—to the death—prove deserving of such trust?

The courtiers accused a man named Phintias, a member of the local Pythagorean community, of plotting against Dionysius. When Dionysius sentenced him to death, Phintias asked for a stay of execution for the remainder of the day, long enough to settle his affairs. It was a Pythagorean custom, established by Pythagoras himself, to keep no private property but own all things in common. Phintias was the oldest among the local brotherhood and chiefly in charge of the management of finances. Dionysius and his court, following their plan, allowed him to send for another Pythagorean, Damon, to remain as hostage until his return. To the astonishment of the court, Damon willingly came to stand as personal surety for Phintias. Phintias departed, and the courtiers—sure they had seen the last of him—sneered at Damon for being such a trusting fool. But the faithful Phintias returned at sunset to face his death rather than leave his friend to be executed in his stead. "All present were astonished and subdued," reported Dionysius, who was so impressed that he embraced the two men and asked to be allowed to join their bond of friendship. Not surprisingly, "they would by no means consent to anything of the kind." What happened to them then is not known. Plato, so often at court in Syracuse, also likely heard about this incident, but he never wrote about it.

Plato's activities in Megale Hellas went beyond learning about Pythagoras and Pythagorean teachings, experiencing day-to-day reality in a tyrant's court, and abortive attempts to tutor Dionysius. He helped Archytas strengthen his position in Tarentum as a minor philosopher king. Archytas went on to play a prominent role in political affairs among the cities of Megale Hellas and Sicily, in accordance with the Pythagorean tradition of wise and able involvement in public service.

IN A SEARCH for the real Pythagoras and the Pythagoreans and what they believed and taught, the information about Archytas, Plato, and Dionysius the Younger provides valuable clues. Most significantly, it reveals a link between Plato and a Pythagorean community that still existed in the fourth century B.C. in the region where Pythagoras and his followers had had their golden age in the late sixth century. Plato knew, and knew of, other fourth-century Pythagoreans, but after his visits to Tarentum, when he thought "Pythagorean" he was probably mostly thinking of Archytas and his associates. When he thought "Pythagorean mathematics and learning" he was thinking of the mathematics and learning of Archytas.

What was he like, this man who was, for Plato, the best available evidence of what it meant to be Pythagorean and what "Pythagorean knowledge" was? What could Plato have learned from him about Pythagoras and what Pythagorean teaching had been more than a century earlier?

Archytas was known to be a mild-mannered man who ruled in Tarentum through a democratic set of laws—information deduced from the news that these were not always obeyed: Though the "laws" said a man should serve no more than one year, the city "elected" Archytas seven times to the office of *strategos*, or ruling general.[2] Aristoxenus wrote that Archytas was never defeated as a general except once, when his political opponents forced him to resign and the enemy immediately captured his men. Archytas, said Aristoxenus—whose father, he claimed, had known the man in person—was "admired for excellence of every sort."

There can be no doubt that as a scholar Archytas lived by the great insight that set the Pythagoreans apart from other ancient thinkers: that numbers and number relationships were the key to vast knowledge about the universe. Archytas was a rigorous mathematician who solved an infamous problem in Greek mathematics known as the Delian problem, or doubling a cube, that is, creating a new cube twice the volume of the first. Archytas' solution was sophisticated, requiring new geometry using three dimensions—"solid" geometry—and involving the idea of movement.[3]*

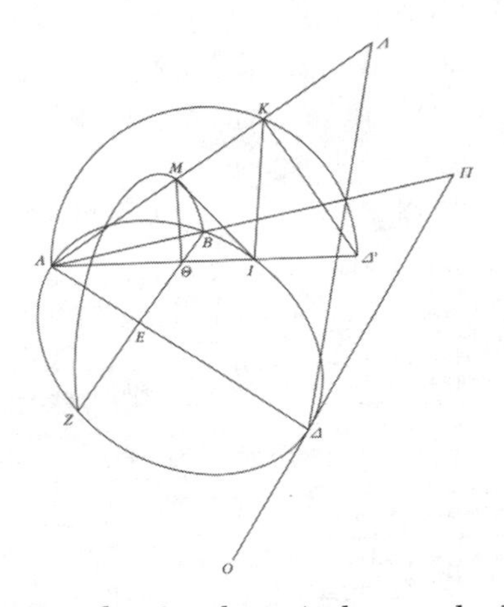

Diagram showing how Archytas solved the Delian problem, evidence of how advanced Pythagorean mathematics and geometry had become in little more than one century.†

Viewing the world through the eyes of his Pythagorean forebears, Archytas could not avoid pondering the possible hidden, underlying numbers and geometry. "Why are the parts of plants and animals (except for the organs) all round?" he asked, "of plants, the stems and branches; of animals, the legs, thighs, arms, thorax? Neither the whole animal nor any part is triangular or polygonal." He suspected there was a "proportion of equality in natural motion, since all things move proportionately, and this is the only motion that returns back to itself, so that when it occurs it produces circles and rounded curves."

* For an example of the use of movement in geometry, take a straight line, fasten down one end of it, and swing the other end about. The result is an arc. Take a right triangle and stand it upright with one of the sides serving as its base; swivel it around the upright leg and the result is a cone. (The ancient scholar Eudemus used this explanation in his description of Archytas' solution.)

† A lengthy text is needed to understand it and is available in S. Cuomo, *Ancient Mathematics*, Routledge, 2001, pp. 58 and 59, and on the Internet at http://mathforum.org/dr.math/faq/davies/cu/bedbl.htm

Later scholars, among them Euclid and Ptolemy, agreed that Archytas' precise work in the mathematics of music was fundamentally linked with the earliest Pythagorean mathematics and music theory.[4] Archytas extended the study of numerical ratios between notes of the scale and showed that if you defined a whole tone as the interval separating the fourth and fifth notes of the scale (such as F and G in a scale beginning with C), as Greek music theorists were doing, then a whole tone could not be divided into two equal halves.* This had dramatic implications, for it was an example of something obviously present in the real world that could not be measured precisely. A different example, discovered in the right triangle, had famously caused the first Pythagoreans to have a devastating crisis of faith in the rationality of the universe, but incommensurability seemed no longer to disturb Pythagoreans like Archytas in the fourth century B.C.

In astronomy, Archytas puzzled over the question of whether the cosmos is infinitely large, and was notorious for asking: "If I come to the limit of the heavens, can I extend my arm or my staff outside, or not?" He replied that whatever the answer—yea or nay—if he were out there performing this experiment, he could not actually be at the limit of the heavens. If he could not extend his arm or staff farther, something beyond the supposed limit had to be stopping him.[5]

In archaic-sounding litanies, the first Pythagoreans had asked "What are the isles of the blessed?" and answered "The Sun and the Moon." Archytas brought this up-to-date in a more sophisticated catechism, asking "What is calm?" and answering as a parent might answer a child, with an example: "What is a man?" "Daddy is a man." Similarly, Archytas' reply to "What is calm?" was "Smoothness of the sea." His catechism, however, implied more than "example answers," for he liked to connect the specific with the general, reflecting the Pythagorean doctrine of the unity of all being, and he enjoyed thinking about the relationship between the whole and the parts or particulars. His questions and answers about the weather and the sea were particular cases of deeper questions about smoothness and motion. The problem of dividing a whole tone into equal halves was a particular case of a mathematical discovery about

* More generally, ratios such as 5:4, or 9:8, in which the larger number is one unit larger than the smaller (mathematicians call these superparticular or epimeric ratios), cannot be divided into two equal parts.

ratios that could not be equally divided. His observations about the roundness in trees, plants, and animals were particular manifestations of a "proportion of equality in natural motion." Archytas was convinced of a tight connection between understanding the universe, or anything else, as a whole and understanding the details. Plato wrote in his *Republic* that this paragraph from Archytas was "the teaching of the Pythagoreans":

> The students of mathematics [by this, Archytas meant students among the Pythagoreans] seem to me to have attained excellent knowledge, and it is not surprising that they have correctly understood how things stand in each matter. For since they have obtained knowledge of the nature of the universe as a whole, they will have come to have a good view of how each thing stands in particular. Concerning the speed and risings and settings of the heavenly bodies they have handed down to us clear knowledge, concerning geometry and numbers, and not least concerning music. For these studies seem to be sisters.

When Archytas wrote about such matters as smoothness and non-smoothness of the sea he was reflecting another Pythagorean traditional favorite—opposites (smoothness/lack of smoothness; motion/lack of motion)—and for him that line of thought inevitably led back to thinking about infinity. Can something be infinitely calm? Or infinitely uncalm? Or infinitely smooth; infinitely rough?

As a politician and general, Archytas was convinced of what he was sure his Pythagorean forebears had demonstrated: The unity of all things had to include ethics and politics. The value of mathematics extended to the political arena. In the following fragment, "reason" could also be translated as "calculation." To a Pythagorean like Archytas, the two meanings were probably synonymous.

> When reason/calculation is discovered, it puts an end to civil strife and reinforces concord. Where this is present, greed disappears and is replaced by fairness. It is by reason/calculation that we are able to come to terms in dealings with one another. By this means do the poor receive from the affluent and the rich give to the needy, both parties convinced that by this they have what is fair.

Plato, of course, could not have agreed more. The ability to use "reason" or "calculation" would make a philosopher king a superbly able ruler.

For Archytas, the concept of unity meant he should also apply a Pythagorean search for deeper levels of mathematical understanding to optics, physical acoustics, and mechanics. His is the earliest surviving explanation of sound by "impact," with stronger impacts giving higher pitches, but he nodded to his Pythagorean forebears by insisting this was a theory that had been handed down to him. By "impact," Archytas meant impact on the air—whipping a stick through the air, playing a high note on a pipe by making the pipe as short as possible (making a stronger pressure on the air, he thought), and the sound of the wind whistling higher pitches as its speed increased, or a "bull-roarer." That last was an instrument used in the mystery religions, a flat piece of wood on the end of a rope. Whirling it around in the air like a giant slingshot produced a fearsome howling sound; the faster the whirling the higher the pitch.[6]

One of the most widely known, influential, and enduring Pythagorean ideas passed down through Archytas to Plato was the concept of the "music of the spheres," the music Archytas and his Pythagorean forebears thought the planets made as they rushed through the heavens. Here is Archytas' explanation for why humans never hear it:

> Many sounds cannot be recognized by our nature, some because of the weakness of the blow (impact), some because of the great distance from us, and some because their magnitude exceeds what can fit into our hearing, as when one pours too much into narrow mouthed vessels and nothing goes in.

According to Pythagorean tradition, only Pythagoras could hear this music.

Archytas was a generous man, kind to slaves and children. He invented toys and gadgets, including a wooden bird (a duck or a dove) that could fly. Aristotle was impressed by "Archytas' rattle," "which they give to children so that by using it they may refrain from breaking things about the house; for young things cannot keep still."[7]

THIS, THEN, WAS the science, mathematics, music theory, and political philosophy that Plato, from Archytas, learned to think of as

Pythagorean. Through Plato, much of the image of Pythagoras and Pythagorean thought in Western civilization is traceable to Archytas' window on Pythagoras.

How unclouded was this window? Archytas regarded himself as an authentic Pythagorean, true to the earliest traditions and teachings. In his era, oral accounts could still be accurate, especially in a continuing community that considered it vitally important to keep an ancient memory alive and clear. In many ways, Archytas was probably a good reflection of what it had meant to be Pythagorean when Pythagoras himself walked the paths of Megale Hellas. However, he was one of the *mathematici*, the school of Pythagoreanism that believed following in Pythagoras' footsteps meant diligently seeking and increasing knowledge. The Pythagorean ideals that underlay Archytas' thought and work led him to newer discoveries. He was among the great scholars and mathematicians of his era, by reputation the teacher of the mathematician Eudoxus. If Archytas had focused only on the knowledge of the first Pythagoreans, this would have been impossible.

Plato himself provided a window through which we view Archytas. No matter how accurately Archytas reflected Pythagoras and Pythagorean thinking, we see him through *Plato's* eyes and with Plato's mind, the eyes and mind of one of the most creative thinkers in all history. It is in the nature of such a man, if he is impressed with an idea, to take the ball and run with it—to say, "This is, of course, what you mean," and restate someone else's good idea with a spin that makes it absolutely brilliant—and his own, not the original. Assuming Archytas was an exemplary Pythagorean, when Plato got the ball to the other end of the field, was it anything like the same ball he had caught in the pass from Archytas? That is one of the most debated questions in all the long history of those who have yearned to know what Pythagoras himself, and Pythagoreans before Plato, really discovered and thought.

On one significant issue, Plato disagreed with Archytas, and that disagreement is a welcome clue, a clear indication of something in pre-Platonic Pythagorean thinking, undiluted by Plato, that differed from Plato. Archytas, Plato complained, was too concerned with what one could see and hear and touch, and with searching for mathematics and numbers to explain it. For Plato, the goal of studying mathematics was to turn away from experience that humans have through their five senses to a search for abstract "form," out of reach of sensory perception.

Numbers and mathematical understanding were a venture into abstract form, but not the same, he thought, as his own concept of the ultimate understanding of "the beautiful and the good." This difference, in Plato's view, made Archytas an inadequate philosopher and himself a better one.

PLATO'S KNOWLEDGE ABOUT Pythagoras and Pythagoreans was not confined to what he learned through Archytas. There is evidence in his dialogues that he heard about them from Socrates; also, Plato and Archytas both knew of Philolaus. If the characters in Plato's dialogue *Phaedo* are not entirely fictional, he was acquainted with contemporaries who were "disciples of Philolaus and Eurytus" in Phlius, a community west of Corinth, as well as with Echecrates, who speaks for them in the dialogue, and Simmias and Cebes. Plato also knew about Lysis of Tarentum who, like Philolaus, had emigrated to Thebes. The Pythagorean community there was apparently still in existence in Plato's time.

Plato could not have avoided also knowing about *acusmatici* Pythagoreans who did not agree that scholars like Archytas were Pythagoreans.[8] The Greek public in the fourth century B.C. generally failed to recognize a distinction between *mathematici* and *acusmatici* and lumped all "Pythagorists" together as an eccentric lot. Athenian comic dramatists lampooned them as unwashed, secretive, arrogant characters who abstained from meat and wine and went about ragged and barefoot. Plato's pupils, educated at his Academy in the "Pythagorean sister sciences"—the quadrivium of arithmetic, geometry, astronomy, and music—spoke of their philosophy and that of Pythagoras as one and the same and featured him in their books. They were certainly more in the *mathematici* tradition than in the *acusmatici*, but they nevertheless were the targets of the same jibes.

The unshakable conviction of the men who inspired the caricature—that they were following in the authentic footsteps of Pythagoras and preserving a precious tradition—caused some of their contemporaries to feel, a bit uncomfortably, that even the most eccentric were favored by the gods and privy to mystical secrets. Antiphanes, in his play *Tarentini* (the title connects it with Tarentum) spoke of Pythagoras himself as "thrice blessed," and Aristophon had one of his characters report:

> He said that he had gone down to visit those below in their daily life, and he had seen all of them and that the Pythagoreans had far the best lot among the dead. For Pluto dined with them alone, because of their piety.

Lest anyone conclude that Aristophon approved of Pythagoreans, another character commented that Pluto had to be a very easygoing god, to dine with such filthy riffraff.[9]

Diodorus of Aspendus, who was not fictional, was described as a vegetarian with long hair, a beard, and a "crazy garment of skins" who with "arrogant presumption" drew followers about him, although "Pythagoreans before him wore shining bright clothes, bathed and anointed themselves, and had their hair cut according to the fashion."[10]

Aristoxenus—who interviewed the tyrant Dionysius in his Corinthian "retirement"—would have none of this. He was effectively a propagandist for the *mathematici*, taking pleasure in contradicting the *acusmatici* by insisting that Pythagoras ate meat and that the aphorisms were ridiculous, and he tried to disassociate "true Pythagoreans" from what he saw as this unsavory, superstitious group who were giving the movement a bad name. He listed the pupils of Philolaus and Eurytus and called them "the last of the Pythagoreans" who "held to their original way of life, and their science, until, not ignobly, they died out." Because these men died a few decades before the comic allusions in the plays, dubbing them "the last of the Pythagoreans" was making the point that the butts of the jokes were only pretending to be Pythagoreans.

Aristoxenus' public relations efforts did not succeed well. Through the fourth century B.C., the popular image of Pythagoreans continued to resemble the *acusmatici* more than the *mathematici*. But after the fourth century, the *acusmatici*, with a few exceptions, dwindled and vanished from notice. Had there been no other Pythagorean tradition than theirs, and if they did indeed represent the truer image of Pythagoras and his earliest followers, it would be almost impossible to explain how such an odd cult figure, not far different from others in antiquity, became so dramatically and rapidly transformed in the minds of intelligent men and women as to inspire deep and effective scientific thinking and seize the imagination of centuries of people to come.

Was it all due to Plato? Did he get so excited about something that was mainly legend that he elaborated on it himself until he had made it hugely significant? At a minimum there had to have been the discovery in music of pattern and rationality underlying nature, and the accessibility of that rationality through numbers—and that was of no small significance. It seems much more reasonable to conclude that Pythagoras, responding to different types of interest and intelligence among his followers, encouraged both kinds of thinking—*acusmatici* for those who needed something naive and more regimented and conservative, and *mathematici* for those with minds eager to grasp difficult, nuanced concepts and explore their implications. He was personally, perhaps, not entirely unlike either group.

However that may be, from the time of Plato, what survived as "Pythagorean" and "Pythagoras" was largely *mathematici*, and that included the conviction that right from the time of Pythagoras himself, and attributable to him, there had been a truly remarkable new approach to numbers, mathematics, philosophy, and nature.

CHAPTER 9

"The ancients, our superiors who dwelt nearer to the gods, have passed this word on to us"

Fourth Century B.C.

NSOCRATES WAS NOT PLATO'S fictional creation. Born about thirty years after the death of Pythagoras, near the time Philolaus was born, he fought in the Peloponnesian Wars and then lived a life of intentional poverty as a teacher in Athens. He wrote nothing, and information about what he taught comes only through Plato's dialogues and similar conversations recorded by Xenophon, another of Socrates' pupils. Socrates' teaching method consisted of asking questions. Plato's dialogues are not word-for-word accounts of real question-and-answer lessons, but are almost certainly faithful to the philosophy as it emerged in conversations like these. When Socrates was seventy, he was accused of impiety and corrupting the youth of Athens. A jury of his fellow citizens sentenced him to death, probably through a dose of hemlock. He died surrounded by his friends and pupils.

In the dialogues Plato wrote, the character Socrates usually directed the discussions, but in Plato's *Timaeus* he relinquished center stage for many pages to a fictional character named Timaeus of Locri. Timaeus was supposed to be a statesman and scientist from southern Italy, and the ideas Plato put in his mouth were heavily indebted to Archytas, with whom he had apparently spent long hours in Tarentum deep in conversation and bent over mathematical diagrams. Perhaps what Timaeus tells

Socrates and his friends in the dialogue is close to what Archytas laid out for Plato. Or maybe Archytas' ideas were only a springboard for Plato. Though many opinions have been expressed about those possibilities, no one knows for certain, and the truth likely lies somewhere in between.

Plato carried forward two great Pythagorean themes: (1) the underlying mathematical structure of the world and the power of mathematics for unlocking its secrets; and (2) the soul's immortality.[1] The stage is set for discussion of the first when Socrates asks Timaeus, an expert on such matters, to "tell the story of the universe till the creation of man."[2] Timaeus' response to this daunting request is a number-haunted, Pythagorean creation story: The mathematical order of the universe was the work of a creative god, whom Plato called the demiurge—not the chief god or the only god, but a figure loosely comparable to Ptah, the Egyptian god at Memphis, or to Jesus acting in the role of the *logos* in the opening of the Gospel of John—"through him all things were created." This craftsman god, says Timaeus, decided that the universe should be a "living being," spherical and moving in "a uniform circular motion on the same spot; unique and alone." Timeaus sets forth a numerical construction of the "world soul":

- First the creator god took his material and "marked off a section of the whole."

- Then he marked off another section "twice the size of the first."

- Next he marked off a third section, "half again the size of the second section and three times the size of the first."

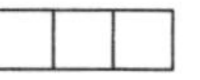

- Next he marked off a fourth section, "twice the size of the second."

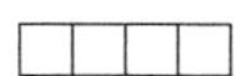

- Next, a section "three times the third."

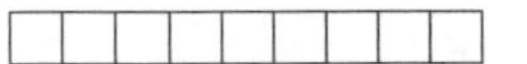

- Next, a section "eight times the first."

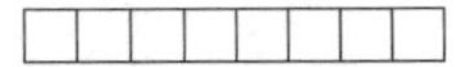

- Last, a section "twenty-seven times the first."

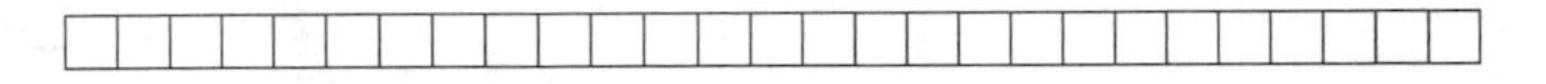

Counting the squares in each line gives 1, 2, 3, 4, 9, 8, 27. The first four of those numbers are the numbers in the *tetractus* and the Pythagorean musical ratios, 2:1, 3:2, 4:3, but it is an interesting challenge to discern a meaningful pattern in the rest of the numbers, and how they could work in a creation scheme. The answer is that if you square each of the first two numbers 2 and 3 (1 not being a "number"), you get 4 and 9. Cube the same two numbers, 2 and 3, and you get 8 and 27. For a Pythagorean it was significant that each pair was an even and an odd number. Plato stopped with the cubes because in the creation of three-dimensional, solid physical reality, only three dimensions are needed.

Next, according to the account Plato put in Timaeus' mouth, the creator divided his material into smaller parts, filling in harmonic and arithmetical means between those numbers and connecting the "world soul" with a diatonic scale in music.* Plato used a scale developed by Philolaus, not the one developed by Archytas.

Astronomy in his *Timaeus* was also worked out in numbers, with the "world soul" cut into two strips bent around to form an "X" at one point, making an inner and outer ring. Two such rings really exist in astronomy, the celestial equator and the ecliptic. The celestial equator is on the plane of Earth's equator and anchors the sphere of the fixed stars that do not change their positions in the sky relative to one another and the celestial equator. This was the ring Timaeus called "the Same." It stays the same and never changes. The ecliptic is the circular path that the Sun appears to follow in its daily round, with the planets appearing to orbit in a band centered on it. This ring was Timaeus's "the Different,"

* "Diatonic" refers to the scales now known as major and minor scales.

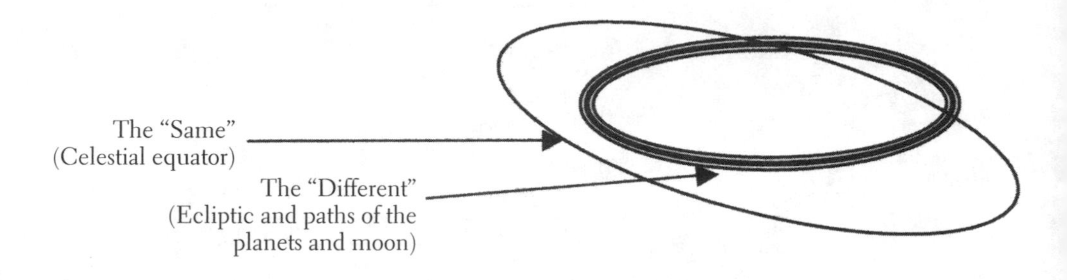

for it changes He called the planets "instruments by which Time can be measured."* The creator cut the Different into seven narrower strips to accommodate Sun, Moon, and five planets, with the radiuses of their orbits proportional to the numbers 1, 2, 3, 4, 8, 9, and 27. Both rings—Same and Different—were in constant motion, which, Plato thought, nothing but a living soul could be, unless something else pushed it. The rings moved in opposite directions, the Same east to west, the Different west to east, and the seven strips of the Different moved at different speeds, corresponding to the speeds of the Sun, the Moon and the planets.

Plato had Timaeus explain that the movement humans see in the sky is the result of this combination: The daily rotation of the Same with the sphere of fixed stars carries everything around with it, east to west, including Sun, Moon, and planets. But the Sun, Moon, and planets—the seven bodies of the Different—have in addition their own contrary west-to-east motion against that background. They "swim upstream," so

* Plato was not the first to think of the planets moving on rings. Anaximander's cosmos involved huge wheels, whose hollow rims were filled with fire. The Sun, Moon, stars and planets were glimpses of this fire, showing through at openings in the wheel rims. Similar ideas had surfaced elsewhere as well. After Plato, the idea was taken up by his pupil Eudoxus, who responded to Plato's challenge to produce an analysis that would account for the appearances in the heavens with an explanation along the lines introduced by the Pythagoreans, involving a combination of movements of the sphere of stars and the planets. Eudoxus did this with a system not of concentric rings but of concentric spheres, and that was adopted by Aristotle and would dominate astronomy until the time of Tycho Brahe and Johannes Kepler.

to speak, against the current of the Same, at varying speeds, and sometimes back up. This, says Timaeus, is because they are souls, and souls exercise independent choices and power of movement. It is believed to be one of the Pythagorean triumphs, showing up in Philolaus' fragments, in more detail in Archytas' work, and then in Plato, to have explained heavenly motion correctly as a combination of opposite movements.

Geometry, Plato had Timaeus explain, had a detailed role in creation when primordial disorder was sorted into four elements—earth, fire, air, and water—and the creator introduced four geometric figures—cube, pyramid or tetrahedron, octahedron, and icosahedron. These "Pythagorean" or "Platonic" solids are four of the five possible solids in which all the edges are the same length and all the faces are the same shape.* Each element—earth, fire, air, and water—was made up of tiny pieces in one of those shapes, too small to be visible to the eye.

Plato had Timaeus continue: The four elements and four solids were not the alphabet of the universe. The solids were constructed of something even more basic, two types of right triangles. Plato, through Timaeus, admitted there was room for argument about which triangles were most basic, but he thought he was correct to choose the isosceles triangle and scalene triangle. Both are right triangles.

The isosceles triangle is made by cutting a square into equal halves on the diagonal. Obviously, two isosceles triangles make a square, and squares make up cubes (one of the solids).

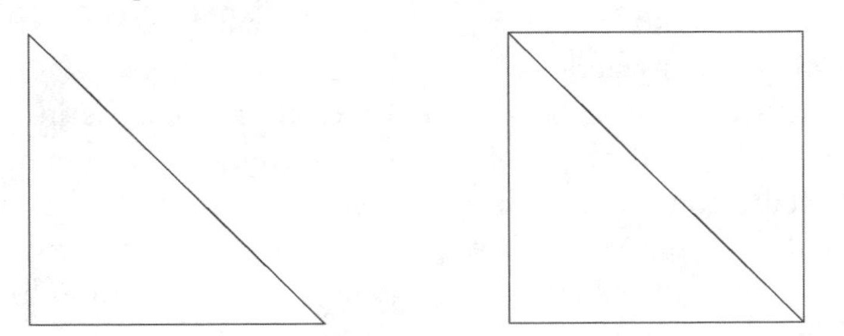

In a scalene triangle, the diagonal is twice as long as the shortest side. Two scalene triangles set back to back create an equilateral triangle—none other than the Pythagorean *tetractus*. The faces of the tetrahedron, octahedron, and icosahedron are equilateral triangles.

* Kepler discovered other regular solids, the "hedgehog," for example, but they did not have all the characteristics of the original five.

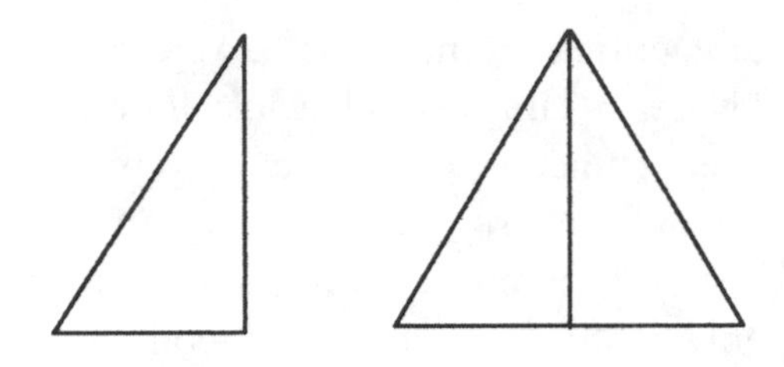

Here is Plato's explanation.

Cube: Fasten together the edges of six squares (each made by pairing two isosceles triangles). The result is a *cube*, the only regular solid that uses the isosceles triangle or square for its construction.

Pyramid or tetrahedron: Fasten together the edges of four equilateral triangles (each made by pairing two right scalene triangles). The result is a *pyramid* or *tetrahedron*.

Octahedron: Fasten together the edges of eight equilateral triangles. The result is an *octahedron*.

Icosahedron: Fasten together the edges of twenty equilateral triangles. The result is an *icosahedron*.

The Pythagoreans and Plato knew the dodecahedron, the only regular solid made of pentagons (12 of them), but Plato did not use it in his scheme.

Beyond those five—cube, pyramid, octahedron, icosahedron, and dodecahedron—there are no other regular solids (polyhedrons). Try to fasten together *any other number* of *any regular figure* (polygon). You get no fit. No wonder the Pythagoreans, Plato, and later Kepler thought these solids were mysterious.

Timaeus explains to Socrates and the other characters in the dialogue that earth is made up of microscopic cubes, fire of tetrahedrons, air of octahedrons, water of icosahedrons. The pairings were based on how easily movable each solid was, how sharp, how penetrating, and on considerations of what qualities it would give an element to be made up of tiny pieces in this shape.

Timaeus pairs the fifth regular solid, the dodecahedron, with "the whole spherical heaven," and in his *Phaedo*, Plato associated it with the spherical Earth, in spite of the fact that in his time most of the Greek world, except for the scattered Pythagorean communities, still assumed the Earth was flat. The dodecahedron comes close to actually being a sphere. In fact, the earliest mention of a dodecahedron was in sports, with twelve pentagonal pieces of cloth sewn together and the result inflated to create a ball. Each of the five solids fits into a sphere with each of its points touching the inner surface of the sphere, and a sphere can be fitted into each of the solids so as to touch the center of each surface, which makes sense of Philolaus' enigmatic (and controversial) fragment: "The bodies in the sphere are five: fire, water, earth, and air, and fifthly the hull of the sphere."

Though the triangles making up the solids in Plato's scheme may have been the basic "alphabet" of creation, he thought they were not the fundamentals or *archai*. In the dialogue *Philebus*, Socrates says knowledge of the principles of *unlimited* and *limiting* is "a gift of the gods to human beings, tossed down from the gods by some Prometheus together with the most brilliant fire. And the ancients, our superiors who dwelt nearer to the gods, have passed this word on to us."[3] Plato's contemporaries and generations of later readers thought that by "some Prometheus," he meant Pythagoras, and that "the ancients, our superiors who dwelt nearer to the gods," were the Pythagoreans, which contributed substantially to the image of Pythagoras as a channel for superhuman knowledge and wisdom. If Plato meant that, he shortchanged Anaximander, who had talked of "unlimited" and "limiting" earlier.

According to Plato, one thing that "some Prometheus" tossed down concerning the unlimited and the limiting was that "all things that are said to be are always derived from One and from Many, having Limit and Unlimitedness inherent in their nature."[4] He explained this in unpublished lectures at his Academy that Aristotle reported firsthand.

Plato chose to transform the concepts of unlimited and limiting into something slightly easier to understand: unity and plurality. He called these "the One" (unity) and the "Indefinite Dyad" (plurality). It is easy enough to grasp what is meant by One, or unity, but the Indefinite Dyad is a more difficult concept. Think of it as more than one, or everything that is *not* One, or—more vaguely, but closer to what Plato apparently meant—something implying the possibility of numbers or a role for numbers (there would be no role for numbers if everything were One), but not implying that numbers actually exist. The Indefinite Dyad also implied the possibility of opposites—large/small; hot/cold—for if everything were One, opposites would not exist.

To start things off, the One acted in some fashion upon the Indefinite Dyad and the result was a definite number, 2. The One went on acting on the Dyad, generating the numbers up to 10. Once they existed, the numbers 2, 3, and 4—numbers in the Pythagorean *tetractus*—predictably played a special role, organizing the Dyad to produce geometry. Plato introduced again the progression point–line–surface–solid, connecting the One and the Indefinite Dyad with the world as humans knew it. The meeting of the One and the Indefinite Dyad had been the flashpoint that brought everything else into existence.

On a more mundane level, Plato connected numbers with ideas about an ideal society and ideal rulers. He had probably only recently returned from his first visit to Syracuse when he wrote his *Gorgias*, his earliest dialogue to deal explicitly with political matters. A character named Callicles in the dialogue lusts for power and luxury, and Socrates admonishes him in words that ring with Pythagorean conviction:

> Wise men say that heaven and earth and gods and men all are held together by community, friendship, orderliness, self-control and justice, which is why they call this universe a kosmos (a world order, or universe)—not disorderliness or intemperance. But, I fear, you ignore them, though you are wise yourself, and fail to see what power is wielded among both gods and men by geometrical equality. Hence your defence of selfish aggrandisement. It arises from your neglect of geometry.[5]

Also in Plato's *Timaeus*, before Socrates relinquishes the floor, he reminds his listeners of two decidedly Pythagorean conditions of "equality" needed for an ideal society: Those whose duty is to defend the community, internally and externally, should hold no private property but own all things in common. Women should share in all occupations, in war and in the rest of life. However, sharing in all occupations did not apparently indicate true equality for women, for later in the same dialogue Timaeus says that if a man fails to live a good life he may be relegated to being a woman in the next.

Archytas had introduced Plato to the Pythagorean quadrivium, the curriculum comprising arithmetic, geometry, astronomy, and music. Plato had Socrates declare, "I think we may say that, in the same way that our eyes are made for astronomy, so our ears are made for harmony, and that the two are, as the Pythagoreans say, sisters of one another, and we agree." That was Plato's only mention of the Pythagoreans by name, but as Socrates continues he clearly is still talking about them: "They gave great attention to these studies, and we should allow ourselves to be taught by them."

Plato was not, however, entirely in agreement with the Pythagorean approach: Studying the stars and their movements was useful insofar as it got one beyond surface appearances to underlying mathematical principles and laws of motion, but, though the stars and their movements illustrated these realities, they never got them precisely right. A philosopher had to go further than what they could show him and attempt to understand "the true realities, which reason and thought can perceive but which are not visible to the eye."[6] Plato was convinced that a new manner of education was needed.

Not long after his first visit to Syracuse, Plato had acquired property near Athens that included an olive grove, a park, and a gymnasium sacred to the legendary hero Academus. In that pleasant setting, he had founded his Academy—the name deriving from the legendary hero. For the rest of his life, except for sojourns abroad, he taught there, lectured, and set problems for his students.* His trainees spent ten years (between the ages of twenty and thirty) mastering the quadrivium, but this was only a preliminary step in Plato's preparation of them to serve as civic leaders who were also philosophers. Education continued in the form of "dialectic." It is not surprising that Dionysius—in the middle of running an impossibly unwieldy tyranny—balked, though this was the training Plato and Dion felt would enable him to rule effectively.

Plato required the dialectic, not merely the quadrivium, because he believed that the world as humans can know it is at best only an imperfect

* "The Academy" also refers to the men associated with this school after Plato's lifetime, including his successors as *scholarch* elected for life by a majority vote of the members. Aristotle was also associated with the Academy, first as a pupil and later as a teacher. In several transformations, still claiming descent from the original, the Academy lasted until the sixth century A.D. as a center of Platonism and neo-Platonism.

Forms, in Plato.

Plato spoke of two levels of reality:

(1) the divine realm of immutable Forms, which is the model for
(2) the realm in which humans live and where everything is continually changing, ruled by the passions, subject to opinion.

likeness to something else—only a flawed copy of a unique, perfect, eternal model that just "is," "always is," "never becomes," and can never change or be destroyed. In the world perceptible to humans, things resembled this higher realm of the "Forms" and had the same names, but they were not perfect and eternal.* They changed—they "became." They began to exist, came to an end, moved about, and were subject to opinions and passions. They were copies or imitations of the Forms; but "never fully real."[7] The realm of the Forms could not be perceived by the human senses, but through reasoning and intelligence humans could come nearer to perceiving it. To stretch toward it, Plato thought, you had to use discussion and debate, hence "dialectic." That was what his characters did in his dialogues—Socrates' question-and-answer lessons—those discussions that never settled anything definitely.

Where did Plato place numbers and mathematics in this picture? Parting company with the Pythagoreans and with Archytas, he thought that although the logic of mathematics and geometry might be part of the universal, immutable truths of the Forms, there was no way humans could find out whether or not they were. Human mathematics was earthbound, deductive reasoning, capable of building only on its own previous knowledge, making the truth of human mathematics only hypothetical, not necessarily Truth with a capital T. In Plato's house, there was no complete staircase from the human-experience level to the level of the Forms. Numbers and mathematics could take you up a

* The writer Richard E. Rubenstein put it succinctly: "Plato did not hate the world, it simply reminded him of a better place" (Richard E. Rubenstein, *Aristotle's Children: How Christians, Muslims, and Jews Rediscovered Ancient Wisdom and Illuminated the Dark Ages* [New York: Harcourt, 2003]).

few flights. By using dialectic, argument, thought, and logic, you could go higher, but those flights also fell short of reaching the top. You could never find out whether what was up there on the unreachable level was mathematical or not. The Pythagorean house, by contrast, had a complete staircase made entirely of numbers and mathematics. Humans could climb it and, reaching the top, would discover that what was up there was also mathematical. Pythagoreans were sure they knew that mathematics and numbers were the rationality of the universe and the key to complete understanding and reunion with the divine level of reality.

"Knowing" in a context like this was problematic for Plato, for it was not compatible with a universe in which the Forms could never be fully known. The Pythagoreans, however, had had an experience that Plato lacked. The discovery that mathematical logic and pattern underlie nature had apparently come as a shocking, intuitive impression for them. Mathematics and numbers were the rational, unconditional principles of the universe, waiting to be discovered, not deduced from things already known. Had they heard Plato speak about a search for "the invisible and incorporeal realm of Form," one of them might well have raised a hand and insisted they had found it. Their experience was that the vein of Truth (call it Forms, if you are Plato), mostly buried deep beyond the reach of the senses, at some rare points lies close enough to the surface to be perceived, like a vein of gold gleaming through a thin layer of dust and rock. The realm of music was one of those thin places.

Plato's pupils, and their pupils, continued to ponder the issues he had wrestled with, including the questions about whether the numbers are Forms. Speusippus allowed numbers and "mathematicals" to take the place of the Forms, while Xenocrates said that the Forms *were* the numbers. Both thought of themselves, and Plato, as Pythagoreans.

Many scientists and mathematicians today still hold to a Pythagorean faith that truth about the universe is inherently mathematical, and that it is possible to grasp at least bits of that truth by using our human level of mathematics. A few insist that mathematics is the only discipline in which some things are unarguably true and not subject to opinion, while others will not grant it that. Still others redefine "complete truth" as "truth that *human beings can discover* through mathematics," stretching the Pythagoreans beyond their own meaning and performing an end run around Plato.

THE SECOND PYTHAGOREAN theme that inspired Plato was the creation and destiny of the soul. He applied the mathematical proportions that went into the creation of the "world soul" also to the human soul, and even described the soul in terms of a version of the Same and Different, reflecting two types of competing judgment—the ability and privilege of a human to say yes or no. For Plato, this free will was the essence of rational thought. But things were not easy for a soul living in a physical body on Earth, the Moon, or one of the planets. At the mercy of all the passions of its body, it inevitably got distorted and stirred up. Proper education could restore it to harmonious equilibrium by reawakening it to its link with the world soul. One way this could happen was through something heard and understood—the musical scale, the proportions of the world soul reproduced in sound.

The Pythagorean belief that a soul could ultimately escape the distorting influences of the world and be reunited with the divine level of the universe fascinated Plato. His ideas about immortality ranged from skepticism in his *Apology* to mystical speculation in his *Gorgias*, where Socrates attributes some of his thoughts about the soul to "some clever Sicilian or Italian"—an allusion to the Pythagoreans and probably to the philosopher Empedocles, who was often included under the Pythagorean banner. The dialogue ended with a myth in which souls witness the horrible punishment of incurable sinners in Hades. Plato, in this dialogue, did not argue for a doctrine of reincarnation, but his myth assumed that reincarnation occurred for those witnesses.

Plato probably wrote his *Meno* after his first visit with Archytas. In it, Socrates speaks of "wise priests and priestesses" whose authority is reliable, who teach about immortality and reincarnation—a bow to the Pythagoreans. Plato had Socrates attribute to those "priests and priestesses" the idea that because of what we have experienced earlier, much of what we know in this present life is "recollection." This does not seem out of line with what Pythagoras claimed for himself, but Plato had something different in mind that he and his pupils thought was compatible with Pythagorean teaching.

As Plato interpreted the Pythagorean concept of the transmigrating soul, the possibility of escape from the parade of reincarnations lay in "becoming just and pious with wisdom," freeing the soul from fear and the passions and pains of the body. The highest goal was "becoming like god," as Plato phrased it in his *Theaetetus*. Several centuries later, the pagan neo-Platonist Porphyry (Pythagoras' biographer) listed Hercules, Pythagoras, and Jesus among those who had succeeded in this ultimate achievement of "becoming like god."

"Recollection," for Plato, did not however mean memories of past lives. Instead, it was the mysterious, innate, a priori knowledge that humans seem to possess, that cannot be explained by what one has learned in one's present life. Plato did not imply that anyone could recall acquiring this knowledge. As a demonstration, he used the geometric exercise that some believe reflected Pythagoras' proof of the Pythagorean theorem.

In this scene from Plato's *Meno*, Socrates and Meno are discussing a figure Socrates has drawn in the sand, a four-foot square. The task is to double the size of the square. Socrates intends to demonstrate that innate knowledge—not of the correct answer but of the underlying geometry that will lead to the correct answer—lies hidden in Meno's slave boy, waiting to be reawakened. Socrates is acting as a sort of midwife.

SOCRATES (to Meno): Now notice what he will discover by seeking the truth in company with me, though I simply ask him questions without teaching him. Be ready to catch me if I give him any instruction or explanation instead of simply interrogating him on his own opinions. *(Socrates rubs out previous figures in the sand and starts again with a four-foot square.)*

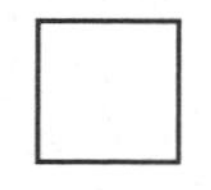

Tell me, boy, is not this our square of four feet? You understand?

BOY: Yes.

SOCRATES: Now we can add another equal to it like this? *(Draws.)*

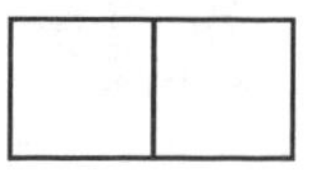

BOY: Yes.

SOCRATES: And a third here, equal to each of the others? *(Draws.)*

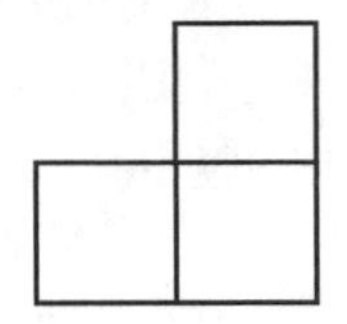

BOY: Yes.

SOCRATES: And then we can fill in this one in the corner? *(Draws.)*

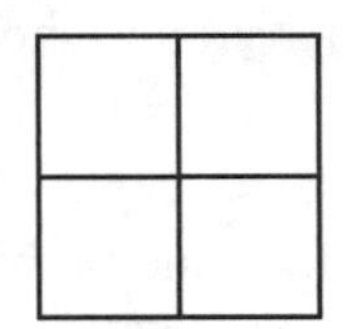

BOY: Yes.

SOCRATES: Then here we have four equal squares?

BOY: Yes.

SOCRATES: And how many times the size of the first square is the whole?

BOY: Four times.

SOCRATES: And we want one double the size. You remember?

BOY: Yes.

SOCRATES: Now, do these lines going from corner to corner cut each of these squares in half? *(Draws.)*

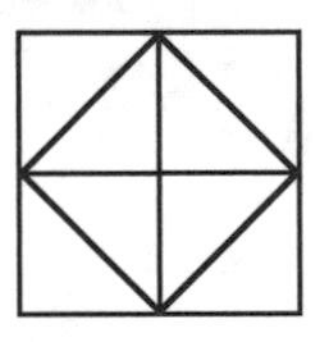

BOY: Yes.

SOCRATES: And these are four equal [diagonal] lines enclosing this [central] area?

BOY: They are.

SOCRATES: Now think, how big is this [central] area?

BOY: I don't understand.

SOCRATES: Here are four squares. Has not each [diagonal] line cut off the inner half of each of them?

BOY: Yes.

SOCRATES: And how many such halves are there in this [central area]?

BOY: Four.

SOCRATES: And how many in [one of the original squares]?

BOY: Two.

SOCRATES: And what is the relation of four to two?

BOY: Double.

SOCRATES: How big is this figure then?

BOY: Eight feet.

SOCRATES: On what base?

BOY: This one. *(Indicates one of the diagonal lines.)*

SOCRATES: The line which goes from corner to corner of the square of four feet?

BOY: Yes.

SOCRATES: The technical name for it is "diagonal"; so if we use that name, it is your personal opinion that the square on the diagonal of the original square is double its area?

BOY: That is so, Socrates.

SOCRATES: What do you think, Meno? Has he answered with any opinions that were not his own?

MENO: No, they were all his.

SOCRATES: Yet he did not know, as we agreed a few minutes ago.

MENO: True.

SOCRATES: But these opinions were somewhere in him, were they not?

MENO: Yes.

SOCRATES: So a man who does not know has in himself true opinions on a subject without having knowledge.

MENO: It would appear so.

SOCRATES: At present these opinions, being newly aroused, have a dream-like quality. But if the same questions are put to him on many occasions and in different ways, you can see that in the end he will have knowledge on the subject as accurate as anybody's.

MENO: Probably.

SOCRATES: This knowledge will not come from teaching but from questioning. He will recover it for himself.

MENO: Yes.

SOCRATES: And the spontaneous recovery of knowledge that is in him is recollection, isn't it?

MENO: Yes.

SOCRATES: Either then he has at some time acquired the knowledge which he now has, or he has always possessed it. If he always possessed it, he must always have known; if on the other hand he acquired it at some previous time, it cannot have been in this life, unless somebody has taught him geometry. He will behave in the same way with all geometrical knowledge, and every other subject. Has anyone taught him all these? You ought to know. He has been brought up in your household.

MENO: Yes, I know that no one has ever taught him.

SOCRATES: And has he these opinions, or hasn't he?

MENO: It seems we can't deny it.

SOCRATES: Then if he did not acquire them in this life, isn't it immediately clear that he possessed and had learned them during some other period?

MENO: It seems so.

SOCRATES: When he was not in human shape?

MENO: Yes.

A modern attorney would probably object that Socrates was "leading the witness." But Plato was not talking about knowledge the boy had hidden somewhere in his mind because he had witnessed it or been taught it in a previous life: the date of an event or the length of a road—knowledge of the changeable world. Plato meant inborn knowledge of truths that do not change—universal and immutable truths of the Forms, in this case truths of geometry. The point of Plato's lesson scene was that at each stage of questioning, the boy knew whether what Socrates was suggesting was correct. Such recollection of the "eternal Forms" came not from past lives at all but from experiences of the disembodied soul.

Many who first encounter proofs in a setting other than a smotheringly dry presentation are struck by this deep, mysterious sense of recognition of something they already knew. Indeed there are truths that have been "rediscovered" time and time again (the Pythagorean theorem may be one of them) by ancient people and by more recent individuals who were unaware they were repeating a former discovery. Socrates' demonstration was an extremely Pythagorean lesson, for it united the two Pythagorean themes: the immortality of the soul and the mathematical structure of the world.

Other dialogues and his *Republic* show that Plato's mind was much taken up with the doctrines of recollection, reincarnation, and immortality. His *Phaedo* ends shortly after Socrates' death, with Phaedo pausing on his journey home from Athens in a Pythagorean community in Phlius to tell Echecrates and other Pythagoreans about the philosopher's last words. In a discussion centering on immortality and reincarnation, Phaedo repeats Socrates' quote from an Orphic poem that Socrates had thought spoke of philosophy's power to raise one to the level of the gods. In his *Phaedrus*, Plato wrote that human "love" was recollection of the experience of Beauty as an eternal Form.

In his "Myth of Er," at the end of *The Republic*, Plato most clearly revealed his belief in reincarnation, although, true to his doctrine that knowledge of ultimate truth is unattainable, he used the term "myth" to indicate that he could not vouch for the absolute truth of the lessons it taught. In the "myth" he imagined what happens when one life has ended and the next has not yet begun: Each soul chooses what it will be in the next life. Choices include "lives of all living creatures, as well as of all conditions of men." Orpheus chooses to be a swan so as not to be born of a woman—for frenzied Bacchic women had torn him apart in a former life—while a soul who has lived previously as a swan chooses to be a man. The harmony of the spheres was also on Plato's mind. The souls see a vision, a magnificent model of the cosmos. On each of the circles in which the planets and other bodies orbit stands "a Siren, who was carried round with its movement, uttering a single sound on one note, so that all the eight made up the concords of a single scale." Though Earth, in Plato's cosmos, sat dead center, and there was no central fire or counter-earth, the "Myth of Er" was suffused with Pythagorean ideas.

When the members of Plato's Academy before and after his death in 348/347 B.C. thought about Pythagoras and called themselves Pythagorean, they had in mind mainly Pythagoras as seen through Plato's eyes. However, to say that Pythagoras was reinvented as a "late Platonist," as some scholars insist, is to be too glib and overconfident about where to draw the lines between original Pythagorean thought, Pythagorean thought shortly after Pythagoras' death, Archytas, Plato, and Plato's pupils, some of whom attributed their own ideas to more ancient Pythagoreans and even to Pythagoras. As time passed, the line between Platonism and what called itself Pythagorean became increasingly difficult to discern. Eventually the two were indistinguishable.

CHAPTER 10

From Aristotle to Euclid

Fourth Century B.C.

NWHILE MOST SCHOLARS WERE content to view Pythagorean teachings through Plato's eyes and not eager to differentiate between Plato's philosophy and the thinking of pre-Platonic Pythagoreans, one person was still curious. That was Aristotle. Born in 384, he was two generations younger than Plato and at age seventeen had come to Athens to study at Plato's Academy. Plato was away at the time, on one of his jaunts to Sicily. Twenty years later, when Plato died at age eighty in 348, Aristotle was only thirty-seven and, perhaps because of his youth, was not chosen to succeed Plato as *scholarch* of the Academy. Instead, though by then hardly anyone failed to recognize that Aristotle was one of the most gifted men around, Plato's nephew Speusippus got the job. Aristotle left Athens and eventually returned to found his own school, the Lyceum. His debt to Plato was clear throughout his work, but so was the fact that the two disagreed in significant ways. Aristotle was not happy with Plato's concept of Forms. Plato thought the world as humans knew it was only an undependable reflection of a real world that humans could never know. Aristotle, by contrast, believed that the world humans perceive is the real world. He highly valued what could be learned about nature through use of the human senses, and what could be extrapolated from those perceptions.

It would not have displeased Aristotle to find that Plato's teachings were at least in part derivative of the Pythagoreans. In his *Metaphysics*, in a passage following his description of Pythagorean philosophies, Aristotle looked down his nose at Plato and invited his readers to do the same: "To the philosophies described, there succeeded the work of Plato, which in most respects followed these men, though it had some features of its own apart from the Italian philosophy."[1]

To make such a statement, Aristotle had to be fairly confident he knew what the "Italian philosophy" was before it fell into Plato's hands. His research was extensive and careful, including the work of Philolaus and Archytas and other sources we know little or nothing about, and he recorded the results in several books.* Unfortunately, those devoted entirely to the person of Pythagoras and Pythagorean teaching are lost, but because he spent so much time and effort on them, and referred elsewhere to his "more exact" discussions in them, there is no doubt Aristotle knew the subject well.† References and quotations from the lost books appear in the writings of authors who lived before the books disappeared, making it possible to peer, indirectly, at a few of the vanished pages.[2] The result is a window into what Pythagoreans were thinking and teaching before Plato, helping, at least a little, to circumvent that frustrating impasse, the question of whether what later generations thought they knew about the Pythagoreans and their doctrine was only a Platonic interpretation.

Aristotle was one of the earliest, most dependable sources used by Iamblichus, Porphyry, and Diogenes Laertius. His information went back to shortly after Pythagoras' death (within about fifty years), but in the books that have survived he never claimed that any specific teaching could or could not be attributed directly to Pythagoras. He also made no distinction between the ideas of Pythagoreans who lived close to the time of Pythagoras and those who lived nearer the time of Plato. He used a Greek form that Burkert says is the equivalent of putting words between quotation marks in modern literature—the

* The classical scholar Walter Burkert thought that the way Aristotle "occasionally plays off the Pythagorean doctrines against the Academy" makes "the conclusion unavoidable that he was using written sources without Academic coloring. Therefore he must have had at least one original Pythagorean document" (Burkert, 47).

† The three surviving books in which he included material about the Pythagoreans are *Metaphysics*, *Physics*, and *On the Heavens*.

"Pythagoreans"—though translating it as "the so-called Pythagoreans" would put too negative a spin on it.

Aristotle wrote that what set both Plato and the Pythagoreans apart from all other thinkers who had lived before Aristotle's own time was their view of numbers as distinct from the everyday perceivable world. However, the Pythagoreans regarded numbers as far *less* independent of the everyday, perceivable world than Plato did. At the same time, for the Pythagoreans, numbers were also more "fundamental." If these distinctions seem confusing, they were, even for Aristotle. His difficulty deciding and explaining what the Pythagoreans thought about numbers was not, at heart, a matter of being unable to find out. Rather, he could not think with their minds. The discussion he was insisting on having—about what was more fundamental, more abstract, or more or less distinct from sensible things—would not have taken place at all among the first Pythagoreans. Whether numbers were independent of physical reality, or how independent, were not questions they would have thought to ask.

In his attempt to squeeze the Pythagoreans into Plato's and his own molds, Aristotle overinterpreted them and became particularly ill at ease with the idea that all things "*are* numbers." The Pythagoreans, he reported with chagrin, believed that numbers were not merely the design of the universe. They were the building blocks, both the "material and formal causes" of things. Physical bodies were *constructed* of numbers. Aristotle threw up his hands: "They appear to be talking about some other universe and other bodies, not those that we perceive."

As Aristotle understood the Pythagorean connection between numbers and creation, for numbers to exist, there first had to be the distinction between even and odd—the "elements" of number. The One had a share in both even and odd and "arose" out of this primal cosmic opposition.* The One was not an abstract concept. It was, physically, everything. Aristotle was puzzled by that idea, and unhappy with it.

Odd was "limited"; even was "unlimited." As the unlimited "penetrated" the limited, the One became a 2 and then a 3 and then larger

* For the ancient Greeks, including the Pythagoreans, 1 was neither even nor odd, and it was not a number. Number implied plurality—more than 1.

numbers.* This emergence of numerical organization resulted in the universe humans know. In Aristotle's words (he was still rankled by the "substance" of the One):

> They say clearly that when the One had been constructed—whether of planes or surface or seed or something they cannot express—then immediately the nearest part of the Unlimited began "to be drawn and limited by the Limited" . . . giving it [the Unlimited] numerical structure.

Aristotle had found that, at least in its broad outlines, the numerical creation of the universe was a pre-Platonic Pythagorean concept. However, he often regarded the Pythagoreans with a frown of frustration, like a professor faced with brilliant students who have disappointed him. Though he was, in fact, not consistent in the way he described Pythagorean ideas about numbers, and was never able to define what he thought "speak like a Pythagorean" and say "the One is substance" meant, it is clear that he feared theirs was a sadly earthbound, material view. "The Pythagoreans introduced principles," said he, that could have led them beyond the perceptible world to the higher realms of Being, but then they only used them for what is perceptible, and "squandered" their principles on the world itself as though nothing else existed besides "what the sky encloses."[3]

His was, in truth, an earthbound interpretation of the Pythagoreans. Their attempt to give numbers a physical role in creation may look as naive to us as it did to him, but they faced difficult questions: What were numbers, really? What was their role—their power—in creating, sustaining, and controlling the physical universe? Those questions have never been answered. Humans have all but given up on them. If numbers underlie, even constrain, physical reality, as the Pythagoreans thought was the case, then where, precisely, is the connection? How do mathematics and geometry exert their grip on the universe? The Pythagoreans tried to find ways to answer such questions, and at the root of their thinking, spanning the years that led to the time of Aristotle, lay

* What emerged as a Platonic idea, the "Indefinite Dyad," was not a Pythagorean concept. Aristotle spoke of no very important role for "Twoness" in Pythagorean doctrine.

that first realization that "what the sky encloses" was much more mysteriously and wondrously interconnected and infused with rationality than anyone had recognized before.[4] A cosmos governed by numbers—no matter how everyday-perceptible it also was, or whether you could figure out how it got built—was a mind-haunted cosmos. Where to go from there, with this treasure that had fallen into their hands? That was new, unknown territory, and the Pythagorean exploration of it was always a work in progress.

In an age when abstract thinking was supposed to be more prevalent than in the sixth century B.C., Aristotle seems, in his interpretation of the Pythagoreans and his frustration with some of their ideas, to have been insisting, for them, that they thought of numbers only as something concrete and physical. He was apparently blind to any other way of interpreting their thoughts and would allow them little sophistication and subtlety. Complicating this issue, the Greeks used the same word for "same" and "similar," making it difficult even to have a meaningful disagreement about whether the Pythagoreans meant a number *was* something or was "something like it" or was a symbol for it.

Aristotle summed up his interpretation of the Pythagorean view of numbers more sympathetically in two statements: "Having been brought up in it [mathematics], they came to believe that its principles are the principles of existing things." And (transmitted through Iamblichus) "Whoever wishes to comprehend the true nature of actual things, should turn his attention to these things, the numbers and proportions, because it is by them that everything is made clear." As Burkert paraphrased Aristotle's *Metaphysics*: "Number is that about things which can, with a claim to truth, be expressed; nothing is known without number."

One approach the Pythagoreans had taken, Aristotle found, was to express the creation process in a "table of opposites."

Limited	Unlimited (recall that the One, when it arises, will have a share in both)
odd	even (recall that the One will be both odd and even)
One	plurality
right	left
male	female

resting	moving
straight	crooked
light	darkness
good	bad
square	oblong

Nothing in the table could be linked with Plato's Indefinite Dyad in a clear way. In Plato's creation scheme the One and the Indefinite Dyad were there first, with limit and unlimited "inherent in their nature." On these points, if Aristotle's interpretation was correct, Plato chose not to follow the Pythagoreans, misunderstood them, or transformed their ideas to suit himself.

The Pythagoreans apparently thought creation had to involve both *drawing together* (of the limiting and the unlimited) and *separation* (as numbers and pairs of opposites arose from the One), and the universe could only exist if things were different from one another—an idea found in many ancient creation accounts. In Genesis, God separated light from darkness, the water above the earth from the water below the earth, and sea from dry land; Adam and Eve ate from the tree of the "knowledge of good and evil." In Aristotle's interpretation of the Pythagoreans, the One was not undifferentiated unity, like the unlimited. It was harmony of many different things whose differences were necessary in order for anything to exist in the way humans experience the world.*

Disappointingly, Aristotle did not really answer the question whether Plato's view of the relationship between the ideal and the material world was derivative of the Pythagoreans, or original, or somewhere in between. What Aristotle concluded has hung in the air for centuries, with the answer depending on what he meant by one ambiguous Greek sentence. Burkert cut to the heart of the matter:

> Again and again it becomes clear that the Pythagorean doctrine cannot be expressed in Aristotle's terminology. Their numbers are "mathematical" and yet, in view of their spatial, concrete

* The table of opposites was probably not meant to imply good (the left column) and evil (the right), though other, later such tables did. For example, for Plato's Academy, "good" led off the left-hand column, and still later, Platonists, neo-Pythagoreans, and pseudo-Pythagorean writers rearranged the columns. Plutarch's table was thoroughly Platonized: "Good" was on top and "Dyad" replaced plurality

> nature, they are not. They "seem" to be conceived as matter and yet they are something like Form. They are, in themselves, Being, and yet are not *quite* so.[5]

Guthrie put it more simply: "By the use of his own terminology, Aristotle imports an unnecessary confusion into the thought of the early Pythagoreans. It is no use his putting the question whether they employ numbers as the 'material' or the 'formal' causes of things, since they were innocent of the distinction."[6]

Aristotle gave what is probably the most reliable description of the pre-Platonic Pythagorean concept of "music of the spheres," grumbling that "it does not contain the truth," though he admitted it was "ingeniously and brilliantly formulated." He explained that the Pythagoreans realized that all harmonious-sounding musical intervals were the result of certain numerical ratios in the tuning of an instrument, so "number" *was* "harmony." The same numerical ratios determined the arrangement of the cosmic bodies, resulting in a "harmony of the spheres." Here, said Aristotle, was "what puzzled the Pythagoreans and made them postulate a musical harmony for the moving bodies":

> It seems that bodies so great must inevitably produce a sound by their movement. Even bodies on Earth do that, although they are not so great in bulk or moving at so high a speed, or so many in number and enormous in size, all moving at a tremendous speed. It is unthinkable that they should fail to produce a noise of surpassing loudness. Taking this as their hypothesis, and also that the speeds of the stars, judging from their distances, are in the ratios of the musical consonances, they affirm that the sound of the stars as they revolve is concordant.

Some heavenly bodies appear to move faster than others. Aristotle wrote that the Pythagoreans had arrived at the idea that the faster the motion, the higher the pitch it produced, and they had taken this into consideration when allowing the ratios of the relative distances between the bodies to correspond to musical intervals. With the full complement of heavenly bodies, the result was a complete octave of the diatonic scale.*

* A modern major or minor scale.

What surprises is that Aristotle or anyone could think the eight notes of the scale heard simultaneously would be harmonious. The sound would not be beautiful. There would be cacophony in the heavens. Humans should be glad they cannot hear it. Pity Pythagoras, who, legend says, could! The explanation cannot be that *harmonia* did not imply audible sound, for Aristotle thought the Pythagoreans believed planetary movement produced actual tones. He never explained how it could be beautiful, but he did give what he thought was the Pythagorean explanation—different from Archytas'—for why ordinary humans do not hear it:

> To solve the difficulty that no one is aware of this sound, they account for it by saying that the sound is with us right from birth and has thus no contrasting silence to show it up; for voice and silence are perceived by contrast with each other, and so all mankind is undergoing an experience like that of a coppersmith, who becomes by long habit indifferent to the din around him.

AT THE TIME of Aristotle and in later antiquity, it was generally assumed that if one mentioned "Pythagorean mathematics," an educated person would know what that meant, but in fact the meaning was vague, apparently referring to a tradition that thought it inspiring to discover hidden, true relationships of the sort that, once found, seemed inevitable. Since the evidence about what sixth- and fifth-century Pythagorean mathematics were like is so sparse, we are at a loss to know how authentically Pythagorean this so-called Pythagorean mathematics was. To modern eyes, its vestiges seem feeble by comparison with Euclid's *Elements*, which appeared around 300 B.C. Did it really reflect a naive mathematics of Pythagoras himself, and his associates? Or was it "a dilute, popularized selection from what had been originally a rigorous mathematical system"?[7] Perhaps it was a hodgepodge of what survived from early, primitive mathematical thought from several sources, mistakenly lumped under the heading "Pythagorean"? Maybe a much more authentically Pythagorean, lively *mathematici* heritage had moved through Archytas to influence Euclid, while this older, calcified, fading mathematics limped alongside, still bearing the name "Pythagorean."

There are also differences of opinion about whether there is valid reason to call the five regular solids that Plato featured "Pythagorean" solids.[8]* The issue is not a simple one, for "knowing about" the solids, or "discovering" them, or "trying to figure them out," are not the same as "giving them a full mathematical description" or being able to prove that they are the only possible perfect solids. It is uncertain which achievement deserves to be rewarded with having one's name attached to it.

Arguing in favor of early Pythagorean knowledge of the solids is the fact that these shapes were familiar in nature and construction. Cubes (and pyramids, for anyone who had been to Egypt) were familiar building shapes, though pyramids often were five-sided including the base, not four-sided tetrahedra. A dodecahedron dating from at least as early as Pythagoras, apparently Etruscan, has been discovered near Padua. Pyrite crystals appear as cubes and also, in southern Italy and on the isle of Elba, in the form of dodecahedra.[9] A fluorite crystal is an octahedron; quartz crystals are pyramids and double pyramids; garnet crystals, dodecahedra. Pythagoras would have known about gems and crystals if his father really was a gem engraver, and someone with a Pythagorean cast of mind would surely have been curious about regular, beautiful shapes that appear without any human intervention. It would have been in keeping for someone obsessed with numbers to try to understand them by means of numbers.

Also favoring an early Pythagorean knowledge of them is that, if the fragment is genuine, fifty to a hundred years after Pythagoras' death Philolaus knew about the five regular solids but was almost certainly not, himself, their discoverer. In the absence of evidence to show who did or did not discover them, it is not far-fetched to think the five regular solids might legitimately be called Pythagorean.

Plato associated four of the five solids with the four elements in his *Timaeus*, as had Philolaus in the fragment that read, "The bodies in the sphere are five: fire, water, earth, and air, and fifthly the hull of the sphere"† But had anyone made that association earlier? The scholar Aëtius, of the second century A.D., thought Pythagoras had:

* Plato did not call them that, though he was using them in the most Pythagorean-inspired of his dialogues.

† Recall that the regular solids each fit neatly into a sphere, and the fifth is close to being a sphere.

> There being five solid figures, called the mathematical solids, Pythagoras says earth is made from the cube, fire from the pyramid, air from the octahedron, and water from the icosahedron, and from the dodecahedron is made the "sphere of the whole."[10]

Since "Pythagoras says" was used for what Pythagoras' followers said, the attribution should probably be read as "the Pythagoreans said." Aëtius got his information from Theophrastus, a pupil of Aristotle who may have been contradicting his teacher, for Aristotle scoffed that the Pythagoreans had "nothing new to add" to knowledge about the elements. However, Aristotle had so little respect for the idea of associating elements with solids that even if irrefutable evidence had existed that the association originated with the Pythagoreans he would still have dismissed it as "nothing new to add." Little survives of Theophrastus' history of philosophy or of the books he wrote about individual philosophers, but more would have been available when Aëtius was doing his research. However, though Philolaus' fragment associated the elements with the solids, and the solids might have been known to earlier Pythagoreans, the identification of the four elements as fire, water, earth, and air did not originate with them. Philolaus was evidently familiar with the idea from his older contemporary, the Sicilian poet-philosopher Empedocles, born ten years after Pythagoras' death.*

The question whether the Pythagoreans thought of a point as having magnitude seems trivial, but it is related to the question of who first knew about the solids. Zeno, one of the Eleatics, reputedly scorned the Pythagoreans for naively thinking that a point had dimensions like a pebble and that two points (pebbles) touching one another made a line, but that way of thinking made the pyramid easy to "discover" by building a little pebble structure. The Pythagorean preoccupation with the numbers 1, 2, 3, and 4 makes it difficult to believe they did not extend their progression past making a triangle with three pebbles to building a little pyramid with four, or better yet a larger one with 10, the perfect number.

Speusippus, Plato's pupil and nephew, attributed the point–line–

* Many called him Empedocles the Pythagorean, but except for agreeing about reincarnation, his ideas ran far from Pythagorean thinking.

surface–solid progression to Pythagoreans before Archytas. It was a more primitive way of arriving at a solid than Archytas' use of "movement." Even the use of movement may have come before Archytas, and leads easily to a square and cube. An example appears in a reference from the Skeptic philosopher Sextus Empiricus, who flourished at the turn of the second to the third century A.D. He called this a "scheme of the Pythagoreans": "Some say that body is formed from one point. This point by flowing produces a line, the line by flowing makes a surface, and this when moved into depth generates a body in three dimensions."[11]

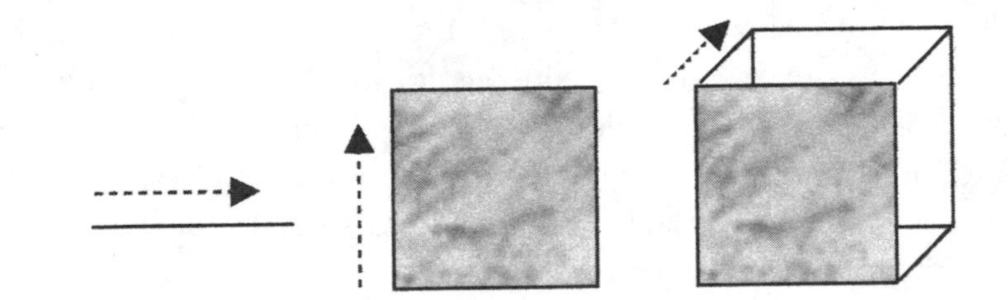

The sophistication of geometry in a Pythagorean community a little more than a century after Pythagoras' death—as witness Archytas' solution for doubling the cube—makes it ludicrous to insist that earlier Pythagoreans could not have discovered the five regular solids. Nevertheless, the man who first arrived at a complete mathematical understanding of them was not a Pythagorean. He was Theaetetus, a friend of Plato who was killed in 369 B.C. Whatever was known about the regular solids earlier, Theaetetus, with his description of the octahedron and the icosahedron, finished the job.

In the end, in spite of differing viewpoints about the solids and the "Pythagorean mathematics" of late antiquity, there is widespread consensus that the first Pythagoreans opened up a new way of thinking about, appreciating, and using numbers, representing a watershed and having very long lasting impact. Their profound musical/mathematical discovery was as modern as tomorrow's science news, as timeless as any discovery ever made, but most of the true mathematical connections and relationships in nature were hidden too deep for them to find. Even Kepler, in the sixteenth century A.D., with a Pythagorean certainty that such relationships existed, spent a good part of his lifetime searching for them on too superficial a level and was surprised when he had to admit that nature followed her own far cleverer mathematics, not his. In spite of the Pythagorean faith in the power of numbers, they had no inkling of how far numbers would lead humankind. Working out the implications of their discovery would take centuries.

ALONG WITH ARISTOTLE, three other authors who lived during the latter part of the fourth century B.C. were the earliest and most reliable sources used by Porphyry and Iamblichus. They were Heracleides Ponticus, of Plato's Academy, and Aristoxenus of Tarentum and Dicaearchus of Messina, both Aristotle's pupils. Heracleides Ponticus, like Plato, wrote dialogues. He used the character "Pythagoras" as a spokesman, telling stories about his former lives and calling himself *philosophos*, lover of wisdom. Other Pythagoreans in the dialogues, Hicetas and Ecphantus, taught that the Earth rotates.* Heracleides believed that the Earth rotates, and that this makes it appear to humans as though the stars are moving.†

Dicaearchus was Porphyry's and Iamblichus' source about Pythagoras' arrival in Croton and his success among the young men, the city rulers, and the women. Dicaearchus claimed that in his own time the memory of the revolts that ended Pythagorean rule was still vivid in Magna Graecia. He revered Pythagoras as a moral teacher and social reformer, but he believed in no sort of immortality and scorned the idea that anyone could remember former lives, joking that Pythagoras had been a beautiful courtesan in one reincarnation. A man of extensive learning and a scientist with an independent turn of mind, an admirer of Pythagoras and yet not an unqualified admirer, Dicaearchus had his ear to the ground at a time when the oral record could be extremely trustworthy, in the region where Pythagoras had lived and flourished—all of which increases the likelihood that what he reported was genuine.

* Scholars such as Kahn think these men were not fictional and that their words reflected a much older line of Pythagorean speculation.

† Copernicus would point to Heracleides Ponticus as an ancient precedent when he presented his hypotheses in the sixteenth century. The Earth also rotated in Plato's *Timaeus*, and the idea was probably not original with either man, for Philolaus and possibly earlier Pythagoreans thought part of the apparent movement of the heavens was caused by the movements of the earth. Copernicus also referred to Hicetas and Ecphantus of Syracuse.

Aristoxenus, like Dicaearchus, did not toe the Pythagorean line precisely. He dismissed the idea of the soul being more than a harmony of the body's various components, and his music theory took a different direction from Archytas'. The information Porphyry and Iamblichus attributed to Aristoxenus probably came from his biography of Pythagoras—thought to have been the first written—but neither Porphyry nor Iamblichus ever actually saw Aristoxenus' and Dicaearchus' books.* The information they took from them came indirectly, through other writers who lived during the centuries in between.

After Aristotle there were no attempts in antiquity to draw a distinction between pre-Platonic Pythagorean doctrine and Plato. Beginning with Plato's pupils Speusippus and Xenocrates, no one for centuries would make a distinction between Platonism and Pythagoreanism at all. Almost without exception, everyone would accept what Plato taught in his *Timaeus* and his "oral doctrine" (reported by Aristotle) as the teaching of the early Pythagoreans. In the eyes of the educated world, Plato *was* a Pythagorean.

BY THE TURN of the century in 300 B.C., the world of classical Greece, of Plato and Aristotle, and of strong and often warring city-states like Athens, Sparta, and Thebes had ended.[12] The rise of a power from the north—the kingdom of Philip the Great of Macedonia—was heralding a new era. Less than forty years after Philip had become king of Macedonia in 359, his son (traditionally Aristotle's pupil) Alexander the Great had conquered not only Greece but also Egypt and the entire Persian empire to the east, as far as present-day Afghanistan, Pakistan, and the Indus River. The culture and learning of Greece and its colonies and of the conquered peoples mixed and, to an impressive extent, enriched one another.

After Alexander died in 323 B.C., though the city-states had not vanished entirely and change was slow in remoter regions such as Magna Graecia, his short-lived and sprawling empire became three "successor states" under his former generals and associates. Mainland Greece became part of Macedon. The Seleucid dynasty controlled Syria. Egypt

* In search of the source of Iamblichus' lists of Pythagoreans, Burkert believed he had narrowed down the possibilties, conclusively, to Aristoxenus (Burkert, p. 105, n. 406).

was ruled by the Ptolemies, the dynasty that would later include Cleopatra. At the time of Alexander's death (and Aristotle's, for he died a year later, in 322), Athens was still the hub of the intellectual world, but Alexandria, with the wealth of the Ptolemys lavished on literature, the arts, mathematics, science, and a library and museum would soon rival and eventually eclipse her.

Around 300 B.C., Euclid, who lived in Alexandria, gave mathematics and geometry a new form of life, surpassing all others in antiquity for putting the power of numbers to use in a truly significant and comprehensive manner. Euclid personified the Pythagorean intellectual and philosophical conviction that mathematics was a precious guide to truth, and he was even known to use a Pythagorean aphorism, but he did not consider himself a Pythagorean nor did he belong to a Pythagorean community.*

Euclid in a sixteenth-century engraving

* When someone asked what the practical use of one theorem was, Euclid turned aside to his slave, sniffed, and muttered, "He wants to profit from learning, give him a penny." The Pythagorean aphorism was "A diagram and a step (an advance in knowledge), not a diagram and penny."

Euclid's *Elements* is one of the premier intellectual achievements of all time, foundational for later mathematicians and geometers. It was both a comprehensive summary and treatment of what had been discovered before him, and wondrously original, and Euclid did not clearly distinguish between what was new and what was old. He knew the Pythagorean theorem and included it in Book I as "Proposition 47," never referring to it as "Pythagorean" but also never claiming it was his own discovery or mentioning another origin. His knowledge of early Pythagorean mathematics and astronomy appears to have come mostly through Archytas, though modern experts who have analyzed the *Elements* believe that many of the results which appear in it[13] predated Archytas, and that some of the material was extremely old.[14] Archytas had previously built on some of this earlier work, and his discoveries, particularly his number theory, were incorporated by Euclid in the *Elements* Book VIII.

By Euclid's standards, a feeling of inevitability and a few examples did not constitute a "proof." The so-called Pythagorean mathematics of his own contemporaries did not fall in happily with his higher abstraction.[15] That tradition nevertheless wheezed along and proved tenacious beyond all expectation. Iamblichus preferred it:

> Pythagorean mathematics is not like the mathematics pursued by the many. For the latter is largely technical and does not have a single goal, or aim at the beautiful and the good, but Pythagorean mathematics is preeminently theoretical; it leads its theorems toward one end, adapting all its assertions to the beautiful and the good, and using them to conduce to being.[16]

Though Euclid was translated into Latin and not unknown in the Middle Ages, the premier mathematical textbook of those later centuries would be in the "Pythagorean" mathematical tradition, not his.* However, and in spite of Iamblichus' opinion, Euclid's *Elements* resonates with joy and appreciation for the beauty of the subject he was exploring

* The *Elements* was translated by Boethius in about A.D. 480, but not until A.D. 1120, when Athelhard of Bath translated it again, this time from Arabic into Latin, did mathematicians begin to appreciate its worth.

as no one had before. Though modern mathematicians still carry forward the ancient Pythagorean/Platonic belief in the beautiful rationality of numbers, and even tend to be suspicious of anything claiming to be mathematical truth that is not beautiful, it is Euclidean technical rigor that guards the gate of beauty.

CHAPTER 11

The Roman Pythagoras

Third, Second, and First Centuries B.C.

IN ROME IN THE SECOND and first centuries B.C. there was a popular legend that Numa, the wisest and most powerful of Rome's ancient kings, had been a disciple of Pythagoras. This was not possible. Dates in the city's early history were under debate, but no amount of fuzziness or fudging could change the fact that Numa died at least 140 years before Pythagoras came to Croton. The Roman lawyer and orator Marcus Tullius Cicero made that clear in his *Republic*:*

> MANILIUS: Is it an authentic tradition, Africanus, that King Numa, was a pupil of Pythagoras, or at least a Pythagorean? This assertion has often been made by our elders, and one gathers that the opinion is widely held. Yet an inspection of the public records shows that it is not properly documented.
>
> SCIPIO: No, Manilius. The whole thing is quite wrong, not only a fabrication but a clumsy and absurd fabrication too (it is

* Cicero's life, and his political life, began when Rome was a republic and ended after the assassination of Julius Caesar and the beginning of the reign of Octavian (Caesar Augustus). He was a strong supporter and defender of the republic and strove on its behalf during the civil wars.

> particularly hard to tolerate the kind of falsehood that is not just untrue but patently impossible). Research has established that it was when Lucius Tarquinius Superbus had been on the throne for over three years that Pythagoras came to Sybaris, Croton and that part of Italy. The Sixty-second Olympiad witnessed both the beginning of Superbus' reign and the arrival of Pythagoras. When the years of the kings have been added up it follows that Pythagoras first reached Italy about a hundred and forty years after Numa's death. No doubt has ever been cast on this conclusion by the experts in chronological research.
>
> MANILIUS: Ye gods! What a gigantic howler![1]

Nevertheless, Numa's discipleship made a good story and represented widespread wishful thinking—that Rome could claim a direct link with Pythagoras. Cicero himself liked the idea:

> For who can think, when Magna Graecia flourished in Italy with most powerful and populous cities, and when in these the name, first of Pythagoras himself, and then of the Pythagoreans afterwards, sounded so high, that the ears of our own countrymen were closed to the most eloquent voice of wisdom? Indeed I think it was because of their admiration for Pythagoras, that Numa the king was reputed to be a Pythagorean by posterity; for, knowing the system and institutions of Pythagoras and having from their ancestors the renown of that king for wisdom and integrity—but ignorant, through distance, of ages and times—they inferred that, because he excelled in wisdom, he was the disciple of Pythagoras.[2]

Cicero was avidly interested in Pythagoras. That a great man of mathematics and philosophy had also reputedly been an effective civic leader—though no specifics were known about his leadership methods or activities—was particularly appealing. Cicero was a prolific author but considered writing a poor second to his active public career.

The connection with Numa was by no means the only bit of fiction and semi-fiction about Pythagoras that was current in Cicero's Rome. The Roman vision of Pythagoras was an amazing mixture of Plato with unfounded legends and assumptions—undergirded by blatant forgeries—and various shades of interpretation and misinterpretation. Pythagoras' name had been familiar to the Roman public at least since the early years of the third century B.C. In the years 298 to 290 B.C., Rome was struggling for the third time to conquer the Samnite tribes in the central and southern Apennine mountains that form the spine of the Italian peninsula. The Samnites were tough warriors desperately defending brutally rugged terrain that was their familiar home ground. When the conflict was going particularly badly for the Romans, they cunningly adopted the military formation that their enemy were using so successfully, a checkerboard pattern in which solid, tight squares of soldiers alternated with square empty spaces.* They also consulted the Oracle at Delphi, which told Rome to honor the wisest and bravest of the Greeks. Responding to this rather insulting order, the Romans chose two figures who were not exactly those a Greek would have chosen: Alcibiades, a notoriously opportunistic military and political genius who had once been a student of Socrates and had often been a thorn in the flesh to the Greeks of his era; and Pythagoras, whom Rome preferred to regard as more Italian than Greek.† The oracle must have been satisfied, for Rome subdued the Samnites. The statue of Pythagoras in the Forum stood for two centuries, until the construction of a new Senate necessitated its removal, probably when Cicero was in his late teens.[3]

By the mid-second century B.C., Rome controlled the entire eastern Mediterranean, and in Greece, Egypt, and Asia Minor, Romans were encountering some of the highest and most ancient cultures in the world. To their credit, for the most part they did not look upon these as the outdated, easily dismissed, quaint cultures of conquered inferiors, but rather chose to regard the older societies as guardians of a valuable legacy to which Rome had now become the heir.

The most significant and long-lasting influence was from the Greeks. The Roman military brought home works of art, slaves who were much better educated than they, and a new thirst for knowledge

* The Romans continued to use this formation effectively through the years of their republic and in the expansion of their Empire.

† Alcibiades' reputation for lack of discipline and unscrupulousness was later used to support the charges brought against Socrates of corrupting the youth of Athens, which resulted in Socrates' death sentence.

and ideas. Before long, upper-class Romans were avidly reading Greek works in translation and even in the original, for many were becoming bilingual. Roman parents sought out educated Greek slaves to tutor their children, and young men traveled to Greece for part of their schooling. Cicero studied philosophy and oratory in Athens and Rhodes. Authors, artists, sculptors, philosophers, and architects who could match the standards of Greek achievements, or at least do a fair job of copying them, were in high demand. Though state business continued to be carried on in Latin, hardly any part of Roman life escaped this peaceful, sophisticated counterconquest. In the midst of what was rapidly becoming not a Roman but a Greco-Roman culture, Pythagoras, an almost homegrown ancient intellectual giant, of mythical stature throughout both the Greek and Italian world, was a Roman treasure. This was "Italian" philosophy. Aristotle himself had called it that.

The poet Ennius—whom later generations would call the father of Latin poetry—also helped provide Rome with a much-needed cultural self-image that involved Pythagoras. One of Ennius' immensely successful poems and dramas was a lengthy historical epic called the *Annales*, purporting to trace Roman history to the fall of Troy. In it, Ennius presented his credentials as the successor to Homer by describing a dream in which that great Greek poet appeared to him on Mount Parnassus and told him that in a former life he, Ennius, had been Homer himself. This dream was symbolic and symptomatic of Rome's vision of herself as the heir to Greek culture, but it did not represent orthodox Roman or Greek doctrine regarding the afterlife. It was instead a nod to Pythagoras and the doctrine of reincarnation. In a satirical poem, *Epicharmus*—the name was that of a Sicilian Pythagorean comic poet—Ennius described another distinctly Pythagorean dream about what would happen after his death, in a place of divine enlightenment.

Ennius was a member of the staff of the Roman consul Marcus Fulvius Nobilior, which gave him yet another Pythagorean connection. Fulvius had returned from military campaigns in the eastern Mediterranean with a passion for Greek culture and laden with captured artistic treasures. He authored a work called *De Fastie* that was probably the original source of a passage claiming to be "what Fulvius reported from Numa," implying something genuinely Pythagorean since Numa, of course, was the early king who was supposed to have studied with Pythagoras.

Fulvius' book in fact owed a great deal to Plato's *Timaeus*, which at the time was almost universally regarded as Pythagorean doctrine.

At the time of Ennius and Fulvius, a cult appears to have existed in Rome and/or Alexandria whose members followed what they believed were the ritual practices and lifestyle of the *acusmatici*. A book had appeared entitled the *Pythagorean Notebooks*, prescribing that lifestyle, and the claim was that Pythagoras had written it himself, though in truth it dated from little earlier than the cult. Nonetheless, Diogenes Laertius later quoted from it in his biography:

> Virtue is harmony, health, universal good, and god, on which account everything owes its existence and preservation to harmony. Friendship is harmonic equality. Honors to gods and heroes should not be equal; gods should be honored at all times with pious silence, clothed in white garments, and keeping one's body chaste; but, to the heroes, such honors should not be paid till after noon. A state of purity is achieved through purifications, washings, ablutions and purifying ones self from all deaths and births and any kind of pollution; by abstaining from all animals that have died, mullet, blacktail fish, eggs and egg-laying animals and from beans and other things forbidden by those who have charge of the mysteries in the sanctuaries.[4]

In second-century-B.C. Rome and Alexandria, many such "pseudo-Pythagorean" books and writings appeared. The semi-historical tradition regarding Pythagoras, fragmentary and confusing as it was already, would be tainted irretrievably by this large body of fiction pretending to be fact.

Cato the Elder, who brought Ennius to Rome and sponsored his introduction to Roman society, read a book called *Pythagoras on the Power of Plants*, a work in the genre of natural and supernatural botany in which he found information about a species of cabbages, *Brassica pythagorea*. Cato included them in his own book *De Agricultura*, a compendium of practical advice for owners of mid-sized agricultural estates, featuring recipes, prescriptions, religious formulae, and high praise for cabbages, especially the Pythagorean variety, leaving little need to grieve for beans. Pliny the Elder, in the next century, like Cato a man of impressive learning and intelligence, nevertheless also failed to discern that *Pythagoras on the Power of Plants* was a forgery and alluded to it in his *Naturalis Historia*, a thirty-seven-volume encyclopedia of every bit of information available to him about animals, vegetables, minerals, and humans.* "Nature, which is to say Life, is my subject," he had declared.[5]

Some authors were meanwhile more focused on attempting to convey authentic Pythagorean doctrine. When Cicero was in Rhodes for part of his education, he sat at the feet of the Stoic philosopher Posidonius, who lived from about 135 to 51 B.C. Many young enthusiasts were seeking out Posidonius as a teacher and role model. Born in Syria, he had traveled widely, and daringly, to Spain, Africa, Italy, Sicily, and what is today France, into regions that were still frontiers, and his accomplishments and physique had earned him the nickname Posidonius the Athlete. Students and contemporaries respected him as one of the most stimulating and learned men of their time.

Only fragments survive of more than twenty books by Posidonius. He apparently discussed what he believed were Pythagorean ideals of good government in a history of the Roman Republic, arguing that Rome's decline in public and political morality was linked to her final defeat of the Carthaginians in 146 B.C. With no enemy on the horizon, Rome had degenerated into a morally weak city, rank with unrestrained behavior and torn by internal political violence and competition for power and wealth.[6] Posidonius treasured Plato's *Timaeus* and attributed part of his own philosophy to the Pythagoreans. According to one of the Posidonius fragments: "Not only Aristotle and Plato held this view about emotion and reason but others even earlier, including Pythagoras, as Posidonius says, who claims that the view was originally that of Pythagoras but Plato developed it and made it more perfect."

Much that is known about Posidonius comes through the Skeptic philosopher and historian Sextus Empiricus, who lived at the turn of the second to third centuries A.D. He apparently took his information from Posidonius when he explained why the Pythagoreans thought that if you claim something is true, mathematical logic is the only standard by which your claim can be judged. "Number" was the principle underlying the structure of the universe: "And this is what the Pythagoreans mean when, in the first place, they are in the habit of saying 'all

* Pliny lost his life when his insatiable curiosity about natural phenomena tempted him too close to the erupting Vesuvius.

things resemble numbers,' and, in the second place, they swear this most naturalistic oath." The oath was the *tetractus* oath.[7] Sextus went on in familiar fashion to point out how the *tetractus* embodied the numbers 1, 2, 3, and 4 that were also in the musical ratios. He listed the four steps, point–line–surface (*tetractus*)–solid (pyramid)—"the first form of a solid body." So "both body and what is incorporeal are conceptualized according to the ratios of these four numbers." To reinforce this idea, Sextus Empiricus gave numerous examples of the ways the numbers and ratios play out in bodily substances, in incorporeal things like time, in everyday life, and in the arts and architecture.

Sextus Empiricus, living at the turn of the second to third centuries A.D., got all this information from an earlier source, but why have scholars concluded it was Posidonius? The clue lies in a sad story set in Posidonius' adopted home, the island of Rhodes. The sculptor Chares of Lindos was engaged to construct an enormous bronze statue, the Colossus at Rhodes. He submitted his estimate of the cost. Then the citizens decided they wanted a statue twice as large. How much would that add to the cost? Chares merely doubled his original estimate—a fatal error. "Twice as large," he remembered too late, did not only mean twice as tall. He had to increase all the dimensions. Chares realized his mistake when all the money was used up on the first phase of the work, and he committed suicide. Sextus included this story in a discussion of numbers and ratios, and scholars see it as Posidonius' fingerprint on Sextus' explanation of Pythagorean theory. The information Sextus preserved was probably what Cicero learned about Pythagoras when he studied with Posidonius.

By the mid-first century B.C., a cultlike group flourished in Rome under the leadership of Nigidius Figulus, a "Pythagorean and magus" in whose Pythagoreanism the line between science and magic grew fuzzy to the point of extinction. Pythagoreanism "for Nigidius and his friends meant primarily a belief in magic," wrote the historian Elizabeth Rawson.[8] Nigidius' reputation for having second sight and occult powers qualified him to work up a birth horoscope of the later-to-be-emperor Augustus, which correctly foretold a brilliant future. Romans of that era did not consider such a scholar out of the mainstream or on the lunatic fringe. Cicero wrote in the introduction to his own translation of Plato's *Timaeus* that Nigidius "arose to revive the teachings of the Pythagoreans which, after having flourished for several centuries in Italy and Sicily, had in some way been extinguished," and that he was "a particularly acute investigator of those matters which nature has made obscure."[9] Nigidius was an educated, prolific author of books on the planets, the zodiac, grammar, natural philosophy, dreams, and theology, with an extensive knowledge of religions and cults from much of the known world.

Romans often invoked Pythagoras' name to represent wisdom and integrity. The scholar and satirist Marcus Terentius Varro, considered by many the most learned Roman of the first century B.C., began his book *Hebdomades* with Pythagorean-sounding praise of the number 7 and a quotation about astronomy from Nigidius. When Varro died he was buried, according to Pliny, in the "Pythagorean mode," in a clay coffin with myrtle, olive, and black poplar leaves.[10] Cicero, for his part, attempted to undermine the credibility of one "Vatinius," a supporter of Julius Caesar, by righteously accusing him of impiety: for he "calls himself a Pythagorean and, with the name of that most thoroughly learned man, tries to shield his monstrous, barbarous behavior."[11] Cicero seems never to have joined a Pythagorean cult, but he made a pilgrimage to Metapontum to visit the house where tradition said Pythagoras died.

Pythagoras made appearances in many of Cicero's works. In a scene from *On the Commonwealth*, set at Scipio Africanus' country estate, Africanus and his nephew Quintus Tubero, the first of several expected visitors to arrive, recline on couches in the Roman fashion, awaiting another guest, Panaetius, who investigates problems of astronomy "with the greatest enthusiasm."[12] In anticipation of his arrival, Scipio mentions a matter that has come up in the Senate about a "second sun,"* then remarks,

> SCIPIO AFRICANUS: I always feel that Socrates was wiser, since he resigned all interests of this sort and declared that problems of natural philosophy either transcended human reason or in no way concerned human life.
>
> TUBERO: I cannot understand, Africanus, how the tradition became established that Socrates rejected all such discussions and investigated only the problems of human life and

* Cicero made several references to this celestial phenomenon that had appeared in the year 129 B.C. The scientific name is parhelion, in the vernacular a mock sun or sun dog. The appearance is of two extra suns, one on each side of the Sun. This happens when the Sun is shining through a thin mist of hexagonal ice crystals falling with their principal axes vertical. If the principal axes are arranged randomly in a plane perpendicular to the Sun's rays, the appearance is of a halo around the Sun.

conduct. Indeed, what more trustworthy authority can we cite than Plato? And Plato, in many passages of his works, even where he represents Socrates as discoursing about ethics and politics, makes him eager to introduce arithmetic, geometry, and harmony, after the manner of Pythagoras.

SCIPIO: What you say is true, but I presume you have heard, Tubero, that after the death of Socrates, Plato went first to Egypt to continue his studies, and later to Italy and Sicily that he might thoroughly master the discoveries of Pythagoras. He was very intimate with Archytas of Tarentum and Timaeus of Locri and acquired the papers of Philolaus.[*] Since at that time the name of Pythagoras was greatly honored in those places, Plato devoted himself to the Pythagoreans and their researches. Thus, as he had been devotedly attached to Socrates and had wished to attribute everything to him, he interwove the charm and argumentative skill of Socrates with the mysticism of Pythagoras and the well-known profundity of his varied lore.

Tubero thinks of Pythagoras in connection with arithmetic, geometry, and harmony. Scipio associates him with mysticism and profound, "varied lore." Later in the same conversation, they invoke his authority on the natural foundation of laws protecting life:

> Pythagoras and Empedocles, men of no ordinary attainments but scholars of the first rank, assert that there is a single legal status belonging to all living creatures. They proclaim moreover, that everlasting punishment awaits those who have wronged anything that lives.[13]

Cicero even weighed in on the bean issue: Pythagoreans avoided them because they cause "considerable flatulence and thus are inimical to those who seek peace of mind."[14]

It was in Cicero's "Dream of Scipio" that he sounded most

* Timaeus of Locri was the central character in Plato's *Timaeus*, but there was no real person by that name. Writings attributed to him cannot be considered examples of Pythagorean doctrine. They are an interpretation of Plato's *Timaeus*, from the first century B.C. or the first century A.D.

Pythagorean—and also much like Plato. The "Dream" concluded Cicero's *De republica*, and in a graceful parallel, he modeled it on the "Myth of Er" that ended Plato's *Republic*. Cicero's "Dream" takes him to a region accessible only to those who through music, learning, genius, and devotion to divine studies have achieved permanent reunion with the highest level of existence. His ears are filled with a sound "strong and sweet," and he asks Scipio what it is. Scipio replies,

> That is a sound which, sundered by unequal intervals, that nevertheless are exactly marked off in due proportion, is produced by the movement and impulse of the orbs themselves, and, commingling high and low tones, causes varying harmonies in uniform degree; for such swift motions cannot be produced in silence, and nature ordains that the extremities sound low at one end, high at the other. Hence the course of the starry heaven at its highest, where the motion is exceedingly rapid, moves with a sharp, quick sound; while the moon in its course (which is the lowest of all) moves with a heavy sound; for earth, the ninth of these bodies, biding immovable in one place, ever holds fast in the center of the universe.[15]

Because Venus and Mercury "are in unison," there are only seven sounds—matching the number of strings on the seven-stringed lyre—"seven distinct tones, with measured intervals between." By imitating this harmony with strings and voices, "skilled men have opened for themselves a way back to this place, as have others who with outstanding genius have all their lives devoted themselves to divine studies."[16] Cicero's metaphor to explain why most humans never hear the celestial music was that their ears are deafened to the sound, just as "where the Nile at the Falls of Catadupa pours down from lofty mountains, the people who live hard by lack the sense of hearing because of the cataract's roar."[17] He gave no indication that he knew Pythagoreans had thought the Earth was not the center of the cosmos. In fact, nowhere in the surviving ancient literature is there a hint of anyone bringing the concept of an audible "music of the spheres" together with the cosmology that included the central fire and the counter-earth, even though the musical ratios had probably played a role in the development of the Pythagorean ten-body model of the cosmos.

In a different realm of scholarship, one extremely successful younger Roman contemporary of Cicero, the architect Marcus Vitruvius Pollio, authored an overview of architecture of his era, *De architectura* or *Ten Books on Architecture*. He recommended Pythagorean ratios and extrapolations on them for the dimensions of rooms, not using any shapes for temples other than one whose length was twice its width (ratio 2:1), or circular. Greek forums were square, but Vitruvius' had a width 2/3 its length, because an audience for gladiatorial combat was better accommodated in that space. For houses, "the length and breadth of courts [atria] are regulated in three ways," two of which employed Pythagorean ratios: "The second, when it is divided into three parts, two are given to the width." The third: "A square being described whose side is equal to the width, a diagonal line is drawn therein, the length of which is to be equal to the length of the atrium."[18] This

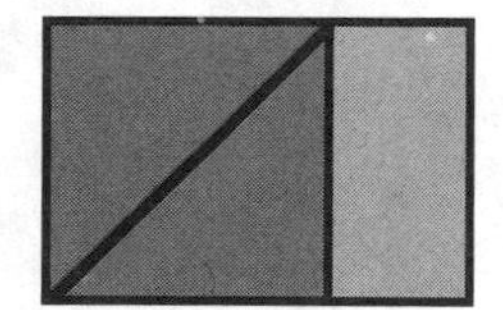

design was based on Socrates' lesson in Plato's *Meno*. "By numbers this cannot be done," wrote Vitruvius. Socrates had used no numbers. The length of that diagonal was incommensurable; so was the length of one side of Vitruvius' room. He frequently mentioned Pythagoras and the Pythagoreans. The Pythagorean theorem was a shortcut in designing staircases, and he unhesitatingly attributed it to Pythagoras.

Vitruvius' books had illustrations, but copies that reached the Renaissance did not. The drawing below, by Cesare Cesariano, is a Renaissance (1521) realization of Vitruvius, who was not easy to interpret. According to the architect Leon Battista Alberti, "Greeks thought he was writing in Latin; Latins thought he was writing in Greek." Nevertheless, this drawing probably faithfully represents his instructions:

> **This proposition is serviceable on many occasions, particularly in measuring [and] setting out the staircases of buildings**

so that each step has its proper height. If the height from the pavement to the floor be divided into three parts, five of them will be the exact length of the inclined line which regulates the blocks of which the steps are formed. Four parts, each equal to one of the three into which the height from the pavement to the floor was divided, are set off from the perpendicular for the position of the first or lower step. Thus the arrangement and ease of the flight of stairs will be obtained, as the figure shows.[19]

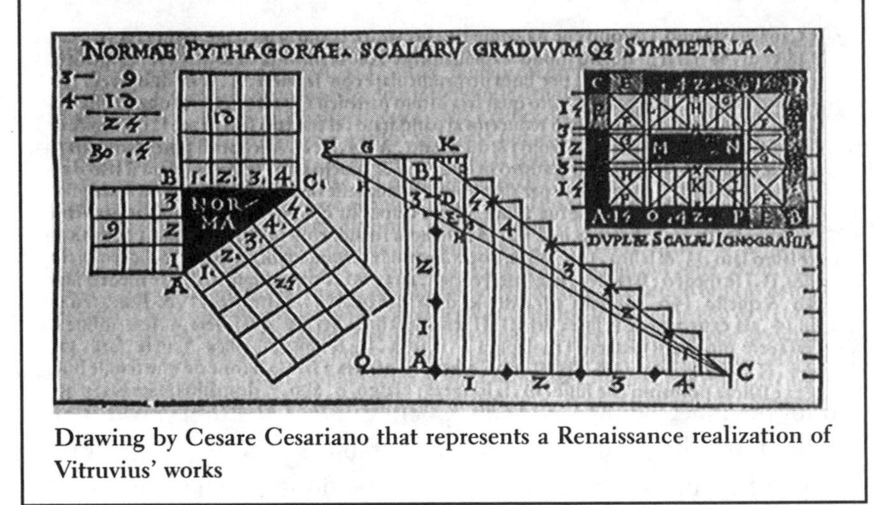

Drawing by Cesare Cesariano that represents a Renaissance realization of Vitruvius' works

Vitruvius' book referred to an unusual application of musical fourths, fifths, and octaves used in an amplification system in Greek theaters. A Roman theater, he pointed out, being made of wood, had good acoustics, but in a Greek theater, made of stone, the voices of the actors needed amplification:

> So [the Greeks placed vessels] in certain recesses under the seats of theatres, fixed and arranged with a due regard to the laws of harmony and physics, their tones being fourths, fifths, and octaves; so that when the voice of the actor is in unison with the pitch of these instruments, its power is increased and mellowed by impinging thereon.[20]

This was by way of demonstrating that an architect must be the master of many subjects—not so difficult as it might seem, thought Vitruvius, for a very Pythagorean reason:

> For the whole circle of learning consists in one harmonious system. . . . The astronomer and musician delight in similar proportions, for the positions of the stars answer to a fourth and fifth in harmony. The same analogy holds in other branches of Greek geometry which the Greeks call λόγος ❑πτικ❑ς: indeed, throughout the whole range of art, there are many incidents common to all.

Music, wrote Vitruvius, assists an architect "in the use of harmonic and mathematical proportion. He would, moreover, be at a loss in constructing hydraulic and other engines, if ignorant of music."[21]

MEANWHILE, THE INSIDIOUS trickle of pseudo-Pythagorean works that had begun in the third century B.C. had become a veritable industry by the first, with publishers and authors trying to meet a continuing demand for books supposedly written by Pythagoras or his earliest followers, or by Philolaus or Archytas. Rome and Alexandria were the places to buy, sell, and collect these scrolls, but those who snapped them up were not only Roman and Alexandrian readers. King Juba II of Numidia, who came to Rome for his schooling, was one of the most avid collectors.[22] The pseudo-Pythagorean books are no help in discovering the real Pythagoras, and would represent unfortunate pitfalls for Pythagoras' biographers, but they are time capsules of what scholars and the public in the third through first centuries, and well beyond, thought Pythagoras had taught and who he had been.

The *Pythagorean Notebooks* were relatively early, from the period when Alexandria was the center of Hellenistic culture and Greco-Roman culture was still largely a thing of the future, and they did not survive long even in complete copies. Their originals are almost as lost in the past as their supposed author. No one knows who wrote them, but it was not Pythagoras, for the author clearly had read the *Timaeus* and was familiar with Plato's unwritten doctrines. In an excerpt preserved by Diogenes Laertius, one of the first sentences mentioned the

Indefinite Dyad.* Traces of pre-Platonic material received an unintentional Platonic update, while passages that depended on later knowledge appear to have been intentionally reworked with an early-Pythagorean twist. Regarding the gestation period of a human embryo: "According to the principles of harmony, it is not perfect till seven, or perhaps nine, or at most ten months." The "harmony" sounded Pythagorean, and "ten months" like a Pythagorean stretch of nature, but other passages having to do with medical matters seem to have mimicked Hippocrates, for whom there was also a large body of "pseudo" literature. A discussion of the significance of opposites in the cosmos rapidly segued into Aristotle, made to sound more "primitive." Aristotle had written that the region below the orbit of the Moon is impure and changeable, but beyond it, all is pure and unchanging, while the *Notebooks* told of the "mortal" area near the earth being stale and "pregnant with disease," and the "upper air" "immortal and on that account divine."† Modern scholarship dates the *Notebooks* to the second or third centuries B.C., not earlier, and certainly not to the sixth century.

Another best-selling pseudo-Pythagorean work was *Lysis' Letter to Hipparchus*, supposedly authored by the Lysis who moved to Thebes after the dispersal of the Pythagoreans in Magna Graecia. Lysis was a real person, teacher of the general Epaminondas, but he did not write this letter. In it, "Lysis" accuses Hipparchus, another Pythagorean, of "philosophizing in public, which Pythagoras deemed unworthy." To prove that Pythagoras frowned on such lack of discretion, the letter writer tells of Damo, "daughter of Pythagoras." Diogenes Laertius quoted:

> When he had entrusted his commentaries to his daughter Damo, he charged her not to divulge them to anyone outside of the house. Though she might have sold his discourses for much money, she did not abandon them; for she thought that obedience to her father's injunctions, even though this entailed

* Diogenes Laertius copied the excerpt not from the original but from an earlier author named Alexander Polyhistor who in turn—this was in the first half of the first century B.C.—copied it from a still older book.

† In view of all the other anachronisms in the *Notebooks*, scholars have ruled out the possibility that they were, after all, authentically early and primitively foreshadowed Aristotle's cosmos.

> poverty, was better than gold, and for all that she was a woman.[23]

Linguistic analysts place the *Letter* in the first century B.C., but some scholars prefer to think it was written at the time of the appearance of the *Pythagorean Notebooks* in order to support their authenticity.[24] The claim would have been that the *Notebooks* were the very discourses that Damo had refused to sell, just recently rediscovered. If the *Letter* was a concoction to support the *Notebooks*, then it was written earlier than 100 B.C. and probably earlier than 200 B.C. But no scholar today believes that *Lysis' Letter to Hipparchus* was written in the fifth century B.C. by the historical Lysis.

The fate of another book, *On the Nature of the Universe* by Occelus of Lucania, is an example of the confusion that occurred even when scholars were well-intentioned. Although Occelus probably lived in the second century B.C., in the early half of the first century A.D. his book was mistakenly regarded as an authentic early Pythagorean text. Occelus and his family considered themselves to be Pythagoreans, but the innocent Occelus had apparently been writing for himself, not trying to pass his book off as something written earlier.[25] However, no less a scholar than Philo of Alexandria, the first-century Grecian-Jewish philosopher, was fooled. Occelus had insisted that the cosmic order was eternal; there was no need for a doctrine of creation. Philo, unaware that Occelus lived after Aristotle, treated his book as evidence that early Pythagoreans, not Aristotle, were the first to introduce the idea that the world is eternal.[26]

By the first century B.C., it had become widely accepted that Pythagoras himself had left no writing, though Diogenes Laertius would later claim otherwise. Works like the *Notebooks* and a three-part book supposedly by Pythagoras (actually from the late third century B.C.) on education, politics, and physics were no longer generally credited, but that did not end the forgeries. It became fashionable to "discover" writings by Pythagoreans like Lysis, the fictional Timaeus, Archytas, and the women Theano and "Phyntis, Daughter of Callicrates." Some offered advice and maxims for daily living. Others claimed to be authentic Pythagorean scientific and philosophical treatises. Many give themselves away today by showing heavy influence from Plato and his pupils, from Aristotle, and from the Stoics, or because their authors made inept attempts to imitate the Doric dialect spoken by the Greeks in Magna Graecia in Pythagoras' time.[27] Even when it was not in "Doric," the writing often had a flowery, pseudo-poetic flavor. (Think of modern attempts to sound like "merrye olde England" and the only slightly more sophisticated efforts of Victorian authors to reproduce medieval speech.) Other Pythagorean forgeries betray themselves simply by their banality; had these been the works of Pythagoras and his followers, the Pythagoreans would hardly have been worth remembering.*

According to one count, at the height of the era of Pythagorean forgeries, there were eighty works "by Pythagoras" in circulation and two hundred purporting to be by his early followers.[28] How could so many readers have been fooled? Not all were. Callimachus, in the mid-third century B.C., declared that a poem Pythagoras was supposed to have written was not authentic. He worked at the Library of Alexandria, and if anyone could spot a forgery, he could. Most readers cannot, however, be seriously blamed for failing to recognize that the pseudo-Pythagorean books were not genuine. The words from the fragment of

* One clue has turned out to be a red herring: the suggestion that inclusion of superstition and "marvelous" events in a work represented more "primitive" thinking and dated the material earlier. Tales about a talking river or being in two places at the same time indicated that what you were reading was authentically early, so it was claimed. However, the late fourth century and the third, second, and first centuries B.C. and the early A.D. centuries were as accepting of magic, marvels, and portents as the fifth and sixth centuries B.C. had been—arguably more so. Such elements were expected in the biography of an important leader. Aristotle wrote during this period, when people may have been more ready to believe in a golden thigh than their forebears would have been at the time of Pythagoras. Clement of Alexandria, an eminent Christian scholar of the second and early third centuries A.D., described a "standard educational curriculum . . . astrology, mathematics, magic, and wizardry"—a quadrivium that would seem appropriate for Harry Potter's Hogwarts School. "The whole of Greece," Clement lamented, "prides itself on these as supreme sciences" (Clement of Alexandria, *Stromateis* 2.1.2. 3–4. Quoted in translation in J. Robert. Wright, ed., *Ancient Christian Commentary on Scripture, Old Testament IX* [Downers Grove, Ill.: Intervarsity Press, p. 18]). For Diogenes Laertius, Porphyry, and Iamblichus, the fact that material included the miraculous did not invalidate the information or call the source into question. There was probably a mystical or magical element to the earliest Pythagoreanism, but late Greek, Alexandrian, and Roman writers were eager to report and exaggerate it. It is difficult to see through the veil of a superstitious age and judge how skeptical an earlier era was, but it is clear that one cannot decide that information was more authentically ancient simply because it included more of the "marvelous."

Posidonius, to the effect that a certain view "was originally that of Pythagoras but Plato developed it and made it more perfect," reflected the assumption that Pythagorean and Platonic doctrines were virtually one and the same—that Plato's philosophy derived from Pythagoras. For readers who believed that, and especially for those who were not aware of how different the philosophies of Plato and Aristotle were from each other, it was an easy step to believe that Aristotle also got his ideas from Pythagoras. So when Platonic and Aristotelian ideas showed up in works claiming to come from before the lifetimes of these two philosophers, why wonder? Was it not from these very documents that Plato and Aristotle had learned?

Pseudo-Pythagorean literature continued to appear for several centuries and was immensely popular. You could pick up a knowledge of "Pythagorean doctrine," unaware or ignoring that it combined some genuinely old material with simplified or summarized Plato and Aristotle, mixed with a good dose of Stoicism, and (in the later books) given a neo-Platonic overcast. You could memorize the maxims of the *Golden Verses of Pythagoras*, or require your children to do so. As was true of *The Prophet* by Khalil Gibran in the twentieth century, you might not notice, or might not care, that what came in the format of authentic ancient wisdom was mostly a contemporary poetic invention and interpretation. The maxims were wise and some of them beautiful. You could find out what "Pythagoras" had recommended regarding the medicinal and magical powers of plants. If it caused you to feel better, this, rather than any scholarly debate, proved the efficacy and authenticity of the book. You could learn what "Archytas" had contributed to knowledge about architecture, agriculture, flutes, ethics, mechanics, wisdom, prosperity, adversity, and "intermediary comfort"—never mind that he had actually had little or nothing to say about some of these subjects. Roman and Hellenistic readers could devour these works, share them, discuss them, make gifts of them, have them read beautifully at weddings and funerals, find themselves uplifted and improved by their high-minded ideas and sometimes enlightened by information that was helpful or challenging no matter where it came from. Romans could feel that they knew something about—and had derived benefit from—their own, magnificent, nearly home-grown sage.

The pseudo-Pythagorean texts outlasted the Roman Empire. *On the World and the Soul*, supposedly by "Timaeus of Locri," was still being recopied in the Middle Ages by scholars who believed this was the early Pythagorean work from which Plato got his cosmology. Copernicus translated *Lysis' Letter to Hipparchus*. One begins to realize the enormous research difficulties, distinguishing Pythagorean fact from fiction, that would confront Diogenes Laertius, Porphyry, and Iamblichus.

CHAPTER 12

Through Neo-Pythagorean and Ptolemaic Eyes

First and Second Centuries A.D.

N Fascination with Pythagoras among Roman and Alexandrian philosophers and scholars of the first century B.C. led to a movement in the first and second centuries A.D. called middle-Platonic or neo-Pythagorean. Books and fragments from men powerfully drawn to what they believed were Pythagorean philosophical and mathematical ideas survive from this period. Some of these writers called themselves Pythagoreans. All regarded Pythagoras as a wellspring, in some cases as the unique wellspring, of a precious intellectual and philosophical heritage that had reached them through Plato.[1] The association of Pythagoras with magic and the occult also continued. Nigidius Figulus' first-century-B.C. version of Pythagoreanism contributed to a growing popular image of Pythagoras—and, oddly, Archytas—as magicians. Nigidius' desire to bring back Pythagoreanism as a way of life and an ongoing approach to the world would attract others in the centuries to follow.

The most important neo-Pythagorean philosophers were, to a man, not from Rome but from other parts of the Empire—Alexandria, predictably, but also from what is now Turkey, from Syria, and even from the Atlantic coast of Spain. The cultlike groups flourished in Rome itself. Information about one of these came through Lucius Annaeus

Seneca, an eminent Roman statesman and orator of the first century A.D. Seneca was a pupil of Sotion, who belonged to a philosophical movement known as the Sextians. The founders, Quintus Sextius and his son, were men of strong moral fiber whose ideal was moral perfection. Theirs was a staunch, Roman approach in which the important thing about a philosophy was how it affected a man's everyday behavior and practical life. Sextians were hard to distinguish from Stoics, but two of their practices were definitely considered to be "Pythagorean": they did not eat the flesh of animals and they performed a self-evaluation at the end of each day, to take stock of personal moral improvement or decline. While no trace of that practice can be found in early Pythagorean communities, it had begun to be associated with "Pythagoreans" in the first century B.C., and Cicero called it a "Pythagorean custom." Seneca described it, as he had learned it from Sotion: A Sextian asked himself, "What bad habit have I cured today?" "What temptation have I resisted?" "In what ways am I a better man?" Similar questions had appeared in the pseudo-Pythagorean booklet called the *Pythagorean Golden Verses*:

> *Never let slumber approach thy wearied eyelids*
> *Ere thrice you review what this day you did:*
> *Wherein have I sinned? What did I? What duty is neglected?*
> *All, from the first to the last, review; and if you have erred, grieve in your spirit, rejoicing for all that was good.*[2]

Sotion had also urged Seneca to adhere to a vegetarian diet, for "souls and animals return in regular cycles. Great men have believed this is so. If these things are true, you avoid guilt by abstaining from meat; if false, you gain in self-control."[3] Seneca's father, who abhorred philosophy, frowned on all this, but Seneca ignored him and avoided meat for more than a year, until under the reign of Tiberius it became dangerous to practice what might be interpreted as a foreign cult.

Another cultlike movement in the mid to late first century A.D. was led by the colorful, eccentric Apollonius of Tyana. Claiming to be the reincarnated Pythagoras, he traveled the Mediterranean world as an itinerant pagan missionary and miracle worker during the reigns of Nero and Vespasian. In a Cilician temple, not far from his birthplace in the Cappadocian region of what is now Turkey, Apollonius established his own "Academy" and "Lyceum," "until every type of philosophy echoed in it."[4] He wrote a biography of Pythagoras, which some have quipped must have been an autobiography, but no one could rival his knowledge of Pythagorean legends and lore from earlier centuries.

More than a hundred years after Apollonius died in 97 A.D., the Roman empress Julia Domna discovered him, probably through a book that she found in the imperial library. This powerful second wife of the emperor Septimus Severus surrounded herself with philosophers and intellectuals; at her request, one of them, Philostratus, agreed to write Apollonius' biography. Julia Domna may have been hoping to undermine the influence of Christianity in the Empire by setting up Apollonius as a competitor to Jesus. Others would put his story to that use.

In Philostratus' book *Life of Apollonius of Tyana*, he had Apollonius retracing Pythagoras' journeys in search of wisdom.[5] In India—not Egypt or Mesopotamia—Apollonius discovers the source of Pythagorean doctrine, including reincarnation with memory of past lives. In other chapters, he is in touch with sacred wisdom closer to home, wrapping himself in his philosopher's cloak and entering a cave shrine in central Greece, announcing "I wish to descend on behalf of philosophy," and emerging after seven days, not there, but at Aulis, clutching a book. He has asked the oracle what is the most complete and pure philosophy and has written down the answer. That book, wrote Philostratus, "contained the views of Pythagoras, since the oracle was in agreement with this type of wisdom." From the time of the emperor Hadrian, the book that Apollonius was supposed to have brought out of the cave was in the imperial library. Many pilgrims and tourists came to look at it in the early third century, around the time of Julia Domna.

According to Philostratus' biography, Apollonius preached abstinence from meat, wine, and sex as necessary for one wishing to draw closer to the spiritual world and see the future. His "Pythagorean" doctrine included supernatural wisdom, universal tolerance, and a way of life dedicated to purification that would eventually release the soul from the prison of the physical body, but no suggestion of witchcraft or magic—extraordinary in an age when hardly anyone discounted them. Philostratus emphasized instead that Apollonius' divine nature allowed him to perform supernatural feats, including escaping persecution by two Roman emperors and reviving a dead girl. Many devotees believed what they read and erected shrines to Apollonius. The emperor Caracalla built a temple to him in Tyana, Apollonius' birthplace. Though he was still venerated as late as Byzantine times, Apollonius did not, eventually, have the staying power of his Christian rival.

Popular interest in Pythagoras was not confined to the Sextians and Apollonius. In the second century A.D., the oracles at Delphi, and at Didyma and Claros on the (now) Turkish coast not far from Samos, adopted a distinctly Pythagorean turn of phrase. The holy man Alexander of Abonuteichos mixed quasi-medical beliefs with his Pythagoreanism.

On a more elevated intellectual level, though "neo-Pythagoreanism" was never a unified philosophy, two themes bound together most of the thinkers grouped under that banner: the old assumption that Plato's philosophy was derived from Pythagoras, and a growing belief that there was one supreme transcendent god. That trend had begun in the second half of the first century B.C., when Eudorus of Alexandria—considered the first important neo-Pythagorean—broke new ground with his own Pythagorean interpretation of Plato, contending that in Pythagorean doctrine the One, the "supreme god," transcended the opposites limited-unlimited and one-plurality. In his table of opposites, One was centered at the very top, not belonging to either column. That alteration would have tremendous importance for philosophy and religion. With Eudorus, "Pythagoras" began to be a code word for a way of thinking in which the One transcended all, something beginning to look like monotheism. Eudorus' interpretation of the Pythagoreans had them believing the invisible supreme god and source of harmony was within reach of human minds. The highest human aspiration was "becoming like god, but Plato had said it more clearly by adding 'as far as possible.'"[6] Eudorus was laying the groundwork for many who would follow him.

The Grecian-Jewish philosopher Philo of Alexandria was younger than Eudorus by about two generations. The Alexandrian Jewish community to which his family belonged was as old as the city, a large, thriving population that had worked hard for more than three centuries to stay on good terms with their Egyptian and Greek neighbors.[7] Under Roman rule, their situation was both helped and hindered by the fact that the Romans gave them special privileges. Roman-Jewish relations were, nevertheless, precarious. Philo served on an Alexandrian/Jewish delegation to Rome that floundered when the emperor Caligula, who thought himself a god, insisted his own statue be erected in the temple in Jerusalem.

Philo was a devout man who made pilgrimages to Jerusalem, where the great temple still stood, but his wealthy, influential parents had made sure he received a thoroughly Hellenistic-Greek education. He was both a devout Jew and a Platonist.

Like Eudorus, Philo interpreted Plato as having taught that one supreme god was primary to everything in the universe, and thought Plato got these transcendental leanings from Pythagoras. Philo quoted Philolaus: "One god, who is forever, is prince and ruler of all things, stable, unmoved, himself similar to himself, different from others."[8] The soul's journey toward God was the ultimate task of life, and, for Philo, the Hebrew Scriptures exemplified that journey. He saw the lives of Moses and Abraham as the pilgrimage of the soul toward God.[9] Adam was intellect; Eve, sensation; Cain and Abel, a soul's being torn in opposite directions of evil and good. Philo's Pythagorean interpretation of Genesis gave special attention to the "fourth day," when God completed the creation of the heavens. The number 4 contained the musical ratios found in the structure of the heavens and represented the four stages in the creation of the planets, point–line–surface–solid. The musical ratios also contained the number 3, representing the three dimensions of created bodies—length, breadth, depth. Numbers were the ideas and the tools of God in creation; they also made it possible for humans to understand the heavens.

While mulling over the issue of whether such a thing as "time" existed before the creation of the universe, Philo got caught up in the question of whether the Pythagoreans or Aristotle had been the first to suggest that the universe is eternal, and mistakenly cited Occelus of Lucania's *On the Nature of the Universe* as evidence that it had been the Pythagoreans. Anticipating the concept of time set forth by the Christian philosopher St. Augustine of Hippo, Philo insisted that "there was no time before the world, but it came to be either with the world or after it."[10]

Some have called Philo a Greek philosopher who remained grounded in his religion; others, a Hebrew mystic who used the tools of Greek thought in the service of religion.[11] He combined the practice (in both Greek and Hebrew traditions) of drawing lessons from Homer or the Hebrew Scriptures with his fine understanding of Greek philosophy and developed a philosophical interpretation of the Scriptures that he hoped would win respect among Greek intellectuals. But his impact on Greek philosophers was not as great as he hoped. No later pagan philosopher appears to have mentioned him directly.[12] Rather, it was the early Christian writers who followed his lead and used the allegorical method as he did for reconciling revealed truth with intellectually worked-out truth. Clement of Alexandria and Origen were admirers (Clement dubbed him Philo the Pythagorean) and generations of early and medieval Christian scholars carefully preserved and copied his work, so that an extraordinary amount of what Philo wrote survives intact.*

The Roman poet Ovid, a contemporary of Philo, captured in his *Metamorphoses* the more popular image of Pythagoras: the all-knowing sage of legendary antiquity with an aura of universal, unworldly wisdom. Ovid revived the old legend about the Roman king Numa and had Pythagoras speak through him, in an oration that stressed the doctrine of reincarnation and abstention from meat, mostly on the grounds of respect and sympathy for animals. Ovid was not attempting to philosophize along Pythagorean lines or argue Pythagorean doctrine. The oration was part of a larger picture he painted in his poem, in which everything is changing, shifting, being transformed, nothing endures, and "Nature, the great inventor, ceaselessly contrives." Hence the title, *Metamorphoses*.

Plutarch, later in the first and second centuries A.D., was more influential than Philo or Ovid in forming the image of Pythagoras for the future. His *Parallel Lives* paired biographies of celebrated Greeks and Romans, part of an effort to find ways of resolving the differences between the Roman culture of power and Greek intellectual culture. How much to trust Plutarch for historical and biographical details has been a frustrating and often unanswerable question, but that has not prevented his *Lives* from being the text on which readers from the Renaissance to the present day have based their understanding and picture of the ancient world. Plutarch was one of Shakespeare's favorite authors.† Copernicus also read Plutarch, and he quoted, in Greek, from

* There was a legend about a Christian Philo, even a Bishop Philo, and a story in which he met the Apostle Peter.

† Shakespeare found the stories of *Antony and Cleopatra*, *Timon of Athens*, and *Coriolanus* in the *Lives*, and sometimes used Plutarch's words virtually verbatim or changed them (as he read them in translation) only as much as was necessary to transform them into verse.

Plutarch, from a sixteenth-century engraving

Plutarch's *Placita*, in a letter to Pope Paul III—a passage in which Plutarch had written that "Philolaus the Pythagorean" claimed the Earth moved around a central fire. Like Plato and Cicero, Plutarch wrote an elaborate myth about the fate of the soul.[13] For his last thirty years he was a priest of Apollo at Delphi, and he came up with a Pythagorean interpretation of the god's name, equating him with the One: *a* meant "not," *pollon* meant "of many." Hence *a-pollon* was One. Given Plutarch's influence, it is significant that he linked Pythagoras with the Pythagorean theorem. It is largely because of Plutarch that nearly everyone believes Pythagoras discovered it.

Of all who thought that Plato derived his ideas from Pythagoras, no other was so convinced and outspoken as Moderatus of Gades, known as the "aggressive Pythagorean." He lived in the second half of the first century, in Gades (later Cádiz) on the west coast of Spain, by then part of the Empire. Moderatus could be called the Pythagorean conspiracy theorist, for he insisted that not only Plato but also Aristotle, Speusippus, Aristoxenus, and Xenocrates were plagiarists who had "taken for themselves the fairest fruit of Pythagorean thought" and designated as "Pythagorean" only the most superficial and trivial aspects of the school, so as to make a mockery of Pythagoras and the Pythagoreans.[14] The aphorisms were among those "superficial and trivial aspects," and Aristotle's listing of them was, to Moderatus, a malicious act. He scorned the *acusmatici*, and his answer to the argument that Aristoxenus also had scorned them was that Aristoxenus' attitude was a particularly subtle part of the propaganda project against the true Pythagoreans.

Moderatus had a fresh take on the Pythagorean use of numbers, which helped explain how he thought the plagiarism had been engineered. The Pythagoreans, "for the sake of a lucid exposition," wisely resorted to "explanation by means of numbers" because it was so difficult to explain the first principles and primary Forms clearly in language.[15] For example, to say the One was above all else did not mean that the number 1 itself was the supreme fundamental of the universe. Instead, One stood for a great unifying principle, implying equality, everything that causes stability and unchangingness, the absence of "otherness." In a "language" of numbers, the Pythagoreans had expressed everything that Plato would later attempt to express in words. Moderatus had got neatly past Aristotle's stumbling block, the notion that numbers were, for the Pythagoreans, both abstract things and the physical building materials of the universe. In fact, he had made a jump into the twentieth century, when some found it easier to describe the quantum world and the origin of the universe in mathematical formulas than in imprecise descriptive words.

There had been hints before of a way of looking at the world in which physical matter was evil or at least negative, and in Moderatus this view came much more to the fore. Physical matter was a "shadow," and that, to him, meant something more negative than it had to Plato. It was "notbeing." But Moderatus thought that everything including matter came from the One, and the One was "the Good." He failed to address the question of how anything wholly good can produce evil, but the problem of the origin of evil was looming on the philosophical horizon.

At the opposite extreme of the Empire, in about A.D. 125, Theon of Smyrna wrote *Mathematics Useful for Understanding Plato*. It included arithmetic, harmonics, astronomy, geometry, the symbolism of the numbers 1 through 10, different forms of the *tetractus*, and the *tetractus*' link with music and the cosmos.[16] "The one who bestowed it was Pythagoras," wrote Theon, "and it has been said that the *tetractus* appears indeed to have been discovered by him." One passage sounded like a collage of almost everything that had been Pythagorean, or been thought to be, up to Theon's time. But if Theon's mathematics were "useful," the mathematics of Nicomachus of Gerasa, a generation later, in the mid-second century, would prove, for better or worse, much more so over a very long period of time. Nicomachus was one of those mathematicians who rejected Euclid's abstract theorems of numbers and their proofs, preferring to stick to what he thought were "Pythagorean mathematics" and to offer only numerical examples. His *Introduction to Arithmetic* was intended to be not an original contribution, but, essentially, a textbook, and that was what it turned out to be for most of Europe for more than a thousand years, until the Renaissance. The opening passages were a paean to Pythagoras, and largely because of that, for centuries—long past the Renaissance, in fact—the Pythagoreans were regarded as the source of Greek mathematics. W. K. C. Guthrie was not overstating the case when he wrote,

> Everyone comes upon the name of Pythagoras for the first time in school mathematics; and this has been true from the earliest stages of the Western cultural tradition. None of the ancient textbooks which formed the basis of the medieval curriculum forgets Pythagoras. . . . the origin of this tradition: Nicomachus.[17]

Nicomachus also wrote a *Handbook of Harmony* that linked the ratios of music with the movements of the heavenly bodies. That book has survived complete, while his two-volume, avowedly Pythagorean *Theology of Numbers* survives in fragments. Nicomachus set up a correlation between the numbers 1 through 10 and the gods of Olympus that Iamblichus and Proclus would later use in a last-ditch defensive effort against Christianity, on behalf of Greek philosophy and pagan religion.[18] Nicomachus' *Life of Pythagoras*, now lost, was a source for the Pythagorean miracle stories.*

* Nicomachus was also intrigued by a pseudo-science called *gematria* that was not Pythagorean but originated with the ancient Babylonians and survived in ancient Greece and the Hellenistic period. In *gematria*, each letter of the alphabet had a numerical value. A word could be spelled in numbers. Sargon II, in the century before Pythagoras' birth, had the wall of Khorsabad built to a measurement that was the numerical equivalent of his name—16,283 cubits. The name for the Gnostic divinity Abraxas had the numerical value of 365, the number of days in a solar year. Nicomachus did not claim that *gematria* was a Pythagorean practice, and it was not.

The neo-Pythagorean philosophical tradition ended on a strong note with an extraordinary writer and thinker named Numenius of Apamea. Born in Syria, he produced his most important work, only fragments of which survive, around A.D. 160. His books were available, however, long enough for Porphyry and others to read and discuss them in the next century when they studied with the philosopher Plotinus.*

Numenius believed that the teaching of Plato's Academy in its purest form came from Pythagoras, but he wanted to know where Pythagoras had, in turn, got his ideas and knowledge. He took at face value all the stories of Pythagoras' travels, and he unearthed what seemed to him "Platonic" philosophy (that Plato got via Pythagoras) among the Egyptians, the ancient peoples of India, and the Magi of Mesopotamia, as well as in the Hebrew Scriptures. His goal was to trace knowledge to the earliest, highest, primal sources, because, in his opinion, it had all been downhill from there. "Who is Plato but Moses speaking Greek?" he asked, and retold the story of Moses and the plagues in Egypt from a more Egyptian point of view, in which Pharaoh's magicians had more success combating the plagues than they did in the Hebrew Scriptures.

According to Numenius' most ambitious work, his six-volume *On the Good*, "the Good" or "the First God" (what other neo-Pythagoreans called the One) was not completely inaccessible. Sense perceptions were not helpful, but a human could work on finding access. In an exquisite passage, Numenius described the degree of solitude necessary for an approach to the Good or the First God:

> Like someone seated in a lookout post, who, straining his eyes, manages to catch a glimpse of one of those little fishing vessels, a one-man skiff all alone, isolated, engulfed in the waves, even so must one remove oneself far from the things of sense, and consort alone with the Good alone, where there is neither human being nor any other living thing, nor any body great or small, but some unspeakable and truly indescribable wondrous solitude—there, are the accustomed places, the haunts and celebrations of the Good, and it itself in peace, in

* Plotinus used and developed Numenius' thoughts so extensively that he was accused of plagiarism. A colleague came to his rescue by writing an entire book to point out the differences between the two.

> benevolence, the tranquil one, the sovereign, mounted graciously upon Being.[19]

Numenius also urged a more active approach: disregarding "sensibles" and devoting oneself enthusiastically to learning the sciences and studying numbers, so as to attain the knowledge of what is Being.[20] He did not regard Plato as a high point in the history of knowledge—rather as part of a downward slide—but he was not unappreciative and gave him a backhanded compliment: "He was not superior to the great Pythagoras, but perhaps not inferior either." He called Socrates a Pythagorean, and Plato a brilliant mediator between Pythagoras and Socrates.

Numenius introduced a doctrine of "three gods" that he called "typically Pythagorean." Though it might have been possible to find hints of the idea in the work of other neo-Pythagoreans (Moderatus, for example, thought the Pythagoreans believed in three "unities"), in Plato, and in the pseudo-Pythagorean literature, "typically Pythagorean" was an overstatement. Numenius' "three-god" passages suggest he was clinging to the theology of Pythagoras/Plato and the polytheism of the pagan world, while at the same time reaching for the concept of the Christian trinity—a prodigious intellectual and theological balancing act. He saw a philosophical need for a trio of roles in the creation and sustenance of the universe and came close to what others would call the Father, Son, and Holy Spirit.

Numenius' First God was intrinsically good, the source of all goodness, and, more than anything else, rational, an intellect—what the earliest Pythagoreans had touched in the discovery of the ratios of musical harmony. This god was "the Good" of Plato, "the One" of the Pythagoreans and neo-Pythagoreans. The "thinking" of this god was the source of life. In the work of Numenius—in a mind deeply informed by Plato but moving beyond him—what had come to the first Pythagoreans as a revelation was, at last, receiving a brilliant philosophical and theological workout.

Numenius's Second God was responsible for the reincarnation of souls and also a mediator between the First God and the material, human, physical world, and had therefore to have two natures in order to understand and focus on both. Numenius gave a great deal of thought to the roles of the Second God and the paradoxes involved. Difficulties that would be the subject of debate in the early Christian church regarding the nature of Christ were already being given deep consideration by the pagan Numenius.

The Third God was either the created cosmos or the world soul. Numenius did not make clear his ideas about which it was and seemed not to think the question was important. But he struck a note firmly in the Judeo-Christian tradition (Numenius loved the Hebrew Scriptures) of creation in the image of God when he wrote that "the nature and Being that possess knowledge is the same in the god who gives and in you and me who receive." He laid all this at the feet of Pythagoras with the words "and that is what Plato meant when he said that wisdom was brought by Prometheus to mankind together with the brightest of fires." Ever since Plato wrote that passage, the intellectual world had thought he was speaking of Pythagoras. Numenius did not disagree.

With him, the problem of the origin of evil at last reared its head in Western philosophy. Numenius wrote that all living things, including the world itself, have two souls. The good soul was the soul he was referring to when he wrote "the nature and Being that possess knowledge is the same in the god who gives and in you and me who receive." The bad soul was made up of primeval matter originating before any god "adorned it with form and order." What was the source of the bad soul? Was there only one source of everything, a good god, who then "withdrew from its own nature," as Numenius put it, to make room for the existence of evil? (Asking the question in Pythagorean words: Did the One have to give up something of itself so that "plurality" and the rest of a table of opposites could exist—including the opposite of good, evil?) Numenius' answer was no. Evil did not emerge from Good or from God. Good or God did not relinquish anything or move aside. Evil was as old as God. There was no One overarching the opposites. Both good and evil were part of primordial reality. Anything else was an incorrect interpretation that had emerged when "some Pythagoreans did not understand this doctrine."[21]

After Numenius, it became impossible to differentiate neo-Pythagoreanism from neo-Platonism.

NEAR THE END of the second century A.D., Ptolemy (or Claudius Ptolemaeus), who did not call himself a Pythagorean and was sometimes critical of the Pythagoreans, picked up strongly on the idea of the harmony of the spheres and gave it a long future. He lived and worked at Alexandria and was interested in a great variety of subjects, including acoustics, music theory, optics, geography, and mapmaking. His most brilliant accomplishment was to draw together, from previous ideas and knowledge and out of his own mathematical genius, the Earth-centered astronomy that would dominate Western thinking about the cosmos for more than a thousand years. Ptolemy's book *Harmonics* also had an impact on the history of science, because Johannes Kepler read it in the seventeenth century. One of Ptolemy's sources was probably Archytas.

Ptolemy knew that harmony in music was based on mathematical proportions showing up in sound, and he agreed with the earliest Pythagoreans that mathematical principles underpin the entire universe, including the movements of the heavens and the makeup of human souls. He devoted nine chapters in *Harmonics* to the harmony of the spheres, applying harmonic theory to planetary motions.

One principle Ptolemy followed was to "save the appearances"—that is, not to make up theories that contradicted what one actually saw happening. He would not have proposed ten heavenly bodies because of the importance of the number 10, if he could not see ten in the sky. His astronomy looks superficially, to modern eyes, as though its inventor made up rules and patterns but never looked up. Indeed, Kepler wrote that "like the Scipio of Cicero he seems to have recited a kind of Pythagorean dream rather than advancing philosophy."[22] But the evidence that seems overpowering now could not be detected in Ptolemy's day. When Aristarchus of Samos proposed a Sun-centered astronomy in the third century B.C., his theory was dismissed on the sound basis that the evidence for it was, simply, not there. For Ptolemy, with musical harmony, "what one actually saw happening" translated to what one actually *heard* happening. The judgment of the human ear about what was pleasing was of first importance when considering theoretical possibilities.

The system of heavenly harmony that Ptolemy worked out was more complicated than previous ones. The early Pythagoreans may have connected the intervals of the octave, fourth, and fifth (rather than a complete scale) to a cosmic arrangement. Or perhaps the ten-body cosmos, with an octave separating the central fire and the outer

Ptolemy, from a medieval book illustration

fire, did constitute a complete scale once all the intervals between were filled in. Plato's "Myth of Er" and Cicero's "Dream of Scipio" proposed cosmic scales of eight or seven notes respectively. Pliny, much more specifically, would have had the cosmos sounding the following scale:

Earth		C
	(whole tone)	
Moon		D
	(half tone)	
Mercury		E flat
	(half tone)	
Venus		E
	(one and a half tones)	
Sun		G
	(whole tone)	
Mars		A
	(half tone)	
Jupiter		B flat
	(half tone)	

Saturn		B
	(one and a half tones)	
Stars		D[23]*

Nicomachus, earlier in Ptolemy's century, had also assigned notes to each of the planets, but in his scale Earth was silent because it was sitting still.

When Ptolemy worked out his own system, he considered it such a significant accomplishment that he had it engraved on a slab of stone at Canopus near Alexandria.[24] He felt that he had made a connection with ancient knowledge, taking the concept of the music of the spheres back to something close to the Pythagorean original. Venus and Mercury shared a note; the stars were in the chorus, with the highest note; the four elements sounded the two lowest notes. For the first time perhaps since the ancient Pythagoreans, the intervals used were larger than tones, half tones, and one and a half tones. Bruce Stephenson, who wrote that Ptolemy's Canopic Inscription is so difficult to interpret that no one can claim to understand it completely and decipher it correctly, nevertheless made the following attempt.[25] The notes (on a piano) are rough equivalents of what they would have been in the tuning in late antiquity:

Fixed stars	D	(a whole tone above Saturn)
Saturn	C	(a fourth above Jupiter)
Jupiter	G	(a fourth above the Sun and a whole tone above Mars)
Mars	F	(a fourth above Venus and Mercury)
Sun	D	(a whole tone above Venus and Mercury)
Venus and Mercury	C	(a fourth above the Moon)
Moon	G	(a fourth above fire, air)
fire, air	D	(a whole tone above water, earth)
water, earth	C	

What might have been the most significant part of Ptolemy's *Harmonics* was lost before the Middle Ages. It is not certain whether some

* This scale adds up to more than an octave, a problem easily corrected by changing the interval between Saturn and the stars to a half tone, as music theorists in later antiquity corrected Pliny. A half tone (half step) is the interval between one key and the next—black or white—on a piano.

text recovered in the fourteenth century by the Byzantine Nikephoros Gregoras is really part of what was missing. In the seventeenth century, Kepler translated the *Harmonics* and attempted to re-create the last three chapters, an exercise that helped him find the way to one of his most important discoveries. In spite of how little is left of the details of Ptolemy's musical theory and the fact that what is left is not fully understood, Stephenson wrote,

> What is clear today—and was clear to Kepler at the beginning of the seventeenth century—is that Ptolemy thought that orderly motion, in the heavens as in music, followed only certain kinds of patterns, so that study of the patterns in one field could in theory elucidate those in the other. Rational motion obeyed the same laws everywhere, in the celestial spheres as in the strings of the *lyra*, not for any mystical reason but precisely because those were the laws of rational motion. . . .
>
> The motions of the planetary spheres could similarly be understood more deeply through an awareness of the principles they shared with musical harmony. In Ptolemy's *Harmonics*, connections such as these were assumed to be rational, although they were not assumed to be understood—yet—in detail.[26]

In Ptolemy, the Pythagorean conviction that the universe is rational and that numerical relationships underpin nature, and the belief in an overall harmony binding creation together—all were here. But humans, if Ptolemy can be taken as an example, though no less obsessive about discovering this harmony, had grown more patient about teasing out examples of it, less prone to force the patterns, a little more willing to be taught by nature itself how the numbers play out.

CHAPTER 13

The Wrap-up of Antiquity

Third–Seventh Centuries A.D.

DIOGENES LAERTIUS, THE earliest author to write a biography of Pythagoras that still survives substantially today, either was born in the town of Laerte in Cilicia—a region that in the days of the Roman Republic was the feared "Pirate Coast" and is now southeastern Turkey—or was a member of a prominent Roman family known as the Laertii and born in Rome. He was probably writing late in the second or early in the third century, during the reigns of the emperors Septimius Severus and his son Caracalla. Not only is nothing biographical known with certainty about him; he also never put his own philosophy in writing.

Reading his *Life of Pythagoras*, however, acquaints one rather well with Diogenes Laertius. His research method set him apart and makes him a delightful writer. Gathering as much information as he could, often in bits and pieces, he put together a rather informally written collection of biographical and bibliographical material, summaries of doctrines, sayings of philosophers, his own poetry about them, and comical and scandalous stories. Most of the time he scrupulously named his sources, and he liked to contrast one piece of information with another, sometimes pausing to try to assess their credibility. The greatest attraction and value of Diogenes Laertius' writing comes from his verbatim quoting (sometimes at length) from authors whose works are otherwise lost. Many of his sources are completely unknown except for his mention and the excerpts in his books. Pythagoras was not the only ancient thinker to interest him. The *Life of Pythagoras* was Book VIII of his ten-volume *Lives of the Philosophers*.

Much more can be said about Porphyry and Iamblichus, both of whom were major neo-Platonic philosophers in their own right. Iamblichus was Porphyry's student, and Porphyry was in turn a disciple of the eminent Roman philosopher Plotinus.

Porphyry lived only a little later than Diogenes Laertius. He was born in about A.D. 233 in Tyre, in Phoenicia (now in southern Lebanon, then part of the Roman Empire), which may explain why he was the only one of the three biographers to link Pythagoras' father with Tyre. "Porphyry" was not the name his parents gave him. He was Malchus, meaning "king," and changed his name when he was in Athens in his early twenties at the suggestion of his teacher, the philosopher Longinus. Longinus knew that the area of Malchus' birth was famous for a purple dye made from the crushed shells of sea snails mixed with honey. The color that resulted was porphyry, so highly prized and expensive that it had come to symbolize royalty. The connection—Malchus/king and Porphyry/royalty—suggested the name that may at first have been only Longinus' nickname for him.

For ten years, Porphyry immersed himself in Platonic and Pythagorean doctrine at the feet of Longinus, whom someone described as "a living library and walking museum," and during this period he published some unofficial oracles, utterances of mediums put into trances at private séances.[1] When he moved to Rome to study with the even more eminent Plotinus, who was carrying forward Pythagorean and Platonic themes and the thoughts of other philosophic predecessors in a creative synthesis of his own, his new teacher turned him into a thoroughgoing advocate of a rational, intellectual approach to truth. However, neither Porphyry nor Plotinus ever entirely discounted magic and the supernatural. As the historian E. R. Dodds wrote of the period, "Could any man of the third century deny it?"[2]

Romans were looking for encouragement wherever they could find it, in the natural or the supernatural.[3] Civil wars were following one after the other. The news of a new emperor hardly had time to spread before it was outdated. Rome was engaged in ruinously expensive conflicts on two fronts—with the Persians in the Orient, with the Goths and other Germanic tribes to the northeast on the European river frontiers and the Black Sea. Septimus Severus—Julia Domna's husband—and his son Caracalla had built the elaborate Baths of Caracalla at the beginning of the century, but most of the decades since had been a period of enforced austerity, with civilians sacrificing nearly all comforts and amenities to make sure the Roman legions could be paid and would remain loyal. Epidemics were rampant; imperial finances a disaster. The government put less and less gold and silver in the coins, and the currency collapsed, with prices rising nearly 1,000 percent between the years 258 and 275. This debacle would have touched Porphyry almost anywhere he might have lived in the Empire, but he was at the center of it, in Rome itself.

In the midst of what for many amounted to sheer misery, Plotinus continued to teach. He led seminars, wrote essays (which Porphyry collected in six books called the *Enneads*) and moved in aristocratic circles that included the court of the emperor Gallienus, who had intellectual and philosophical pretensions of his own and apparently enjoyed the philosopher's company. Edward Gibbon candidly described Gallienus as

> a master of several curious but useless sciences, a ready orator and an elegant poet, a skillful gardener, an excellent cook, and a most contemptible prince. When the great emergencies of the State required his presence and attention, he was engaged in conversation with the philosopher Plotinus, wasting his time in trifling or licentious pleasures, preparing his initiation to the Grecian mysteries, or soliciting a place in the Areopagus of Athens.[4]

Gallienus became enthusiastic about Plotinus' plan to create Plato's Republic in the countryside near Rome. When the emperor's interest waned, the idea was abandoned, and the city of Platonopolis was never built.

Porphyry's mentor had a high regard for the Pythagorean and neo-Pythagorean concept of the One. It was, for Plotinus, the First Principle, transcending all else. The One, the Spirit, and the Soul were his trinity. Porphyry heard him teach that it was not possible even to think about the One, much less to define it. There was no movement or number to it. It was just One, unity, absolute pure reality and goodness, never changed or diminished.[5] Plotinus' One was close to the concept of God in Christianity, except that it never intervened in the world. It remained external, outside all orders of being. Yet, in the words of the historian Michael Grant, Plotinus thought the One "pours itself out in an eternal downward rush of generation which brings into being all the different, ordered levels of the world as we know it, in a majestic, spontaneous surge of living forms."[6] This meant that all levels of the cosmos, all levels of existence, all living beings, were linked. Mortal bodies were base and degraded, but every soul had the potential to rise to reunion with the One by means of intellectual work and discipline, and life itself implied a longing for that reunion. The quest required banishing space, time, and body into a nothingness even more profound than the solitude of Numenius' little "one-man skiff, all alone." Plotinus claimed he had experienced this mystic union himself and there was nothing "magic" about it, "except the true magic which is the sum of love and hatred in the universe."[7]

This philosophy that Porphyry was studying was not depressing, but—perhaps because of conditions in Rome or for personal reasons—he sank into melancholy and considered taking his own life. Plotinus prescribed travel, so he went to Sicily. In 269 or 270, Plotinus died shortly after retiring to the country and Porphyry returned to become the head of his school.

Perhaps Dodd's description of Porphyry as "an honest, learned, and lovable man, but no consistent or creative thinker"[8] was correct, but he was a prolific writer. In addition to publishing his teacher's essays, he wrote more than seventy books on metaphysics, literary criticism, history, and the allegorical interpretation of myth, as well as his short *Life of Pythagoras*. He classed Pythagoras with Orpheus, Herakles, and Jesus, as "divine heroes" who led exemplary, devout lives and became immortal, but like many other neo-Platonists of his generation he saw Christianity as a dire threat to the Platonic tradition. Porphyry expressed his fears in a famous, now lost book called *Against the Christians* and in letters to the soon-to-be wife of his old age, Marcella. Though Christianity was still struggling, it was nearing the political triumph that would occur not long after Porphyry's death, and its adherents thought of Jesus not as a divine hero but as equal to or one with God. Drawing believers and potential believers away from Jesus was not the only motivation for Porphyry's biography of Pythagoras—he intended it as a popular introduction to Platonic philosophy—but he did hope that getting it before the public would provide competition for the Christian Gospels. Readers then, like modern ones, were more drawn to a personality than to a collection of philosophical ideas, and, in Khan's words, "among the Neoplatonists, it was Porphyry who reinstated Pythagoras as the patron saint of Platonic philosophy, in the tradition of Nicomachus and Numenius."[9]

Porphyry shared Numenius' interest in ancient sources of Pythagoras' knowledge. He concluded that the Egyptians, the Hebrews, and the ancient peoples of India and Mesopotomia had possessed not only invaluable but identical primordial wisdom, and that Pythagoras was the earliest to have this wisdom among the Greeks, with Plato later putting it most fully into words. True to his teacher Plotinus, Porphyry preferred a rational, intellectual approach to truth, but his enthusiasm for the richness and mystery of ancient sources of wisdom, and the fact that he lived in highly superstitious times, prevented him from dismissing miraculous reports about Pythagoras. He also had what Dodds described as "an incurable weakness for oracles,"[10] for which the rationalism of Plotinus had not succeeded in providing a permanent remedy. Porphyry died in A.D. 305, when he was seventy years old, soon after taking Marcella, to whom he had written many eloquent letters having to do with Platonic thought, as his young bride.

Iamblichus wrote the third and longest of the three biographies of Pythagoras. He was Porphyry's pupil, but also a rival. Porphyry may have had an incurable weakness for oracles and not denied the existence of supernatural experience, but Plotinus had convinced him that the approach to ultimate truth and reunion with the divine was through the intellect, the rational, and had nothing to do with magic. Not so Iamblichus. For him, that reunion could only be achieved through ritual and magical evocation. His treatise *De Mysteriis* has been dubbed a "manifesto of irrationalism."[11]

Iamblichus was born in about A.D. 260 in Chalcis, in Syria, and was a wealthy man who owned slaves and suburban villas, but he dedicated his life to contemplation, teaching, and writing, had many devoted disciples, and became widely renowned as the "divine" Iamblichus. The emperor Julian, in the next century, exclaimed that Iamblichus was "posterior indeed in time but not in genius to Plato."

Iamblichus was much more focused on Pythagoras than was Porphyry, whose *Life of Pythagoras* was one volume out of a ten-volume *Lives of the Philosophers*. Iamblichus' *On the Pythagorean Way of Life* (his biography of Pythagoras) was the introductory book to a work of nine or ten volumes with the collective title *On the Pythagorean School*, all of which were about Pythagoras and the Pythagoreans. Iamblichus attempted to put into this compendium everything that was known about Pythagoras and Pythagorean doctrine and philosophy, reflecting the view that Plato got most of his ideas from Pythagoras. However, in spite of his focus, to Iamblichus a mastery of philosophy meant more than knowledge of everything Pythagorean. It meant an understanding of Aristotelian logic and Plato's dialogues. This was, in its way, Iamblichus' attempt to provide what Plato had recommended, a thorough grounding in everything one could know about and through numbers, followed by dialectic, and Iamblichus' contemporaries held him in high regard for his ability to reduce Plato's and Aristotle's thoughts to a more manageable form. He tried to make converts for Platonism by using logical arguments and warnings of fearful consequences for those who failed to pursue a philosophical life. Iamblichus died in about A.D. 330, during the reign of the emperor Constantine.

From that time on, neo-Platonic philosophy would include a Pythagorean emphasis on mathematics and numbers, adding some numerology that had no Pythagorean roots. Neo-Platonists would assume that Plato's *Timaeus* derived from Pythagoras via a real Timaeus of Locri who wrote *On the Nature of the Cosmos and the Soul*—which was actually one of the pseudo-Pythagorean books. In the fifth century A.D., the philosopher Proclus would open his *Commentary on the Timaeus*: "It is agreed by all that, since he acquired the book which the Pythagorean Timaeus composed *On the Universe*, Plato undertook to write the *Timaeus* in the Pythagorean manner." "Pythagorean thought" had assumed the form in which it was going to survive for a thousand years and reach Copernicus.

BY A.D. 400, efforts to block the impact of the Christian Gospels in the Greco-Roman world had failed. Proclus, the last major Greek philosopher—born in Constantinople, educated in Alexandria and Athens, and later head of what remained of Plato's Academy—would go on for much of the next century opposing Christianity, but that was an exercise in futility, with unintended consequences. He and those who read him were largely responsible for a great spread of neo-Platonism throughout the Roman, Byzantine, and, later, Islamic regions of the world, but, more than that—and it would surely have caused him chagrin to learn of it—Proclus' work, thanks to a mistaken identity, became a major influence in Christian theology. His philosophy was adapted by a writer of his own century who was for a long time confused with the New Testament figure Dionysius the Areopagite, a first-century convert of the Apostle Paul. In the writing of this "pseudo-Dionysius," the philosophy of Proclus passed into Christian thought.

The advance of Christianity in the Empire took place on several levels. In at-home, family piety, it replaced the old household gods. House churches, with a few families and individuals (before there were larger Christian communities) competed with pagan cults all over the Empire with steadily increasing success. Among intellectuals, Christian authors co-opted excerpts from pagan books when the wording seemed equally applicable to a Christian context, or when they hoped to undergird Christian teaching by pointing out that the pagan material represented an independent witness to truths now more fully explained. Some of these fragments survived without much change at all. In fifth-century Christian sermons and books, phraseology and imagery appeared that clearly had originally come from a pagan oracle.[12] During its first three centuries, Christianity had very little interest in or influence on politics, but that changed with the conversion in 312 of the emperor Constantine, who ruled first in Rome and finally from Byzantium over both the eastern and western halves of the Empire. Christianity became the official religion, and after that the Empire was mostly ruled by Christians, though some of them followed Arian teachings and did not accept the full divinity of Jesus.

Christianity's victory would have been much more difficult had it not been able, to such a large extent, to accept and assimilate the great pagan intellectual and philosophical traditions. It might have been as short-lived and ineffectual as most of the mystery cults if its leaders had listened to men and women who advised complete rejection of pagan Greek and Roman culture and learning. However, beginning with the Apostle Paul, many Christian theologians and writers were educated men with tremendous respect—indeed, great love—for this heritage. When Paul first arrived in Athens, he expected that this city, which felt like his intellectual home and obviously treasured learning and wisdom, would welcome with open arms and minds the new knowledge he brought. At first it seemed he was right, and the Athenians' eventual rejection of his views was one of the lowest points of Paul's life.

To many early Christian intellectuals it seemed, as it had to Paul, that most of the older culture, wisdom, and knowledge of their intellectual world, which had so informed and inspired them, was consistent with the greater truth they believed they now knew. All had been, in a way, an intellectual preparation. They felt compelled to bring the pagan philosophical heritage into the embrace of Christian thought, make it part of Christian education, and show that there was a continuum. The early church became a guardian of the treasures of classical literature, both in its actual preservation of the books and in the way many key ideas were assimilated into Christian writing and thinking. The Gospel of John in the New Testament opens with the words "In the beginning was the Word." In the original Greek, "Word" is *logos*, and probably the best translation of *logos* is not "word" but "reason" or "rationality." With that in mind, it is possible to read John's words with Platonic and Pythagorean eyes:

> In the beginning was Reason. And Reason was with God, and Reason *was* God. Reason was with God in the beginning. Through Reason all things were made; without Reason nothing was made that has been made. In Reason was life, and that life was the light of men. . . . Reason became flesh and lived among us. We have seen his glory, the glory of the One, who came from the Father, full of grace and truth.[13]

A Pythagorean mathematical interpretation of nature presented no conflict with Christian doctrine. St. Augustine, one of those who strove most successfully to bring Christian belief and pagan philosophy into harmony, mentioned the importance of numbers in his *City of God*: "Not without reason has it been said in praise of God: 'Thou hast ordered all things in measure, and number, and weight.'"[14] Though the doctrine of reincarnation was discarded in favor of Christian immortality, the image of the body being a tomb or prison for the soul was retained. Clement of Alexandria called it a Pythagorean doctrine conveyed through Philolaus.[15]

The prodigious efforts of intellectuals like Augustine to seize upon similarities and work through conflicts in hope of finding deeper levels of agreement encouraged the Christian church to undertake a mission to preserve classical Greek and Roman literature in a more deliberate fashion. In Italy at Monte Cassino, a monastic center of scholarship called St. Benedict's was established in A.D. 529, and other centers soon followed, particularly in the sixth and seventh centuries, when Irish missionaries reached England and then the Rhine, and Gothic missionaries reached the Danube. Because hardly anyone on the mission frontiers spoke Greek, Greek writings failed to spread much beyond the borders of the late Roman Empire, but even remote monasteries were able to preserve some Latin works while one horde of invaders after another swept across Europe—Goths, Vandals, Franks, and later Norsemen. The treasured, scattered works would be the only ancient classical literature known in Latin Europe for centuries. It is difficult in the twenty-first century—because so much ancient literature has been recovered—to realize how dim and fragmentary knowledge of the past became, how pitifully little was remembered, how completely civilization had to start over in the Middle Ages.

Of the few works that survived, one was a Latin translation by a fourth-century Greek Christian scholar, Chalcidius, of the first fifty-three chapters of Plato's *Timaeus*. Others were a fragment of Cicero's translation of the same dialogue, the works of the "encyclopedist" Macrobius, Nicomachus' *Introduction to Arithmetic* in a slightly revised Latin translation, and *De institutione musica*, possibly also a paraphrase of Nicomachus. The latter two were produced by Boethius, a Roman of the late fifth and early sixth centuries. Why these and not others? Much had to do with the language in which a work was written or into which it was translated. Now we return to Rome as her great empire began to disintegrate.

From the close of the third century and the reign of the emperor Diocletian, the Roman Empire had no longer been ruled by one emperor. Sometimes there had been two, sometimes more. Though the administrative division line between the Empire in the east and the Empire in the west was not deliberately drawn along a language frontier between areas that spoke Greek and those that spoke Latin, over time it began to seem so, as Greek and Latin came to dominate in their respective regions. With the institution of bishoprics, a parallel division took place in the church, with Christians in the East regarding the patriarch of Constantinople as their religious authority while those in the West followed the bishop of Rome, the pope.

It was a dangerous, uncertain time in both parts of the Empire, with barbarian tribes pushing one another around the European map and not stopping when they reached borders that had been secure for centuries, close to Rome and Constantinople. On New Year's Eve of 406, the Rhine froze, rendering the river useless as a natural boundary between Roman Gaul and tribes on the other side. The Roman legions previously guarding the Rhine had been recalled to hold defenses against barbarians nearer to home, and Vandals, Alans, and Sueves poured into Gaul, moving inexorably south and west across the countryside, marauding, plundering, and burning, meeting virtually no resistance. Feeling pressure on too many frontiers, and with her former lands in Gaul now "like an enormous funeral pyre," Rome was nearing the end of her tether, but, though her Empire was divided, the city herself had remained unconquered for more than a thousand years, and it was unthinkable that this could change.[16] Then in A.D. 410 the Visigoths sacked Rome. It would be more than another half century before the last emperor of the western Empire ceased to rule, but this date marked a loss of morale and political identity that would never be recovered. Some thought the Christian God had failed the city.

During this period, a writer named Ambrobius Theodosius Macrobius pulled together, as Pliny had done, a body of knowledge from a vast variety of sources with little or no attempt to discriminate between what was authentic and what legend, forgery, reinterpretation, and inaccurate retelling.[17] He said he was not a Roman, though he served Rome in official capacities at home and in Spain and Africa. He also seems not to have been a Christian, in an era when most offices as highly ranked as those he held were filled by Christians. But whatever else he was or wasn't, Macrobius was indeed an "encyclopedist," and prone to expanding on the original. Largely thanks to him, Cicero's "Dream of Scipio" would be popular in the Middle Ages, but when Macrobius wrote about the "Dream," what Cicero had covered in a few pages took him nearly sixteen times that many, for he added commentary and inserted the opinions of other authors. The result was not very original, but for scholars in the Middle Ages it would be a treasure.

Macrobius preserved, in Latin, much that would otherwise not have survived, particularly from the neo-Pythagoreans and neo-Platonists, whom he knew primarily through the writings of Porphyry. Medieval scholars would learn from him that Pythagoras discovered the ratios of musical harmony, and would read the story that this happened in a blacksmith shop. They would know Macrobius' quotation from Cicero about the harmony of the spheres and would glean from him ideas about connecting the musical ratios and the planetary distances, but nothing of a link between specific notes and specific planets, though both Nicomachus and Ptolemy had worked these out before Macrobius' time. The Pythagorean view of numbers underlying everything in the universe, and the way it was exemplified by linking the harmonic ratios with the arrangement of the cosmic bodies, would become standard in medieval writing on music theory, but not so well formulated as it had been by ancient scholars, getting vaguer and less well understood as time passed.

The Visigoths who sacked Rome in 410 did not consolidate their victory by establishing a new government. For a while the imperial government limped along, sometimes surprisingly effectively, though virtually all of the western part of the Empire was now overrun by Germanic tribes who continued to make and break agreements and alliances with Roman and local authorities and to war among themselves. German settlers in Italy rather quickly converted to Christianity, many of them at first to the form known as Arianism which had been declared a heresy by the Council of Nicaea in 325; but eventually they entered Roman Catholicism.

In 429, the Vandals accomplished an end run through Spain, invaded North Africa, and became a new pirate threat in the Mediterranean. A quarter century later, it was their turn to sack Rome. Again, the invaders did not stay, but when they left they carried a former empress and her daughters back with them to Africa. On August 23, 476, the German troops, who by then actually made up most of the Roman army in Italy, elected their general Odoacer as king and overthrew the last Roman emperor, Romulus Augustus. The Roman Empire in the West, long in its death throes, at last expired. In theory, the emperor of the eastern Roman Empire, Zeno, now ruled the entire Empire, but Odoacer, though in no position to reestablish anything like the former Empire, was, in effect, an independent ruler, while various German factions in Italy could only war uselessly among themselves and with other tribes who continued to appear on the horizon. In the western

Empire, including its former vast holdings to the north in Europe, one might assume that the Dark Ages had begun. They had not, quite.

Boethius, born in 480, after the overthrow of the last Roman emperor, was a Roman aristocrat in an era when conventional wisdom would seem to indicate there should no longer have been such a thing. Roman life as usual had not, however, completely ended in the city and its environs. The Roman civil service continued to operate, courts administered Roman law, Roman and Gothic landholders were paying their taxes, and learning and culture had not disappeared. The Roman Senate was still meeting, and Boethius became a Senator. He was also a philosopher, theologian, poet, mathematician, and astronomer—one of the last generation to study at what was still calling itself the Academy in Athens—and he was deeply troubled to see his contemporaries losing the ability to read Greek, which had for centuries been part of Roman education.* No longer could they experience Plato, Aristotle, the neo-Platonists, or many of the Christian church fathers in their original language. Boethius vowed to remedy this potentially disastrous loss: "I will translate into Latin every work of Aristotle that comes into my hands, and all the dialogues of Plato."[18] Much else also, it turned out.

Boethius accomplished an astounding amount of translation before he made an unfortunate career decision. Rome was far from a dead city, but it was no longer the city from which Italy was ruled. The Ostrogoth leader Theodoric, in Ravenna, had taken over from the Visigoths in 493, when Boethius was thirteen, murdering (some said personally) the king, Odoacer, whom the German army had elected. Having grown up in Constantinople and married a Byzantine princess, Theodoric admired classical culture and liked to surround himself with intellectuals. Boethius was particularly attracted to him because Theodoric hoped to reconcile the Romans and the Goths, and that shared goal also drew Theodoric to Boethius. Boethius decided to attach himself to Theodoric's court in Ravenna.

He did not last long there. In 523, he was falsely accused of treason and the use of magic. Theodoric imprisoned him and executed him the next year, but not before Boethius in his prison cell had done his

* Along with all other pagan schools, the Academy would close in 526, two years after Boethius died, by order of the Byzantine emperor Justinian.

Boethius, probably a late medieval representation

deepest thinking and most eloquent writing—a book redolent of Platonic ideas, called *Consolations of Philosophy*. Throughout the Middle Ages, Boethius was considered on the level of one of the church fathers, if not exactly one of them, but in this book he wrote not of the Christian command, but of the Pythagorean command, to "follow God."

For centuries, medieval scholars in Latin Europe would know the Greek authors through the Latin translations and commentaries Boethius had written. In large part thanks to him and Macrobius, the flame of classical Greek philosophy was kept burning in the monasteries of the Middle Ages. Also because of Boethius, Nicomachus' version of neo-Pythagorean mathematics became the bane of every student's existence.

The impact in the Middle Ages of another of Boethius' books, a multivolume work called *De institutione musica*, was almost as great as that of his *De institutione arithmetic*, which preserved Nicomachus' mathematics. Islamic philosophers would refer to it when they wrote about music from a Pythagorean point of view, and it would become a staple when medieval educators adopted the Pythagorean quadrivium. The first three volumes of *De institutione musica* were probably a translation or close paraphrase of Nicomachus' *Introduction to Music*, since lost. The approach was Pythagorean, emphasizing the importance of the musical ratios, linking specific notes of the scale to the Sun, Moon, and planets, and referring frequently to Pythagoras. Boethius divided "music" into three subjects: *musica mundana* was the harmony of the spheres, *musica humana* the relationship of music to the human soul, and *musica instrumentalis* what we normally think of as music.

In the sixth century, the old Roman Empire in the East had a new name, the Byzantine Empire, and was still alive and flourishing brilliantly. Greco-Roman civilization had certainly not died there. Alexandria was a wealthy, thriving city, as were Jerusalem and Antioch; Constantinople had replaced Rome as the capital of the civilized world, and its emperor was for all intents and purposes the head of the Christian church. Before mid-century, the Byzantine general Belisarius drove the Vandals out of North Africa, conquered the southern part of Spain, and retook Rome. The Byzantine Empire soon held Ravenna—ending Theodoric's brief golden age there—as well as Genoa, most of Sicily, and southern Italy, including Calabria (the old Magna Graecia) which would not be lost until the middle of the eleventh century. However, the reconquest of Italy, rather than restoring prosperity there, destroyed what little was left. It was in the Near and Middle East and North Africa that the old traditions of teaching and learning continued, and where Christian scholars were carefully preserving ancient texts and knowledge of the ancient Greek language.

The preservation and treasuring of classical philosophy and learning would continue in those regions for many centuries, but not under the aegis of the Christian Byzantine Empire. In the seventh century, followers of Mohammed poured out of the east. In Syria and Egypt there was scarce resistance, and the great cities surrendered quickly with little damage when the conquerors assured the Jewish and Christian populations that they could continue as usual with their beliefs and worship. This was fortunate for still-existing ancient texts, which came into Islamic hands and were regarded as a precious heritage. By 718, the Arabs held all of Spain, where they would continue as a small but powerful elite, ruling in a manner that was astoundingly tolerant in religious matters and open to cultural influences from all over the Mediterranean and Islamic world.

In the monasteries of Christian Latin Europe, scholars eked out a meager intellectual living on the works of Macrobius and Boethius and a few other classical Latin authors, copying and preserving them with excruciating care, occasionally hearing and hardly believing rumors that the lost literary and philosophical treasures of Greece and Rome still existed in a far-off place. But it was under the rule of Islam in the Middle East, North Africa, and Moorish Spain that most of the preservation of ancient knowledge and writings, and the development of newer mathematics and astronomy based on them, moved forward from the eighth century to the eleventh.

PART III

Eighth–Twenty-first Centuries A.D.

CHAPTER 14

"Dwarfs on the shoulders of giants": Pythagoras in the Middle Ages

Eighth–Fourteenth Centuries

By the eighth century, the book destined to be Ptolemy's most celebrated work had reached Baghdad. Islamic scholars translated it into Arabic, and *Almagest*, "The Greatest," was its ninth-century Arabic title. While Islamic mathematicians and astronomers were advancing beyond the methods and models of Hellenistic scholars, no one apparently questioned the Earth-centered model of the cosmos or seems to have been aware of Philolaus' Pythagorean ten-body model with the central fire and counter-earth. Al Fargani, a brilliant ninth-century Arab astronomer, estimated the sizes of the spheres in which the planets move and worked out relationships among their distances, but musical ratios were not part of his calculations or those of other Islamic astronomers. Those with an ear for Pythagorean harmony of the spheres in the Islamic world were men concerned with the effects of instrumental and vocal music on the health of the body and the well-being and morality of the human soul. They were following the lead of a ninth-century writer named Honein Ibn Ishak al-'Ibadi, or Hunayn.[1]

The mission of the Bayt al-hikma, or House of Learning, an academy in ninth-century Baghdad, was to retrieve the knowledge of antiquity and make it available to readers in Arabic. Baghdad was a cosmopolitan city where ideas flowed freely and minority religions were regarded as no serious threat. Hunayn, though not a Muslim but a Nestorian Christian, was both a member of this academy and the chief court physician to the caliph.*

Hunayn's fluent Greek made him useful for more than his medical expertise. He translated books from Greek into Arabic for Islamic patrons and into Syriac for Christians, and he also produced Arabic translations of the works of his ancient medical predecessor Galen, and of the Hebrew Scriptures (from a Greek version).

Curious about the way music affects the human body and psyche, Hunayn wrote the first known treatise on music in Islamic literature, full of Pythagorean/Platonic themes of unity, harmony, Forms, the estrangement of the soul from the divine, and the possibility of their eventual reunion. "Living in solitude, the soul sings plaintive melodies, whereby it reminds itself of its own superior world," he wrote, and described how life often works to seduce the soul away from this superior world. For him, music, rather than numbers, was the great underlying connector: "The excellence of music is evident by the fact that it appertains to every profession, like a man of understanding who associates himself with everybody." Hunayn compiled a collection of aphorisms, anecdotes, letters, and excerpts from a variety of Greek sources that he titled *Maxims of the Philosophers (Nawadir al-falasifa)*.[2] He took these from compilers like Plutarch, not from the originals, but the excerpts frequently began with "Plato used to say," or "Aristotle said," or "Alexander asked Aristotle," invoking Archytas and Euclid as well. Numbers were involved in some of the excerpts, but no real mathematics.†

Hunayn had a sense of humor:

> Once a philosopher went out for a walk accompanied by his disciple. They heard a voice and a guitar. The philosopher said to

* Nestorian Christians were a group that originated in Asia Minor and Syria in the fifth century A.D. and stressed the human nature of Christ. There are still many thousands of them; today called the Church of the East, the Persian Church, or the Assyrian or Nestorian Church. Most Nestorians live in Iraq, Syria, and Iran.

† It is indicative of the cosmopolitan mix of religions and ideas in the Middle Ages in Islamic regions of the world that Hunayn's writing, reflecting ancient pagan ideas and coming from a Christian who lived and worked in Islamic Baghdad, survived mainly because of a twelfth/thirteenth-century Hebrew translation by Judah al-Harizi.

> his disciple: "Let us approach the guitar; perhaps we can learn some sublime Form." But as they came closer to the guitar, they heard a bad tone and an inartistic song. The philosopher then said to his disciple: "The magicians and astrologers assert that the voice of an owl indicates death for man. Were this true, the voice of this man should indicate death for an owl."[3]

Some of Hunayn's collected aphorisms were later incorporated into a mammoth Islamic encyclopedia that appeared about a hundred years after his own lifetime, produced by a tenth-century community known as the Ikhwan al-Safa', or Brethren of Purity, in Basra in southeastern Iraq. Like the scholars of Hunayn's House of Learning, the Brethren attempted to preserve all they could of the ancient scientific and philosophical material that had come into Islamic hands. Their chief undertaking was an encyclopedia called the *Rasa'il,* in fifty-two volumes. Its purpose was to cover human knowledge in its entirety. Some ancient books were paraphrased, but few passages were taken verbatim in translation. Instead, the *Rasa'il* was an extravagant re-envisioning of earlier doctrines, an example of a second phase of the work to which so many Islamic scholars were devoted. One of the greatest, Al-Kindi, described it as striving to "complete what the ancients have not fully expressed, and this according to the usage of our Arabic language, the customs of our age, and our own ability."[4]

The Brethren of Purity viewed all knowledge as a continuum of revelation taking place in all times and places and among all races and religions. Pythagoras, Plato, Abraham, Jesus, Mohammed, and the imams who succeeded Mohammed were all part of it. The Brethren put together a cosmology of their own having many interlinked levels of being, with "One God" whose holiness lived in all things. The highest destiny of a human was to rejoin his or her inner holiness to this One God. The *Rasa'il* wove together many aspects of the world—music, numbers, medicine, theology, astronomy, and other areas. The interlinking was based in a precise manner on numbers and music and conjured up a "unity," not as the Pythagoreans had done it but very much in the Pythagorean spirit and with them in mind. Referring to ancient "musician-philosophers" who had drawn a connection between the four elements—fire, air, water, and earth—the Brethren linked these with human health, the arrangement of the cosmos, and the four strings of an instrument called the *oud*:

> This we have expounded in the treatise on arithmetic. In effect, the first string is comparable to the element of fire, and its sonority corresponds to the heat and its intensity. The second string is comparable to the element of air and its sonority corresponds to the softness of air and its gentleness. The third string is comparable to the element of water and its freshness. The fourth string is comparable to the element of earth and its sonority corresponds to the heaviness of earth and its density.[5]

The link with the human body and physical and mental health, echoing Hunayn, had the sounds of the different strings producing different effects in those who heard them: "The sonority of the first string reinforces the humor of yellow bile, augments its vigor and its effect; it possesses a nature opposed to that of the humor of phlegm," and so forth.[6] Unlike Hunayn, however, the Brethren included specific mathematics. In their arrangement of the cosmos, each of the four elements, plus something called frigidity, predominated in one of a set of nested spheres with the Earth in the center. The size of each sphere, in relation to the next, was in the ratio of 4:3. Beyond the orbit of the Moon there was a "harmonious proportion that exists between the diameters of the spheres in which the planets move and those of earth and air."[7]

The Ikhwan al-Safa' or Brethren of Purity arrangement of the cosmos beneath the orbit of the Moon. This drawing shows the spheres, but not the exact proportions.

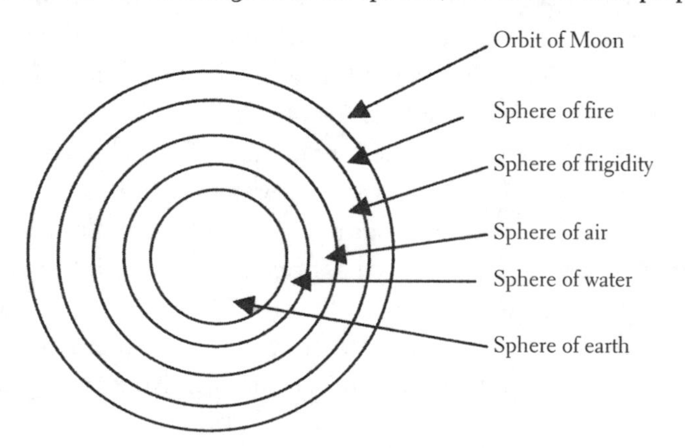

The Brethren made connections among the cube, the notes of the strings on the *oud*, and the relationships between the notes: The ancients, they said—continuing their reinterpretation of Pythagorean thinking and working Euclid into it as well—had a preference for the octave, because 8 was the first cube number (2×2×2). A cube has six sides and 6 is a perfect number.* All of a cube's sides are equal and all of its angles are equal, and

> we have said that the more the created thing possesses the property of equality, the greater its eminence. It is for this reason that it was said in Euclid's last treatise that the form of the earth is probably cubic and that of the celestial sphere probably a dodecahedron defined by twelve pentagons.[8]

The preference for "equality" calls to mind Archytas.

Also in the tenth century, when the Brethren of Purity were compiling their encyclopedia, a Shiite *katib* (secretary) in Syria wrote a treatise in the same tradition. What seemed most significant to Al-Hasan al-Katib was the Pythagorean insight that numbers and the relationships between them were the key to human understanding of the universe. He reformulated this doctrine and applied it both to the human body and soul and to the cosmos: Three was the number of the simple consonances in music (fourth, fifth, octave) and also the number of the divisions of the soul (rational, sensible or sensitive, and natural or vegetative). Seven was the number of notes in the scale short of the octave and seven was also the number of elements of the rational soul: comprehension, intelligence, memory, deliberation, estimation, syllogism, knowledge. Different modes in music were equivalent to different virtues of the soul: justice—the mode of the index finger on the second string; good understanding—the mode of the open third string; purity—the middle finger on the third string; and so forth. In Al-Hasan's scheme, the movements and positions of the celestial spheres also had their equivalents in music, and the origin of the zodiac could

* By "perfect number" they did not mean what the Pythagoreans had meant when they identified 10 as the perfect number. A perfect number by more modern standards (found already in Nicomachus) is a number the sum of whose divisors equals the number. The number 6 is the smallest perfect number: 1+2+3=6.

be explained in similar terms. He wrote that he was indebted to Nicomachus of Gerasa for these ideas.

> The sphere of the Zodiac is divided into twelve parts which represent the houses of the Zodiac. We believe that this division was established [or defined] thus because the number 12 is divisible into halves, thirds, and quarters. These are the elements which are found in the division of the complete system, because the last note of the octave is half the first (2:1), the note of the fifth is in the ratio of one and a half to it (3:2), the note of the fourth is in the ratio of one and a third to it (4:3).[9]

IN THE PARTS of Europe not under Islamic rule, populations in different areas spoke a great variety of vernacular languages and dialects, but Latin was the lingua franca that united scholars, who were almost without exception the only people who could read. As Boethius had feared, Greek had disappeared almost entirely. Little remained in Europe to be read in Greek anyway, since Greek manuscripts that at first had been preserved by Christians were now nearly all far away in Islamic lands.

However, Latin Europe did not languish in total intellectual darkness. Even in the ninth and tenth centuries, when Viking, Magyar, and Saracen invasions repeatedly wreaked havoc, scholars were carrying forward Pythagorean/Platonic ideas about numbers and music. Taking them in a different direction from their Islamic contemporaries, they explored the links between music and astronomy and inventively manipulated the numbers and mathematics of music and the cosmos.

Aurelian of Réôme (now Moutiers-Saint-Jean) was a contemporary of Hunayn; most of his writing dates from the decade 840 to 850.[10] The earliest medieval music treatise that has survived was his *Musica disciplina*, based in part on Boethius. Aurelian did know Greek and some astronomy and was knowledgeable about the movements of the planets and their periods.* He observed that the eight musical modes seemed to copy eight kinds of celestial motion. He wrote of instances when angelic music was audible on Earth, and about the Muses and the zodiac, though he had to be inventive to link eight musical modes with nine Muses. Aurelian followed the Pythagorean lead in more ways than his

* A planet's period is the time it takes to complete one orbit.

interest in the harmony of the spheres: He knew of the quadrivium that Plato learned from Archytas, and he was convinced that the truth of the universe lay in numbers.

> The motions of the stars are eight, seven of the planets and one of that which is called the Zodiac [the sphere of fixed stars], which all say make the sweetest harmony of song; that is, consonance. Even the Lord, in the reply that he made out of the whirlwind to Holy Job, called this the harmony of heaven.[*]
>
> There are other things that writers about this art have discovered. They say that the whole theory of the art of music consists of numbers.
>
> The natural discipline is given over to four sciences, namely, arithmetic, geometry, music, and astronomy. In these, numbers, the measurements of the earth, sounds, and the positions of the stars are examined; but their essence and their whole origin is in mathematics.[11]

John Scotus Eriugena also lived in the ninth century, at about the same time as Hunayn and Aurelian.[12] Nothing is known about his early life, though his name must indicate he was from Ireland ("Eriugena" can be translated as "Erin born"), which escaped the barbarian invasions overrunning most of Europe until the Danes arrived later in the century. Eriugena may have started out as a scholar connected with one of the great Irish monasteries, but when he was in his thirties, Charles the Bald invited him to France to be head of his court school. He is also thought to have traveled to Greece and Italy, studied Greek, Arabic, and Chaldean, later moved to Oxford at the invitation of Alfred the Great of England, and finally taught at the Abbey of Malmesbury. Eriugena was the scholar who translated the works of "pseudo-Dionysius" into Latin.

The idiosyncratic cosmological scheme that Eriugena developed made him one of the most remarkable scholars of his era—in fact, of all who predated Latin Europe's tenth–twelfth century rediscovery of the classical literature. In his cosmos, the stars, Moon, Sun, and Saturn orbited the Earth, but Mercury, Venus, Mars, and Jupiter orbited the

* Aurelian was reading from a mistranslation of the Book of Job.

Sun. This arrangement was not at all harebrained; in fact, it was an insightful step in the direction of Copernican astronomy. However, it presented a challenge to the harmony of the spheres. In Eriugena's cosmos, the four planets with Sun-centered orbits were of course continually changing their distances from Earth. A change of distance meant a change in the musical pitch of a planet, and so a theory of cosmic harmony had to allow for varying pitch. That was an idea that no one but Eriugena would explore until Giorgio Anselmi in the early fifteenth century and Johannes Kepler in the late sixteenth and early seventeenth—Kepler, at last, with a correct understanding of the solar system and much more fruitful results.

Eriugena worked out the problem of the varying pitches of the heavenly bodies in his own way in an elaborate system involving numbers, ratios, and musical intervals, drawing examples from organ pipes and stringed instruments: "Here one must admire the wonderful virtue of Nature; for what anyone can accomplish on a four-stringed lyre is achieved in the eight celestial sounds. But the method by which it is done must be sought out with diligent investigation."[13]

He explained some of the results of this investigation in language that he tried to make reader-friendly:

> As you see, the sounds do not always relate by the same intervals, but according to the altitude of their orbits. No wonder, then, that the Sun sounds an octave with Saturn when it is running at the greatest distance from it, but when it begins to approach it, it will sound a fifth and when it gets closest, a fourth. Considered in this manner, I think it will not disturb you when we say that Mars is distant from the Sun sometimes by a tone, sometimes by a semitone.[14]

Eriugena also urged his readers to keep in mind that when comparing the distances of planets, one was talking about the ratios and relationships of the distances between the planets, not the absolute distances in *stadia* (or, in modern terms, in miles or parsecs).

He had his own take on the agreement between neo-Pythagoreanism/Platonism and Christianity: Everything in creation derived from the One, and the One was the same thing as God. From this One, who was universal, all-containing, infinite, and incomprehensible, emanated the realm of Plato's Forms. Under the influence of the Holy Spirit, the Forms manifested themselves in created things. All creatures would eventually be drawn back to reunion with the divine level of being from which they had fallen. God was both "the source of all things and the final end of all things."

For Eriugena, *all* was "fairest harmony," including not only the heavenly spheres but "even the sounds that will arise from the punishment of evil, for punishments are good when they are just, and so are rewards when they are more in the nature of gifts than payments for what is earned." The result of punishment and reward would be a final purification and redemption, even of animals and devils, and reunification into the divine One, with full knowledge of God, seen "face to face," as St. Paul had written. For Eriugena, the great harmony of creation was a "combination of low, high, and intermediate sounds making a certain symphony between them through their proportions and proportionalities."[15]

A younger contemporary of Aurelian and John Scotus Eriugena, Regino of Prüm, referred to the Pythagoreans in the introduction to a book he wrote about the plainsong melodies used in Trier, his native town. Claiming that he got his information from Boethius' *De musica*, he offered what he believed was the Pythagorean argument for the existence of heavenly music.*

> The Pythagoreans argue the presence of music in the heavenly motions thus: how, they say, could the heavenly apparatus, so rapid in its course, move in silence? Even though it does not reach our ears, it is still quite impossible that such headlong speed should lack sound, especially since the course of the stars are arranged in so convenient and well-adapted a way that nothing so enmeshed and conjoined can be imagined. Some are higher, others lower, yet all are turned with an equal impulse so that their unequal and disparate orbits fall into a determined order. From this it is argued that there is a harmonious arrangement in the heavenly motion.[16]

* Regino's description sounds very much like Aristotle's, which means he must indeed have gotten it through Boethius. Regino lived before the reintroduction of Aristotle to Latin Europe.

The more important issue for Regino was not whether the heavenly motions produced a sound but whether they took place in a "harmonious arrangement." For him, "harmony" was a beautiful scheme of numbers and number relationships pervading the universe, underlying both music and the arrangement and movements of the planets and stars.

Regino was a musician, not an astronomer, but before moving on from the introduction of his book to its main subject matter—plainsong—he wrote a paragraph that sounded like Archytas' idea about the connection between pitch and fast and slow motion and the beating of the air, described the connections between the planets and strings or cords on a lyre, and paid tribute to Cicero's "Dream of Scipio." To cover all bases, Regino closed his introduction with the words "We would just add that not only the heathen philosophers but also vigorous commenders of the Christian faith give their assent to this heavenly harmony."[17]

IN THE ELEVENTH century, a Europe that had been relentlessly tormented for two hundred years by waves of marauding invaders experienced an era of relative peace and optimism. The pace of life quickened, and populations and trade increased, including trade with regions under Islamic rule.

Teaching and study in Europe during the centuries of upheaval had never come to a halt, and monks had gone on preserving ancient writings and copying and illuminating manuscripts. However, in the eleventh century new centers of learning started to appear not in the monasteries but in the cathedral precincts and in the larger medieval cities.[18] At first, these amounted to no more than one or a few learned men with a huddle of students gathered around them, and most of the teaching was oral. In the twelfth and thirteenth centuries, these groups became formalized, with better defined roles and obligations for students and teachers, better established relations with local populations and governments (often a touchy matter), and student lodgings resembling the colleges of Cambridge and Oxford. Universities on this model were an authentically European development.

Among the earlier gatherings of teachers and scholars, and later in the universities, a critical and discursive ("combative," Thomas Kuhn called it) tradition emerged that became known as scholasticism. Many giants of medieval thought who engaged in these combats are still familiar names today—Thomas Aquinas, Peter Abelard, Anselm of Canterbury, to name only a few. A primary goal of scholasticism was to integrate classical Greek ideas and learning with Christian belief. With the reintroduction to Latin Europe of the works of Aristotle, translated into Latin in the twelfth century but not immediately available to all scholars, this became a much greater and more complicated undertaking. Scripture was given a more metaphorical, less literal reading, and Aristotle came to be considered, after Scripture, the supreme authority. Scholars revered him not just as a philosopher, but as "the Philosopher," no other identification required.

The groundwork for the rediscovery of classical literature had been laid in the tenth century, when Christian knights gradually began to take over what is now Spain and Portugal from the Muslims who had ruled there for more than three hundred years. The culture of the Iberian peninsula was one of the highest on Earth, the best of both Jewish and Muslim. Because the population of Christian Europe—from which the conquering knights came—was, on the whole, much rougher and less civilized and literate, the situation somewhat resembled the Roman conquest of the Greeks many centuries before. This newer "conquest" was glacially slow, allowing time for a remarkable intermingling of the three different faiths and cultures. Eventually, in 1492, the Christians would drive the Muslims out of Spain, but for centuries before that the Christians who came there found themselves in the presence of, and mingling with, a long-established, intellectually confident, highly cultivated Muslim and Jewish society.

Clergy who accompanied or followed the knights were awed by the beauty of the cities, the architecture and gardens, the peace in which minority communities coexisted, and the level of learned discussion and scholarship—but most of all by the libraries of Cordoba, Toledo, Segovia, and Lisbon. As long as any cleric in Latin Europe could remember, there had been rumors that priceless manuscripts and books, containing the lost knowledge of the ancients, still existed somewhere in the Muslim countries. The old rumors turned out to be true to a degree beyond their dreams. Here in Spain was the fabled material—much of it translated into Arabic—that had been in the repositories of Christian scholars before the Muslims had taken over most of the former Roman Empire in the seventh century. Since then, Muslim and Jewish translators and scholars had treasured and preserved these works.

By 1100, Christians controlled Toledo and Lisbon. Archbishop Raymund of Toledo invited the cream of the scholarly world to join in an effort to translate a vast collection of ancient writings into Latin. The first translators were representatives of the three faiths, Christian, Jewish, and Muslim, who were already living in Spain, but soon scholars joined them from all over—Christian clergy from Latin Europe and England, Jews and Muslims, Latin, Greek, and Slavic scholars—to work with no censorship, no banning of any book, no rewording to give a Christian spin to pagan words. Some of the translators were not just bilingual but multilingual. Michael Scot, from England, knew some Arabic and was fluent in Latin, Greek, Hebrew, Syriac, Chaldean, and several other languages. When not enough men could be found who had mastered both Arabic and Latin, two translators with a common language worked together. The effort continued for years. One particularly prolific translator, Gerard of Cremona, translated seventy or eighty books in all, including Ptolemy's *Almagest* and Euclid's *Elements*.[19]

Similar work was going on in Palermo, Sicily, under the patronage of the Norman King Roger. Known for an opulent court befitting an Eastern potentate, Roger considered it essential to surround himself with intellectuals, and he was patron to a number of them—Roman Catholic, Byzantine Christian, Jewish and Muslim alike. The translation in Palermo was more often directly from ancient Greek manuscripts into Latin, rather than via Arabic translations, for Sicily in the time of Pythagoras had been a Greek colony and had retained the Greek language through the Roman and Byzantine eras. Roger's retinue included a number of Greek-speaking scholars. Plato's *Meno* and *Phaedo* were, fittingly, first translated into Latin there on the island where one of the earliest Pythagorean communities had existed and where Plato himself had dabbled in court politics and almost lost his life.

With no printing presses yet in existence, copyists devoted long hours to reproducing the translations. The dissemination of the new books was slow, but for the first time in many centuries, scholars in Latin Europe were reading the ancient Greeks, and in the universities Aristotle joined Plato.

The basic medieval curriculum had begun as none other than the Pythagorean quadrivium of Archytas, and students also had to master dialectic, as Plato had required. But when Aristotle's works began to influence university education, they became the foundation of philosophical and theological studies in a "trivium" that followed after the quadrivium. The seven subjects of the combined *quadrivium* and *trivium*—arithmetic, geometry, music, astronomy, grammar, rhetoric, and dialectic—became known as the Seven Liberal Arts.[20]

The standard arithmetic text was the old, familiar *Introduction to Arithmetic* by Nicomachus, the second-century neo-Pythagorean who had clung doggedly to "Pythagorean mathematics" and identified himself as a Pythagorean. Boethius' slightly reworked Latin version of his book had been in the libraries of Latin Europe for centuries. Now, thanks to the translation projects in Spain and Palermo, scholars and students were able to circumvent Boethius' rewrite and read Nicomachus in direct translation from the original. Whichever version they read—Boethius' *De institutione arithmetic* from the early sixth century, or Nicomachus' original *Introduction to Arithmetic* from the second—they encountered Pythagoras before they encountered any arithmetic, for the opening passages lauded him. Medieval students thus learned their arithmetic in the neo-Pythagorean form, which they took to be *the* form; and almost entirely through this one book, the Pythagorean faith in the power of numbers to unlock the secrets of nature and the universe was conveyed to the Middle Ages and beyond. It was a tremendously significant channeling of thought. The image of Pythagoras as the creator of Greek mathematics became entrenched.

In the twelfth century, in many universities, the geometry section of the quadrivium was taught from a much better book, Euclid's *Elements*. Though translated into Latin earlier, it had never caught on or become widely available. Now there were fresh translations from the Arabic by Gerard of Cremona and Abelard of Bath, another of Archbishop Raymund's translators. Early in the century, Abelard had journeyed the whole length of the Mediterranean collecting ancient texts.

In the third section of the quadrivium, music, Boethius' *De institutione musica* was the text. Through Boethius' music books, again probably taken originally from Nicomachus, the "Scale of Timaeus" had already become a significant part of medieval music theory. There is good evidence that this scale did not, in fact, originate with Plato but was used by Philolaus and perhaps earlier, so medieval scholars were dealing with something of impressively ancient origin.[21] They accepted

Boethius' divisions of music into *musica mundana* (harmony of the spheres), *humana* (relationship of music to the human soul), and *instrumentalis* (what we normally think of as music), and most agreed that all three were essential parts of their subject.

In spite of a few doubters such as the Florentine Coluccio Salutati, who insisted that the motions of the heavenly bodies could not possibly produce sound, the idea of *musica mundana* was still favored in the fifteenth and sixteenth centuries in Italy, when Franchino Gaffurio, the most important music theorist of his time, made every attempt to be a true Pythagorean. He refused to consider any but the intervals approved by Boethius as consonant intervals, which made him something of a throwback. Boethius had not regarded major thirds and sixths as consonant, and musicians among Gaffurio's contemporaries certainly did. Tradition had it that only Pythagoras himself could hear the music of the spheres, but Gaffurio amended that slightly to insist that only men of significantly great virtue could hear it.

As for the fourth part of the quadrivium, astronomy, the stationary-Earth-centered systems of Aristotle and Ptolemy prevailed unchallenged and unquestioned until one tentatively raised hand in the fourteenth century. It belonged to a Parisian, Nicole d'Oresme, who went only so far as to argue that Aristotle had fallen short of proving Earth does *not* move. Otherwise, no one anywhere in the Middle Ages and until the fifteenth century took seriously the Pythagorean suggestion mentioned by Philolaus that the Earth does not stand still, or even that it rotates. When a more aggressive challenge eventually came, in the fifteenth century, it would be from a man with a decidedly Pythagorean cast of mind: Nicholas of Cusa.

THE INFLUENCE OF Pythagoras and the Pythagoreans was not confined to the universities during the Middle Ages.[22] Freemasons included Pythagoras among their *ars geometriae*. Gerbert of Aurillac, who became Pope Sylvester II, in the tenth century referred to Pythagoras in his geometry. Gobar numerals—direct ancestors of modern Arabic numerals—were widely believed to have been the invention of Pythagoras.* A work

* In the mid-twentieth century, there was still one expert, Vincenzo Capparelli, who was convinced that Pythagoras invented Arabic numerals (Vincenzo Capparelli, *La sapienza di Pitagora* [Padua: CEDAM, 1941]).

supposedly (though not really) by Boethius included a method called *mensa Pythagorea* for calculating with these numbers on an abacus.[23]* In truth they were originally Hindu and were transmitted to the West through Islamic countries and Spain, with Arabic numerals first appearing in a Latin manuscript in 976.[24] For Nicomachus, the neo-Pythagorean numerology in his book had been even more significant than the arithmetic, and this numerology too continued to be important in the Middle Ages, for had not even St. Augustine himself taken enthusiastically to the Pythagorean-like idea of the allegorical interpretation of numbers in the Bible? Like Philo of Alexandria, Augustine had written about the six days of creation in the Genesis account and pointed out that six was a perfect number.

Whoever chose what to celebrate in the sculptures adorning the doors of the cathedral at Chartres, one of the architectural wonders of the Middle Ages, decided to include a series of statues representing the Seven Liberal Arts and selected Pythagoras to symbolize music. The sculptor made him long-haired and bearded, hands and face middle-aged at least, seated and clothed in a beautifully adorned robe as he bent intensely over his work. At the cathedral school in Chartres, in the twelfth century, the scholastic movement's long endeavor to bring together Platonic and scriptural narratives and concepts, including giving the Genesis account of creation a more Greek (in modern terms "scientific") interpretation, reached its zenith. John of Salisbury called Bernard of Chartres, head of the Chartres school in the first part of the century, "the finest Platonist of his time." The Platonism of Bernard and his fellows was based mainly on Augustine and other early Christian scholars, the writings of Boethius, Macrobius' commentary on Cicero's "Dream of Scipio," and Plato's *Timaeus* in a translation by Chalcidius. The Chartres scholars saw *Timaeus* as an explication of Genesis. Bernard had Pythagoras and Plato in mind when he praised the ancients in words usually attributed to Isaac Newton five centuries later:

> We are dwarfs perched on the shoulders of giants. Although we may see more and further than they, it is not because our sight

* Most who used an abacus were still using Roman numerals, the English exchequer as late as the sixteenth century! (H. G. Koenigsberger, *Medieval Europe, 400–1500* [Harlow, England: Longman Group, 1987], p. 202.)

Pythagoras depicted in a frieze of the Seven Liberal Arts on the western front of the Cathédrale Notre-Dame de Chartres

is keener or our stature greater, but because they bear us up and add their gigantic stature to our height.[25]*

The scholars of the Chartres school were addressing an old question: What is the best guide on the journey toward God, or (if one wished to use more Pythagorean/Platonic language) toward reunion with the divine? Was it "reason" or "faith"? Is it not best that the two work together? Boethius had written, "As far as you are able, join faith to reason," and that was the goal of the scholastics. The hope at Chartres was to stake out intellectual and spiritual ground where one could accept what God had revealed but still strive for more comprehensive knowledge of truth. Faithful to their Platonism, and also to their Christianity (St. Paul had said that humans could only see "through a glass, darkly"), these scholars accepted that full knowledge could not be had in this life. Nevertheless, they thought it essential, insofar as humanly

* T. S. Eliot echoed those sentiments when he suggested that to those who say we shouldn't read the old authors since we know so much more than they did, we should answer, "And they are what we know."

possible, not only to believe but also to understand what one was believing. Plato's *Timaeus* seemed a splendid example of this effort and this understanding, albeit from a pagan philosopher. Not surprisingly, these ideas offended some who accused the Chartres scholars of undervaluing religious revelation and mocking simple faith.[26]

The masters at Chartres influenced thinkers in Paris in the following century, when scholarship took a far more Aristotelian turn, prioritizing sense perceptions, experience, and experiment in the pursuit of knowledge. The church continued to sound much more like Plato—for whom the "Forms" were real and the sense-perceived world a shifting illusion—by encouraging rejection of the perceptible, sin-ridden world.

Though in the late Middle Ages and early Renaissance, scholastic and humanist scholars continued to have success meeting the challenges of new translations, broadening knowledge and reconciling Greco-Roman and Christian thought, even as late as the seventeenth and eighteenth centuries sporadic resistance to their efforts would continue. Some still pointed to doctrines they felt had entered early church thinking as a "pagan corruption" from the philosophy of Plato. This resistance did not come from ignorant people. Isaac Newton dismissed the doctrine of the Trinity on those grounds.[27] So did the late eighteenth century English Unitarian religious dissenter Joseph Priestley, who thought the dualism between matter and spirit was not inherent in the Gospels but had entered the early church through Greek philosophy.[28]

CHAPTER 15

"Wherein Nature shows herself most excellent and complete"

Fourteenth–Sixteenth Centuries

In the fourteenth century, most educated people in Europe regarded foreign languages as completely impenetrable and unlearnable, so the author Francesco Petrarca (Petrarch) was being venturesome when he decided to learn Greek. He engaged a teacher, a monk named Barlaam of Seminara, but the project was not a success and Petrarch was fated to go on lamenting that he would never arrive at the best understanding of philosophy because his Greek was not good enough.

He was disarmingly modest. Perhaps he did, as he claimed, merely chuckle when he was an old man and heard the news—it was being repeated all over Venice and beyond—that four young aristocrats, who had dined and drunk exceedingly well, had off-handedly dismissed him as "certainly a good man but a scholar of poor merit." In a letter written just a few years before that Venetian slight, Petrarch described himself:

> Let me tell you, my friend, how far I fall short of your estimation. This is not my opinion only; it is a fact: I am nothing of what you attribute to me. What am I then? I am a fellow who has never quit school, and not even that, but a backwoodsman who is roaming around through the lofty beech trees all alone,

> humming to himself some silly little tune, and—the very peak of presumption and assurance—dipping his shaky pen into his inkstand while sitting under a bitter laurel tree. I am not so fortunate in what I achieve as I am passionate in my work, being much more a lover of learning than a man who has got much of it. I am striving for truth. Truth is difficult to discover, and, being the most humble and feeble of all those who try to find it, I lose confidence in myself often enough.[1]

Some of the "lofty beech trees" among whom Petrarch hummed his tune were Augustine and Cicero, Aristotle and Plato (he read them in Latin translations), and Pythagoras, whom he knew through those other authors.

Collecting works from the classical period, tracking down manuscripts and early copies, had become the fashion among those sufficiently educated and wealthy, and the acquisition of something interesting was a matter of great excitement to share with like-minded friends. Petrarch's own large library reflected that fashion and his love of learning, but, for all his modesty, the library he stored in his head was vaster than most men's collections. He read more than anyone else, remembered most of it verbatim, and had a habit of imagining himself personally involved in history and literature. As one commentator wrote,

> Since he was such a keen observer of actual life and so lovingly devoted to the investigation of the human heart, all the records of the past became a living reality to him, and he felt himself sharing in the drama as if he had an active part in the cast. It was not just a whim that he, the untiring letter writer, started to "correspond" with characters of ancient times, as if they could answer him. When he read their works, he almost forgot that they were long since dead.[2]

No wonder Shakespeare so often found inspiration and material for his plays in Petrarch. Through Shakespeare and others who read Petrarch, he played an influential role in shaping future culture.

Petrarch was no fan of the Pythagorean doctrine of reincarnation, which he thought was an example of the way a wise and brilliant man can be perfectly capable of coming up with nonsense. "Who does not know," he wrote, "that Pythagoras was a man of exalted genius? However, we also know his Metempsychosis. I am amazed beyond belief that this idea could spring up in the brain, not of a philosopher, but even of any human being." Pythagoras' claim to have been Euphorbus in an earlier life was "an empty lie" and "deceitful pretense." But then Petrarch also scorned Democritus' suggestion that "heaven and earth, and all things in general, consist of atoms."[3]

Petrarch, as imagined by engraver Rob Hart, 1835

A few pages after his disparaging words, Petrarch turned around and referred to Pythagoras in reverential tones as "the most ancient of all natural philosophers." No one knows where he got the quotation that he attributed to Pythagoras and used to defend not only the Christian faith but also Plato and Moses from those who "blind and deaf as they are, do not even listen to Pythagoras, who asserts that 'it is the virtue and power of God alone to achieve easily what Nature cannot, since He is more potent and efficient than any virtue or power, and since it is from Him that Nature borrows her powers.'"[4] Petrarch did not believe that Pythagoras had actually written this, or, indeed, anything, but he thought that others had written down "what he expounded in his conversations."

Petrarch is often called the first humanist. He trusted God so devoutly and completely that he felt free to leave the deepest religious issues alone and concentrate instead on philosophy, which he preferred to define as the study of the art of happiness and living well.[5] Pythagoras, Plato, and Christianity seemed a natural, logical continuum to him.

In the middle of the next century, the fifteenth, no less a personage than Lorenzo de Medici lent his patronage to an attempt to re-create Plato's Academy at the villa of his acquaintance Marsilio Ficino, near Florence. The Accademia Platonica was Ficino's brainchild and dream. He translated all of Plato's works into Latin directly from the Greek, wrote commentaries on them, and gathered a group of writers, thinkers, and artists to study them in a congenial setting. When Ficino had also finished translating Porphyry, Iamblichus, Proclus, and Plotinus, those who knew no Greek could read nearly the entire surviving output of the Platonic and neo-Platonic writers in Latin. It is a pity that Petrarch had lived a century too early to enjoy all these works in translation!

One of Ficino's Academy members was the artist Botticelli, whose painting *Primavera* was supposed to be a visual metaphor for the music of the spheres, relating mythological creatures to planetary orbits and the notes of an octave in music. Ficino himself developed an elaborate system of heavenly music. He was also interested in the early church fathers and, like Petrarch, thought that Platonic doctrine and reasoning (which he thought were divinely inspired) were in harmony with Christianity, having particular value in that they could provide independent confirmation of Christian beliefs in a manner that would satisfy those among Ficino's contemporaries who were of a skeptical and even atheistic frame of mind. He gave a Pythagorean/Platonic spin to his treatment of the fall and salvation of man, referring to the belief that the earthly existence of the soul is an exile from its divine home. The Pythagoreans and Platonists agreed, he wrote, that "because of a certain old disease of the human mind, everything that is very unhealthy and difficult befalls us; but, if anyone should restore the soul to its previous condition, then immediately all will be set in order." To Ficino, that sounded like humanity in its fallen state looking toward the salvation of Jesus, in Christian doctrine. A yearning to turn back to God was built into human nature:

> Just as [according to Aristotle] when an element is situated outside its proper location, its power and natural inclination toward that natural place are preserved together with its nature, in so far as it is able at some time to return to its own region; so, they [the Pythagoreans and Platonists] think, even after man has wandered from the right way, the natural power remains to him of returning first to the path, then to the end.[6]

Ficino agreed with those neo-Pythagoreans who had concluded that the same primordial wisdom had emerged in different ages and cultures. The truth of philosophy, religion, and natural science, in all times and places, was, at some deep, so far unplumbed level, one consistent truth. This, Ficino thought, was a manifestation of the "unity" that the Pythagoreans had held so in awe.

In the city of Parma during this same period, the musician and physician Giorgio Anselmi (some thought he was also a magician) developed the first system since Eriugena's to take into account the fact that the planets change their distances from Earth. In Anselmi's cosmic musical plan, a planet produced not one tone but many different notes as its distance changed, so that each planet sang its own song. All the planet songs together produced magnificent counterpoint and harmony. Though no music of his time went beyond a three-octave range, Anselmi's planetary scale, calculated from the planets' periods, was eight octaves long from the stars to the Moon.

Ficino's younger Florentine friend Giovanni Pico, Count of Mirandola (known as Pico della Mirandola), was fond of using the phrase, the "ancient theology of Pythagoras." He regarded Pythagoras as no less than a Christian sage and connected the peace promised by Jesus—"Come unto me, ye who have labored, and I will give you peace, which the world and nature cannot give"—with a Pythagorean peace in which

> all rational souls not only shall come into harmony in the one mind which is above all minds but shall in some ineffable way become altogether one. That is the friendship which the Pythagoreans say is the end of all philosophy. This is that peace which the [Christmas] angels descending to earth proclaimed to men of good will, that through it men might ascend to heaven and become angels.[7]

Until that time, "Let us wish for this peace for our friends, for our century . . . for every home into which we go," he wrote.

Pico did not always write so clearly and simply. One of his more impenetrable documents was "Fourteen Conclusions after Pythagorean Mathematics,"[8] which arose out of his fascination with "the method of philosophizing through numbers" as it was taught by "Pythagoras, Philolaus, Plato, and the first Platonists."[9] Aristotle would have summoned his Delian diver!

1. Unity, duality, and that which is, are the causes of numbers: One, of unitary numbers; two, of generative ones; that which is, of substantial ones.
2. In participated numbers some are species of numbers, others unions of species.
3. Where the unity of the point proceeds to the alterity of the binary, there the triangle first exists.
4. Whoever knows the series of 1, 2, 3, 4, 5, 12, will possess precisely the distribution of providence.
5. By 1, 3, and 7 we understand the unification of the separate in Pallas: the causative and beatifying power of the intellect.
6. The threefold proportion—Arithmetical, Geometrical, and Harmonic—represents to us the three daughters of Themis, being the symbols of judgment, justice, and peace.
7. By the secret of straight, reflected, and refracted lines in the science of perspective we are reminded of the triple nature: intellectual, animal, and corporeal.
8. Reason is in the proportion of an octave to the concupiscent nature.
9. The irascible nature is in the proportion of a fifth to the concupiscent.
10. Reason is in the proportion of a fourth to anger.
11. In music the judgment of the sense is not to be heeded: only that of the intellect.
12. In numbering forms we should not exceed 40.
13. Any equilateral plane number may symbolize the soul.
14. Any linear number may symbolize the gods.

Not surprisingly, when the twenty-three-year-old Pico went to Rome and offered to debate another of his lists, *Nine Hundred Conclusions*, there were no takers. Like the "Fourteen Conclusions," the *Nine Hundred* were short sentences, covering the subjects of scholastic and earlier theology, Arabic and Platonic philosophy, the Chaldean Oracles, the Zoroastrian Magi, and Orphic doctrines.* All, Pico insisted, were reconcilable with one another, and he was prepared to debate anyone who disagreed. Truth was universal. What might seem to be opposing schools of thought and doctrine really were all the same primordial wisdom of humankind, sharing a common truth.

Pico's interest was piqued by the Jewish Cabalistic literature, in which words and numbers serve as a form of mystical code. Cabala is a form of Jewish mysticism that, though it had roots as early as the first century A.D., fully emerged in the twelfth century. Though a text of Merkava mysticism (a precursor of Cabala) had included a creation story with ten divine numbers, and one of the most important Cabalistic texts, the twelfth-century *Sefer ha-bahir* ("Book of Brightness"), introduced into Judaism the idea of the transmigration of souls, in neither case was there a known link with Pythagoras. But another man who immersed himself in the Cabala at about the same time as Pico, insisted there was a connection. Johann Reuchlin, a German humanist, set out to combine the study of Hebrew, Greek, theology, philosophy, and the Cabala, and to link it all with the name of Pythagoras. He wrote to Pope Leo X that, just as Ficino had so admirably done for Plato in Italy, he would "complete the work with the rebirth of Pythagoras in Germany." He rationalized the connection with the Cabala by drawing attention to the (questionable) fact that "the philosophy of Pythagoras was drawn from the teachings of Chaldean science."[10†]

IN THE SAME century when Ficino set up his Florentine academy and Pico issued his intellectual challenges, their older contemporary Leon Battista Alberti, inspired by the work of the ancient Roman Vitruvius,

* The "Chaldean Oracles," written in verse in the second century A.D. by a man named Julianus the Theurgist and his son, combined Babylonian and Persian beliefs with Platonic and neo-Platonic philosophy and became an important religious book for neo-Platonists.

† "Chaldean" in this case meaning Babylonian.

was insisting on beautiful proportions in buildings and applying Pythagorean principles to architecture. Books on architecture seemed to come in sets of four or ten volumes—two good Pythagorean choices. Vitruvius had written his "Ten" in the first century B.C., and, Alberti produced his "Ten" in 1485.* They were translated from Latin into Italian in the mid-sixteenth century. Alberti liked to use what he thought were Pythagorean ideas and extend them in ways of his own:

> I am every day more and more convinced of the truth of the Pythagorean saying, that Nature is sure to act consistently, and with a constant analogy in all her operations. From whence I conclude that the numbers by means of which the agreement of sounds affects our ears with delight, are the very same which please our eyes and mind. We shall therefore borrow all our rules for the finishing of our proportions from the musicians, who are the greatest masters of this sort of numbers, and from those things wherein nature shows herself most excellent and complete.[11]

Alberti divided the kinds of areas to be measured in an architectural design into three categories: short, medium, and long. The Pythagorean ratios were the only ones that he applied to the "short" or "simple" areas: The shortest was a square; the next an area that started with a square and then added on a third again as much space, making a ratio of 3 to 4 between the square and the total area.

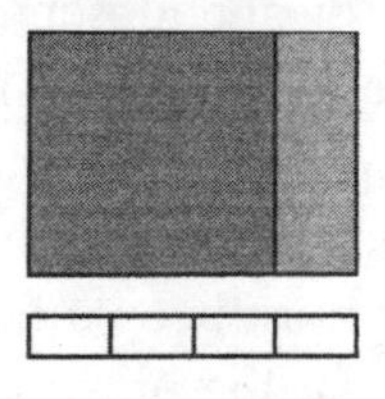

The last also started with a square and added on half again as much space, making a ratio of 2 to 3 between the square and the total area.

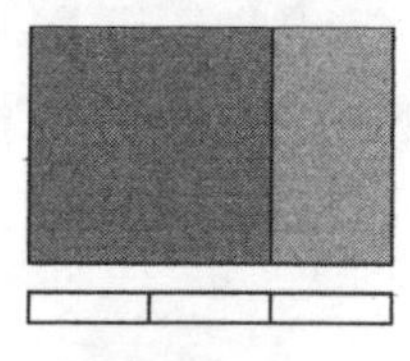

* The great Andrea Palladio was to write four.

For larger areas, Alberti used proportions that went beyond these ratios, but all could, in one way or another, be linked to them.

Though Alberti was one of the most important theorists of architecture in the Renaissance and also one of that era's greatest practitioners, his achievements were by no means confined to architecture. He was truly a "Renaissance man"—a moral philosopher, a major contributor to the techniques of surveying and mapping, a pioneer in cryptography, and the first to systematize and set down the rules for drawing a three-dimensional picture on a two-dimensional surface, establishing principles that would underlie perspective drawing from that time forward. Nevertheless, it was arguably in architecture that he had his most lasting impact, not only because of the splendid buildings he designed, but also because his Ten Books, with their Pythagorean principles, were read and studied by all Renaissance architects after him, including Andrea Palladio, perhaps the most influential architect of all time.*

IN THE EARLIER part of Alberti's century, Nicholas of Cusa, born in 1401, had been considering a startling, fresh approach to structure on a much larger scale: the entire cosmos. Though his name sounds Italian, Nicholas was the son of a boatman on the Mosel River. He received his religious training with a devotional group of laymen in the Netherlands and his university education at Heidelberg and Cologne. Later, as a university scholar and a cardinal of the Catholic church, Nicholas not only found Christian faith and classical philosophy compatible, but that compatibility became for him a fertile ground from which to begin innovative thinking in other areas of knowledge. He decided that God was infinite, and the universe had no limit other than God . . . so the universe was infinite too. Contrary to what most people believed (they had learned it from Aristotle), he insisted that the universe was not made of different types of substance at different levels, such as the impure region near Earth and the pure region of the celestial spheres. The universe was homogeneous. The stars were "each like the world we live in, each a particular area in one universe, which contains as many such areas as there are

* Alberti's most important buildings included, in Florence, the Palazzo Rucellai, the Rucellai Chapel, the Annunziata, and the façade of the Maria Novella church; in Rimini, the Tempio Malatestiano; and in Mantua, the churches of San Sebastiano and San Andrea.

uncountable stars."[12] Nicholas was sure that Earth was a star like the Sun and the other stars, and it moved. This was not the orthodox, Ptolemaic/Aristotelian stationary-Earth-centered astronomy that was being taught in the universities! Nicholas worked his ideas up in a highly original, mathematics-based system. He did not suggest another body to usurp the importance of the Earth, but even without nominating a competitor for "center of the universe," his proposal was a huge demotion.

Nicholas believed the human mind had innate power to know things and to acquire knowledge, and, like Aristotle, he thought that knowledge had to be acquired directly from nature and experience. He also believed that learning about nature and the universe required the use of numbers and the study of numerical proportion and ratios. He was fond of the Pythagorean practice of applying numbers to many aspects of life. In his treatise "On Catholic Concordance" he used the order of the heavens as a model for harmony in the church; and in his book *Of Learned Ignorance* he drew a parallel between the search for truth and converting a square to a circle.

Nicholas, like Alberti, was a Renaissance man. He drew up a map of Europe and was the first to prove that air has weight. He apparently never worried whether his ideas about the arrangement of the cosmos might conflict with church doctrine. It seems he had no reason for concern. The church never condemned or criticized him.

Astronomy was about to take an even more decidedly Pythagorean turn. In 1495, twenty-two-year-old Nicolaus Copernicus and his older brother Andreas journeyed south from their native Poland and "walked across the Alps"—their destination Bologna, seat of Italy's oldest university. Nicolaus had completed four years at the Jagiellonian University in Kraków, which was renowned for its astronomy. If a student intended to continue his education after he had finished the quadrivium and the trivium, he chose an area of study and went to a university that specialized in that. Nicolaus' uncle and guardian, an influential man who became bishop of Warmia, was apparently worried that his nephew was developing a keen interest in astronomy. Hoping that the Italian sunshine and the stimulating intellectual community of the University of Bologna would turn the young man's interest in a better direction, he insisted Nicolaus go to Bologna, famous for its law faculty. (Copernicus did eventually receive a doctorate in canon law, the law of the church, although not from Bologna.)

While studying in Bologna, Copernicus met the university's leading scholars and teachers of astronomy and astrology, and also a mathematician named Maria de Novara, whose influence was probably the most valuable of all that Copernicus carried away with him from these years. Novara was a neo-Platonist and a close younger associate of the men of Ficino's academy in Florence. His neo-Platonism was decidedly Pythagorean. He fervently believed in the need to uncover the simple mathematical and geometric reality that underlies the apparent complexity of nature, and he insisted that nothing so complicated and cumbersome as Ptolemaic astronomy could possibly be a correct representation of the cosmos. His young friend Copernicus came to agree.

No new astronomical discovery, nor any better or more accurate observations of the heavens, caused Copernicus to discard Ptolemaic Earth-centered astronomy and replace it with a system in which the Sun was at the center. Though over the long passage of years the errors produced by the Ptolemaic system had made it less and less accurate in predicting planetary positions, no observational instrument during Copernicus' lifetime was accurate enough to show whether the Copernican system solved this problem. The telescope would not appear until early in the seventeenth century, and the astronomical observations that Copernicus made himself were often less accurate than those of Hellenistic and Islamic astronomers centuries before him.

The early Pythagoreans, in the wake of their discovery of the ratios of musical harmony, had gone off in wild and misguided directions to decide there had to be ten bodies in the cosmos, disregarding the fact that there was no evidence of that number's correctness, running ahead of nature, and arriving at the wrong conclusions. And here was Copernicus, doing something of the same kind, for when he decided that Ptolemaic astronomy could not be correct, he did so largely for reasons other than physical evidence. The beginning of the scientific revolution was perhaps not so scientific—not in the way we most commonly think of "scientific."

Copernicus translated at least two Greek texts into Latin, unaware that one of them, *Lysis' Letter to Hipparchus*, was a forgery. That he knew of the *Letter* at all was symptomatic of his intense interest in

Pythagoras and the Pythagoreans. He even originally named his new system not the "Copernican system," but the Astronomia Pythagorica or Astronomia Philolaica, and he considered adopting the Pythagorean practice of secrecy. In the prefatory letter dedicating his *De revolutionibus* to Pope Paul III, he defended his long delay in publishing this masterwork by pointing to the example of Pythagoras and the Pythagoreans.

> Thinking therefore within myself that to ascribe movement to the Earth must indeed seem an absurd performance on my part to those who know that many centuries have consented to the establishment of the contrary judgment, namely that the Earth is placed immovably as the central point in the middle of the Universe, I hesitated long whether, on the one hand, I should give to the light these my Commentaries written to prove the Earth's motion, or whether, on the other hand, it were better to follow the example of the Pythagoreans and others who were wont to impart their philosophic mysteries only to intimates and friends, and then not in writing but by word of mouth, as the letter of Lysis to Hipparchus witnesses.[13]

Nicolaus Copernicus

Copernicus had thought of including the *Letter* in *De revolutionibus*, but decided not to. Having defended his long period of secrecy, however, he went on in the same preface letter to point to the Pythagoreans as an ancient precedent for his own ideas. Because of his dissatisfaction with the Ptolemaic accounts of the heaven's motions, he said, he had begun to search in "the works of all the philosophers on whom I could lay hand." He had discovered some influential figures who had not, after all, agreed with the overwhelming consensus. Aristarchus in the third century B.C. had moved the Sun to the center in his remarkable cosmology. Cicero had mentioned Hicetas' suggestion that the Earth moved. Even better, Plutarch had written in his *Placita* (Copernicus quoted in Greek),

> The rest hold the Earth to be stationary, but Philolaus the Pythagorean says that she moves around the fire on an oblique circle like the Sun and Moon. Heraclides of Pontus and Ecphantus the Pythagorean also make the Earth to move, not indeed through space but by rotating round her own center as a wheel on an axle, from West to East.[14]*

The philosopher Paul Feyerabend observed that when Copernicus decided to order the heavens he did not consult his "scientific forebears," but instead cited a "crazy Pythagorean."[15]

In Chapter 10 of Book One of *De revolutionibus*, Copernicus illustrated most fully the new and aesthetically beautiful harmony of his system, revealing in the process how well he knew his Plato, a wealth of other classical literature, and even the work of the Islamic astronomers. Calling attention to the simplicity of the new system, he wrote:

> I think it easier to believe this than to confuse the issue by assuming a vast number of spheres, which those who keep Earth at the center must do. We thus rather follow Nature, who, producing nothing vain or superfluous, often prefers to endow one cause with many effects. . . . So we find underlying this ordination an

* Ecphantus the Pythagorean lived in the fourth century B.C. There is some suspicion that he may have been only a fictional character in one of Heracleides' dialogues, but Copernicus thought he was a historical person, and most modern scholars tend to agree.

> admirable symmetry in the universe, and a clear bond of harmony in the motion and magnitude of the spheres such as can be discovered in no other wise.[16]

The Pythagorean insight, from the sixth century B.C., that harmony and simple pattern expressed in numbers underlie nature clearly was for Copernicus a persuasively strong point in favor of his rearrangement of the cosmos. The potential of numbers, in combination with a preference for harmony and simplicity, to lead to a truer understanding of the universe—a potential that had been poorly exploited by the early Pythagoreans and reinterpreted in many ways, some of them admittedly quite strange, by a great many people since—was finally about to be realized.

Copernicus would not live to see the result of his own Pythagorean dream in print. If he saw a printed copy of *De revolutionibus* at all it was on his deathbed, for he had followed the Pythagorean example of secrecy for years before deciding, finally, to publish. The astronomy he was able to devise in the book turned out to be, in its details, almost as complicated as Ptolemy's, but those few who read it carefully and recognized that Copernicus meant his revolutionary Sun-centered suggestion to be taken seriously, found their minds set on a fresh path indeed. The great Pythagorean insight that had led Copernicus was about to lead younger men out of the Middle Ages and into the modern world.

Others in the sixteenth century were captivated by the ideas of the Pythagoreans for what might on the surface seem to be entirely different reasons. However, there were deep connections having to do with harmony and numbers.

Architectural trends begun by Vitruvius in antiquity and continued by Alberti in the fifteenth century were brought to their zenith in the sixteenth in the work of one of the most gifted architectural geniuses of all time, Andrea Palladio, whose rise from stonemason to educated architect occurred thanks to an "academy" like the one that Marsilio Ficino had established and Lorenzo de Medici had patronized in the fifteenth century at Ficino's villa near Florence. Following Ficino's model, academies had become a part of life in northern Italy; nearly every important town had one. Conscious attempts to re-create Plato's original, they were a combination of boarding school, lecture center, and attractive location for scholars, intellectuals, and lovers of learning to meet and discuss literature, philosophy, mathematics, and music. The activities often included physical exercise and musical performances. To a surprising extent, social rank was disregarded and talented or clever men of no social standing rubbed shoulders with wealthy aristocrats.

When Palladio was in his early twenties, in the early 1530s, he was hired as a stonemason for a building project near Vicenza. Count Gian Giorgio Trissino, a wealthy humanist scholar and poet, was rebuilding his villa in the classical style to house an academy. Trissino had designed the new buildings himself, and he thought of his design as an interpretation of the work of Vitruvius. Keeping an eye on the progress of the construction, Trissino watched Palladio at work, made a point of getting to know him, and decided that the young man deserved to have a humanist education.

In his famous *I quattro libri dell' architettura*, published in 1570, Palladio would make a deliberate connection with the Pythagorean discovery that certain ratios in music produced sounds that were pleasant to human ears regardless of whether the hearer knew the underlying numbers. "Just as the proportions of voices are harmony to the ears," he wrote, "so those of measurement are harmony to the eyes, which according to their habit delight in them to a great degree, without it being known why, save by those who study to know the reasons of things."[17] For him, the "preferred" numbers that would produce such spontaneous delight for the beholder of a building were those based on the same sequences the Pythagoreans had discovered in the ratios of musical harmony: 1 to 2, 2 to 3, and 3 to 4.

In Book I of *I quattro libri*, Palladio chose "seven sets of the most beautiful and harmonious proportions to be used in the construction of rooms." Of course the circle and square were among them. Four others were derived from the Pythagorean musical ratios, and the remaining one was the same room Vitruvius had designed based on Socrates' lesson in Plato's *Meno*, with one dimension of the room being incommensurable. Palladio's seven shapes and proportions were a circle, a square (1:1), a room whose length was the same as the diagonal of the square (1:1.414 . . . etc.), a square plus a third (3:4), a square plus a half (2:3), a square plus two thirds, and a double square (1:2).

Though Palladio devoted only one chapter in the second book of *I quattro libri* to harmonic proportions, and other authors who wrote about him later were more concerned than he, the craftsman, for the theoretical aspects of his work, these Pythagorean proportions were abundantly evident in his drawings.[18] It seemed not to bother Palladio that there were differences between the drawings of buildings and the actual buildings that resulted. If one believed Plato, the Forms were never perfectly realized in the material world.

Andrea Palladio

I quattro libri was probably the most influential book ever written about architecture. Palladio wrote it in Italian for a lay audience, and Daniele Barbaro, an architectural expert in his own right, for whom Palladio designed the Villa Barbaro in the Veneto near Venice, aptly described it as a complete guide to building from the foundation to the roof. Soon after publication in 1570, the book and its drawings became the rage throughout mainland Europe and, early in the next century, Inigo Jones returned from a trip to Italy and introduced Palladian design to England. Following this "first great English Palladian," whose surviving buildings include the Queen's Chapel at St. James's Palace and the Banqueting House at Whitehall, many of England's large country houses were soon being built, or rebuilt, along Palladian lines.

Lord Burlington constructed the Assembly Rooms at York on Palladio's designs and fashioned his own home, Chiswick House, after Palladio's Villa Rotonda. Around 1800, Thomas Jefferson designed his Palladian Monticello in Virginia, and numerous American churches of many denominations, university buildings, and official structures and memorials and monuments in Washington, D.C., were following suit, for the feeling was that there was a link between Palladian principles of architecture, with their Pythagorean proportions, and the education, enhancement, and wise governance of society. Palladian design spread to Germany, Russia, Poland, back to Italy, and to Scandinavia.

One of the most unusual houses built using Palladian proportions was the palace-observatory that Tycho Brahe, the finest pre-telescope astronomer, constructed in the latter part of the sixteenth century on the island of Hven in Denmark. As a young aristocrat traveling in Europe, Tycho had visited Venice and the Veneto during the years when Palladio himself was building there, and had probably also seen *I quattro libri*, for he had a connoisseur's appreciation of fine books. Perhaps Tycho was also aware of Palladio's humble origins as a stonemason because, for his own project, he hired a stonemason named Hans van Steenwinkel and raised him to the rank of master builder.

Not everyone who would build in the "Palladian" style would pay mind to Pythagorean or Palladian proportions, but Tycho Brahe did. When his "Uraniborg" was finished, although it looked at first glance anything but Palladian, the Pythagorean musical ratios were all there and the symmetry extended into the landscape, just as Palladio advised. The portal towers on the east and west sides of the house were each fifteen Danish feet wide and fifteen feet long; the height of the façade was thirty feet, the peak of the roof forty-five feet, the side of the central block sixty feet, giving the ratio 1:2:3:4. The same ratios underlay the dimensions of Tycho's rooms and other elements of the structure. The perimeter wall around Tycho's garden enclosed a square divided by avenues on the diagonal, just as Socrates had divided the squares in Plato's *Meno*. Someone unaware of Tycho's intentions, and not steeped in the architecture of Palladio or on the lookout for Pythagorean ratios, would not have noticed these mathematical and musical subtleties, but Tycho was sure this harmony would make his home and gardens satisfying to the eye and soul, encouraging peaceful, intelligent work and inspiring any sensitive person. Tycho designed and built Uraniborg to be both a palace home and an observatory, all for the purpose of better scrutinizing the heavens where the Pythagorean harmony of the spheres—the musical ratios, or perhaps even some deeper harmony—might be discovered. Nowhere else was the Pythagorean and Palladian ideal of proportion so literally, and so idiosyncratically, realized, as in Uraniborg.[19]

For complicated reasons involving Danish politics and personal issues, Tycho Brahe eventually abandoned this remarkable, beloved palace, and Denmark, and went into exile—the exile that made it possible for him to meet Johannes Kepler.

CHAPTER 16

"While the morning stars sang together": Johannes Kepler

Sixteenth and Seventeenth Centuries

NAS THE LAST DECADE OF THE sixteenth century began, the two-thousand-year-old Pythagorean dream of rationality, unity, and the power of numbers was about to be given a serious test. Pythagoras and his followers had been sure they had caught a glimpse, as through a crack or a keyhole, of truth based on numbers that lay beyond the façade of nature. Johannes Kepler would force the door wide open, once and for all. After him, ironically, and though Kepler did not intend it to be so, the Pythagorean concept of the music of the spheres would survive only in poetic imagery. Yet in a profound and magnificent way, the faith embodied in that concept—faith in a wondrously rational and ordered universe—tempered by Kepler's imaginative genius and rigorous mathematics, would finally place real examples of that music under the feet of science.

The higher seminary at Maulbronn, which Kepler attended in the 1580s as a troubled but exuberantly intellectual and religious teenager, taught "spherics" and arithmetic, but it was not until he enrolled at the University of Tübingen that he encountered astronomy. The mission of the Stift at the university where Kepler studied and had his lodgings was to prepare young men for careers of service to the Duke of Württemberg or for the Lutheran clergy, but the course of study was broadly focused. The conviction that there was a unity to all knowledge lived on in the "Philippist" curriculum at the great Lutheran universities after the Reformation, as it had in the classical and medieval quadrivium and trivium and in humanist thinking. "Philippist" referred to the educational philosophy of Martin Luther's disciple and friend Philipp Melanchthon, who had insisted that one could not truly comprehend and master any part of knowledge unless one comprehended and mastered the whole of it—a sentiment the Pythagorean Archytas would have applauded. Melanchthon felt the church could not succeed in teaching the path to salvation unless it produced a well-read scholarly clergy thoroughly grounded in the liberal arts. Reading the Scriptures, the church fathers, and the classical philosophers required facility in Hebrew, Latin, and Greek. Arithmetic and geometry were necessary for comprehension of both the secular and the sacred aspects of the world, and astronomy was the most heavenly of the sciences. Philippist philosophy also held that since the cosmos was orderly and harmonious, one could, and should, not only observe and record things but also hypothesize about them.

Early in his university career, Kepler realized that theology, mathematics, and astronomy would all be essential in his personal search for truth. He never ceased to be a devoutly religious man, but, as he later wrote, he believed that "God also wants to be known through the Book of Nature." Perhaps it was in that interest (Kepler would have thought so) that God had placed a superb professor of mathematics and astronomy at the University of Tübingen: Michael Mästlin.

When Kepler first arrived there in 1589, forty-six years had passed since the death of Copernicus and the publication of his *De revolutionibus* in 1543. Many scholars were finding Copernicus' grasp of celestial mechanics and his mathematics invaluable, while choosing to ignore his rearrangement of the cosmos. The University of Tübingen still officially taught Ptolemaic astronomy, and Michael Mästlin made sure his pupils had a good grounding in that, for which Kepler would later be grateful when he sought to overturn it. But Mästlin believed that Copernicus' system had to be taken literally and that the planets and the Earth do, indeed, orbit the Sun. Kepler also read Nicholas of Cusa and was soon writing: "I have by degrees—partly out of Mästlin's lectures, partly out of myself—collected all the mathematical advantages which Copernicus has over Ptolemy." In a letter

Johannes Kepler

he wrote later to Mästlin, Kepler called Pythagoras the "grandfather of all Copernicans."[1]

During his university years, Kepler rapidly became well-read in the classics and also encountered neo-Platonic/Pythagorean thinkers of his own era. He gave all of it a religious and Pythagorean spin of his own: A universe created by God must surely be the perfect expression of a profound hidden order, harmony, simplicity, and symmetry, no matter how complicated and confusing it might appear to people who, like himself, were only beginning to understand it. This was the conviction that set fire to his spiritual and scientific imagination, and that flame would last him a lifetime. He was about to pin this idea to the wall using more precise observations of the heavens and his innate genius for rigorous mathematics: a potent combination.

While still a student at Tübingen, Kepler openly defended Copernican astronomy in two formal debates, arguing that the planets' periods and their distances from the Sun made far better sense in the Copernican system; and that if the Sun was indeed (like the Creator) the source of all change and motion, then it might follow that the closer a planet was to the Sun, the faster it would travel. He worked busily and happily on astronomical questions and wrote a piece about how the movements of the heavens would appear to someone on the Moon. Despite all that, it seems not to have occurred to him that he might pursue any career other than as a clergyman.

Near the end of his fifth university year, Kepler learned that his time at Tübingen was to end immediately, and not in the way he had planned. A Protestant school in southern Austria appealed to the university for a teacher, mainly for mathematics but also with knowledge of history and Greek. Tübingen had decided to send Kepler. Sorely discouraged and frustrated, he made the move to Graz. It was there that, about a year after his arrival, while drawing a diagram on the board for his pupils, he made the startling discovery that a triangle seemed somehow to be dictating the distance between the orbits of Jupiter and Saturn. The triangle was the Pythagorean *tetractus*.

The date was January 19, 1595, and Kepler was lecturing about the Great Conjunctions that occur when Jupiter and Saturn, as viewed from the Earth, appear to pass each other. This does not happen often in anyone's lifetime, for Jupiter overtakes Saturn only approximately every twenty years. Imagine the two planets moving on a great circular belt around the Earth. During the twenty-year interval between two Great Conjunctions, Saturn moves about two thirds of the way around the belt, while Jupiter makes one complete revolution and two thirds of another. The locations of the Great Conjunctions leap forward on the belt by two thirds of the circle every twenty years.

Kepler had drawn a circle on the chalkboard to represent the great circle of the zodiac belt, and then marked the points in the zodiac where the successive Great Conjunctions occurred, viewed from Earth. If one plotted only three Great Conjunctions, those points were very near to being the corners of an equilateral triangle, but not quite. Beginning another triangle where the first ended (plotting the next conjunctions), the new triangle did not precisely retrace the first one. For example, the fourth conjunction in Kepler's drawing (the conjunction that occurred in the year 1643) happened at almost the same point as the first (in 1583), and the fifth at almost the same point as the second. Draw lines connecting them and you almost have an equilateral triangle . . . but, again, not quite, and you have not retraced the first triangle. So the triangle "rotates," as Kepler's diagram shows. The result is

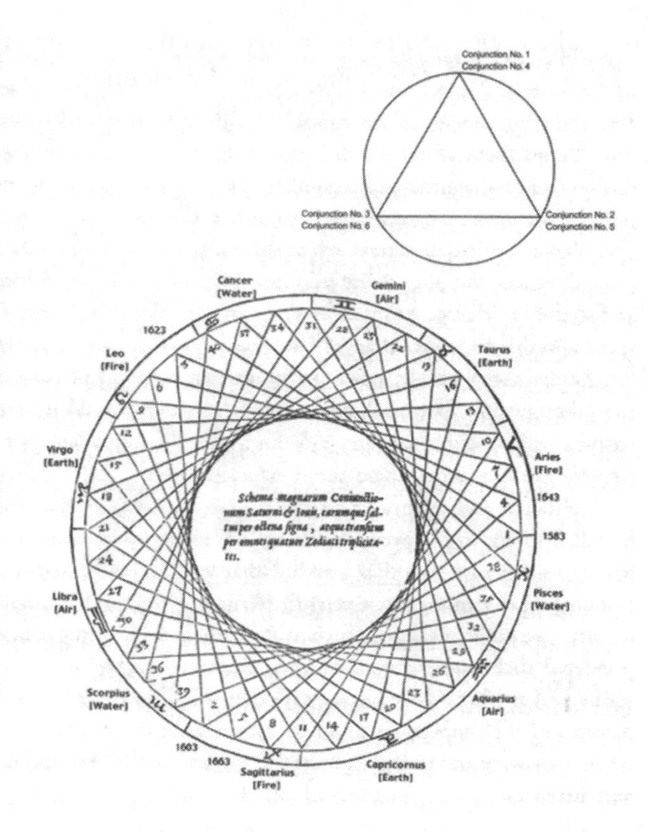

Drawing from Kepler's *Mysterium cosmographicum* depicting the pattern of Jupiter-Saturn conjunctions and where they happened in the zodiac. The conjunction in 1583 (right) occurred when the two planets were in Aries/Pisces. The conjunction in 1603 (lower left) was in Sagittarius, in 1623 in Leo, in 1643 in Aries, in 1663 in Sagittarius, and so on. If the conjunctions occurred repeatedly in the same positions in the zodiac, Kepler's drawing would have looked like the insert (upper right). Instead they "progress," as represented in the central figure.

two circles, outer and inner, with the distance between them set by the rotating triangle. Thus, Kepler's triangle seemed to be mysteriously dictating the distance between the orbits of the first two planets. Interestingly, the radius of the inner circle looked as though it were half that of the outer circle, and observations of the heavens showed that the radius of Jupiter's orbit was approximately half the radius of Saturn's.

An amazed Kepler decided immediately to try the next regular polygon—the square (the triangle has three sides, the square four)—to see whether it would serve similarly for the separation between the orbits of Jupiter and Mars.* If it did, he planned to try a pentagon (five sides) for the separation between the orbits of Mars and Earth, a hexagon for Earth and Venus, and so forth. He hoped the arrangement of the cosmos would resemble this diagram, with the triangle, then the square, then the pentagon, then the hexagon, and so forth, all nested between the separate planetary orbits. The idea failed on the first try, when the square would not work for the known separation between the orbits of Jupiter and Mars.

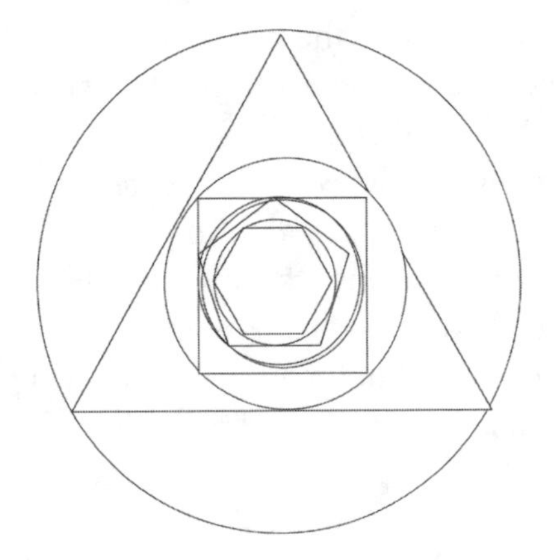

Kepler experimented with other regular polygons, searching for a fit, but he realized that given the infinite number of polygons available, success was assured. To the early Pythagoreans, this might have seemed adequate. Not to Kepler, for the question remained, why—among all the possibilities—*these* particular polygons worked and not others. Why had God chosen to construct the universe in this way and not in some other?

Though many of his contemporaries considered questions like these naive, they bothered Kepler, who had already been focusing his thinking along two lines of investigation: what reasoning God was using when he made things the way they are; and the physical reasons why the universe operates as it does. Clearly, for Kepler, shuffling through all the polygons and finding five that fit neatly between the six planetary orbits was not satisfactory. Since there were regular polygons to fit any planetary distances one might find, he felt there had to be a scheme that would *limit* the

* A regular polygon is a flat shape in which all edges are the same length. For example: the triangle, square, pentagon, hexagon, etc. ad infinitum.

actual, possible ratios (Saturn to Jupiter, Jupiter to Mars, Mars to Earth, Earth to Venus, Venus to Mercury), accounting for why some ratios, not others, existed in the heavens and there were only six planets.

It occurred to Kepler that he was making a mistake in trying to apply two-dimensional, flat figures (polygons) to a three-dimensional universe, and he decided to experiment instead with solid figures, the regular polyhedra.* That thought was a Pythagorean knockout. There were, after all, only five regular polyhedra (the Pythagorean or Platonic solids) not an infinite number of possibilities. To Kepler's immense satisfaction, he found he could fit the five polyhedra into a nested arrangement that quite nicely coincided with the known separations between the six "spheres" in which the planets orbit.†

Any of the five regular polyhedra—cube, tetrahedron (pyramid), octahedron, icosahedron, and dodecahedron—can be set inside a sphere so that each of its points touches the sphere; and a smaller sphere can be set inside any polyhedron so that it touches the center of each side of the polyhedron. This was almost certainly what was meant by the cryptic words of the Philolaus fragment: "The bodies in the sphere are five." So Kepler pictured the solids nesting among the planetary spheres, giving the separation between them just as the triangle had seemed to give the separation between the orbits of Jupiter and Saturn in his drawing on the chalkboard. This "polyhedral theory" appears to have been completely original with Kepler.

Despite Kepler's conviction that there were deep, harmonious connections in nature, and his hope that he had found a stunning example, there was a side to his intellectual makeup that set him apart from the Pythagoreans who had decided there must be ten bodies in the cosmos. He did not merely assume that the universe must surely fit his beautiful

* A regular polyhedron is a solid shape in which all the edges have the same length and all the faces the same shape. The Pythagorean or Platonic solids are the regular polyhedra.

† When astronomers of Kepler's time and earlier spoke of the "spheres," they did not mean the planets. The Ptolemaic view of the cosmos had the planets traveling in transparent "crystalline spheres," nested within one another like the layers of an onion and centered on the Earth. Though Kepler and Mästlin discussed spheres in their correspondence about Kepler's new idea, Kepler (like his predecessor Tycho Brahe) did not believe there were actual glasslike spheres that one could crash through in a space vehicle. Thinking about them in a geometrical sense, not as physical reality, was nevertheless helpful in visualizing the movements of the planets.

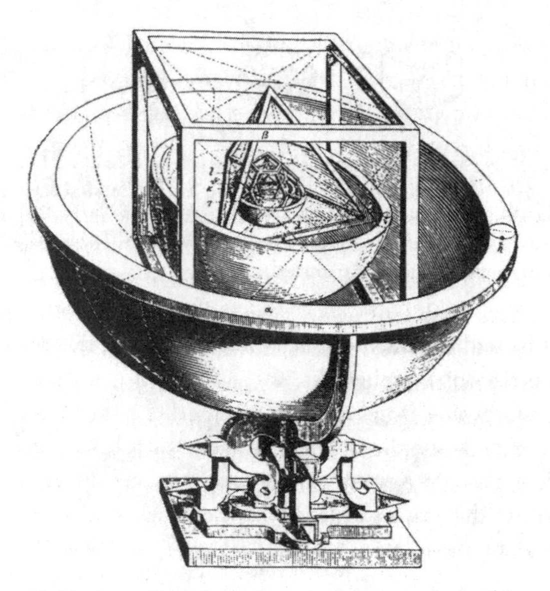

Kepler's polyhedral theory drawing, from his *Mysterium cosmographicum*, showing his nesting arrangement of the six planetary spheres and the five Pythagorean/Platonic solids

geometrical scheme without testing it against Copernican theory and the available observational records "to see whether this idea would agree with the Copernican orbits, or if my happiness would be carried away by the wind." Though the need for such a testing procedure seems obvious today, it did not to those who studied nature in the sixteenth century. Kepler was, in fact, feeling his way into the process that would later be dubbed the scientific method.

Given that there are eight or nine planets orbiting the Sun,* not only the six known in Kepler's time, and that his polyhedral theory has turned out to be quite off the wall, it is astonishing to read Kepler's exclamation that "within a few days everything worked, and I watched as one body after another fit precisely into its place among the planets." He knew, however, that there were better observations than those he was using. Those undertaken by the older Danish astronomer Tycho Brahe were far more precise. Tycho was a reputedly arrogant, supremely talented man whose

* Depending on one's definition of "planet," Pluto and some other bodies that orbit the Sun may or may not have that status. Hence "eight or nine."

nose, it was widely known, had been partially hacked off in a youthful duel and reconstructed out of gold and silver. Unfortunately for Kepler, Tycho was behaving like a new Pythagoras or Copernicus: He was keeping his findings to himself and refusing to publish them.

Kepler finished writing *Mysterium cosmographicum*, his book about the polyhedral theory, in the winter of 1595–96, and it came off the press in 1597. When he was an old man, Kepler would reminisce that this small volume with the long title (the complete title required about six lines of print) was the point of departure for the path his life would take from that time forward. He could, with some justice, have said the same with regard to its watershed significance for the whole of science. As the eminent historian of science Owen Gingerich has commented, "Seldom in history has so wrong a book been so seminal in directing the future course of science."

The game was now afoot in earnest. The polyhedral theory was not a dead end, and the reason was that Kepler—Platonist and Pythagorean when it came to his faith in harmony and symmetry—was a thoroughgoing Aristotelian when it came to his respect for down-to-earth, or at least visible-from-Earth, observational data. He first tried to approach Tycho soon after *Mysterium*'s publication, but it was a truly labyrinthine four-year trail of events that finally brought him to the moment when Tycho's logbooks lay open before him. Meanwhile Kepler set off in another even more Pythagorean direction.

KEPLER MENTIONED MUSIC in *Mysterium* only once, noting that just as there were five regular solids in geometry, so there were five harmonic intervals in music. He was counting more than the octave, fourth, and fifth of the Pythagoreans. Kepler had begun to give himself a thorough grounding in music theory. In his own musical calculations, he decided to use what is known as "just" tuning, rather than "Pythagorean" tuning. Pythagorean ideas about harmony, based on the ratios among the numbers 1, 2, 3, and 4, considered only the intervals of the octave, fifth, and fourth as being consonant. In "just" tuning, more commonly used in Kepler's day, as it still is, major and minor thirds and sixths were also recognized as pleasing to the ear.* Kepler thought the

* An example of a third on the piano is the interval from C to E (major third) or C to E-flat (minor third). An example of a sixth is the interval from C to A (major sixth) or C to A-flat (minor sixth). These are intervals that modern ears are most likely to hear as "beautiful" and easy to listen to.

addition of these intervals was a great improvement over the music of the ancients, and it was a rare musician or listener of his era (or later) who would not have agreed.

His comment was no musical theory of astronomy, but in 1599, two years after the publication of *Mysterium*, Kepler mentioned some ideas about the harmony of the spheres in letters he wrote, first to an Englishman in Padua who he hoped would pass his idea along to Galileo, and then to his patron Herwart von Hohenburg and his old mentor Michael Mästlin in Tübingen. Kepler's proposal was not exactly the same in all three letters, for his thoughts were developing rapidly regarding a question he had raised in *Mysterium*: Why does each planet take the time it does to orbit the Sun? Sure that planets farther from the Sun actually move more slowly and are not merely handicapped by being assigned an outer lane in the race, he was pondering what logic might lie behind the planets' different distances from the Sun and their different velocities.

Kepler thought, as the Pythagoreans and others had before him, that the planets, moving through something like air, must produce a sound, just as the strings of a musical instrument would if hung in a breeze, and he believed the sound was harmonious. Only two people had anticipated Kepler's precise linking of music and planetary *movement*: John Scotus Eriugena, in the ninth century, and Giorgio Anselmi of Parma, in the early fifteenth.[2] Eriugena had recognized that if you associated a musical pitch with a planet, and if pitch depended on the planet's distance from the Earth, then you had to include in your theory the way the pitch would change as a planet moved and changed its distance—which planets clearly did, especially in Eriugena's arrangement with some of the planets orbiting the Sun. Anselmi had not imagined each planet as having an individual tone but rather as singing its own melody in counterpoint with the others. In devising his eight-octave planetary scale, he had taken into consideration the planets' orbital periods. The result was a great cosmic symphony.

In 1599, Kepler was considering the possibility that the velocities (in his word, the "vigor") of the six planets might be related to one another in the same relationships that would produce a harmonious chord if translated into lengths of strings on a musical instrument. For example, a relationship of 3:4 between the velocities of Saturn and Jupiter, used as the relationship between two string lengths, would produce the interval of the fourth. So one could think of the "interval" between Saturn and Jupiter as a musical fourth. Kepler calculated the proportions of the velocities of the planets as 3:4 for Saturn to Jupiter, 4:8 (1:2) for Jupiter to Mars, 8:10 (4:5) for Mars to Earth, 10:12 (5:6) for Earth to Venus, and 12:16 (3:4) (Venus and Mercury). Translating those ratios into musical intervals, he worked out a chord composed of (starting from the lowest note) intervals of a fourth, an octave, a major third, a minor third, and another fourth. In modern notation, an example of this chord would be:

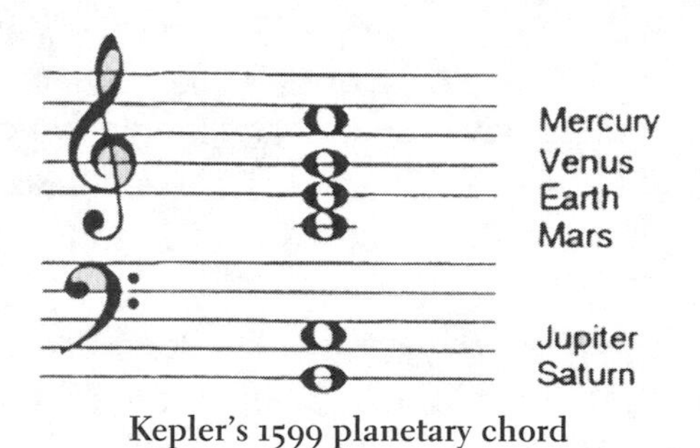

Kepler's 1599 planetary chord

Kepler had chosen the velocities with the aim of having a harmonious chord, and now he found that by doing so he had produced musical intervals that were close to the spatial intervals between the planets in his polyhedral theory. The planetary orbital periods had been well known since antiquity, so he was able to proceed to calculate how large the different orbits had to be in relation to one another if the planets, with these known periods, were traveling at the velocities his musical intervals predicted. He compared the results with the orbital sizes calculated from Copernican theory and found that his harmonic theory was in somewhat better agreement than his polyhedral theory.

Kepler's own summation of what he had learned from each theory was that with the harmonic theory he could calculate the planets' distances from the Sun, relative to one another; and with the polyhedral theory he could calculate the thickness of the empty spaces between the spheres in which the planets orbited.

Kepler wrote to Mästlin that he had found a clever way to connect his polyhedral theory to three of the five intervals in his chord. The cube was the polyhedron that separated the orbits of Saturn and Jupiter. Three flat squares meet at each corner of a cube, and the corner of each of the three squares is a 90-degree angle. Add those three 90-degree angles together and you get 270 degrees. The ratio between 270 and 360 (the number of degrees in a complete circle) is 3:4. It seemed appropriate that the musical interval (the fourth) that required the ratio of string lengths 3:4 was the one that defined the space interval between Saturn and Jupiter. Kepler found similar relationships working for the intervals between Jupiter and Mars, and Earth and Venus.

Kepler thought that he had made good progress with his harmonic theory, and that the harmony he was discovering reflected the mind of the Creator and was surely carried out in the cosmos. He confided to Mästlin and von Hohenburg, late in the summer of 1599, that he felt as though he had "a bird under a bucket." He was soon writing to von Hohenburg that he was planning a work titled *Harmonice mundi*.

When von Hohenburg wrote expressing concern that the numbers were not really a fit, and that the theory was perhaps based on suspicion and not really demonstrated, Kepler replied:

> First, I think that aside from a few propositions, I have proposed not an ironclad demonstration but one which nevertheless stands, in the absence of contrary argument. Second, the suspicion is not entirely false. For man is the image of the Creator, and it may be that in certain matters pertaining to the adornment of the world the same things appear to man as to God.

His own ideas of harmony, he believed, were in synchrony with the Creator's; the remaining difficulties in his theories would not be difficult to overcome, and he would soon have them solved. He had little idea of the arduous intellectual journey that lay before him, or that twenty years would pass before he arrived at the great "harmonic theory" that would continue to be regarded as correct in the twenty-first century.

In August 1599, von Hohenburg mentioned in a letter to Kepler an opinion of Ptolemy on the number of consonant intervals there were in music. Kepler eagerly wrote back to say that if Ptolemy's book would not overburden the messenger, he very much hoped von Hohenburg would send it. In two more letters he continued to ask, and in July 1600, von Hohenberg finally complied. The book was a poor Latin translation of Ptolemy's *Harmonics*, and Kepler later complained that he could hardly make sense of it. Nevertheless, it amazed him that Ptolemy's speculations were not far different from his own, though "to be sure, much was still lacking in the astronomy of that age; and Ptolemy, having begun badly, could plead desperation. Like the Scipio of Cicero, he seems to have recited a kind of Pythagorean dream rather than advancing philosophy."[3]

The late summer and autumn of 1600, when Kepler was first reading the *Harmonics*, was not a convenient moment to consider harmonic theories more deeply. The previous winter, Kepler had joined Tycho Brahe at Benatky Castle near Prague, where the imperious astronomer was then in residence under the patronage of the Holy Roman Emperor Rudolph II. Kepler had arrived anticipating a fruitful collaboration and thinking that his hopes of being able to consult Tycho's phenomenal astronomical data were about to be realized. Instead he had found himself having to cope with a difficult, paranoid, secretive old man who treated him more like an untrustworthy and unpaid servant than a collaborator and would allow him only tantalizing, inadequate glimpses of the precious data. Tycho's longing to gain immortality with his own Tychonic system made him highly suspicious of young Kepler, who openly preferred the Copernican system.* Kepler's hope of improving his financial situation had sent him on a fruitless journey back to Graz, seeking a continuance of his salary as District Mathematician (in absentia) there. But in the summer Kepler found himself not better off but worse. His health was failing, and a drastic turn in the Counter-Reformation in Catholic Graz suddenly made him a penniless Protestant refugee. Reluctantly, when all other possibilities failed (including an appeal to his old mentor Michael Mästlin for a job at Tübingen) he settled with his frightened family in Prague, even more at the mercy of Tycho Brahe than before. Kepler set aside Ptolemy's book and his own ideas about harmony, but only temporarily.

* The Tychonic system had the Sun and the Moon orbiting the Earth, and all the other planets orbiting the Sun. It was the geometric equivalent of he Copernican system, but retained the unmoving Earth.

He had meanwhile not by any means abandoned his polyhedral theory. For him, that theory, his studies of the harmony of the spheres, his great revision, later, of all of astronomy in the light of Tycho's observations—and much else—were not isolated, disconnected efforts. They were all part of a unity of thought and work.

By January 1607, the wheel of fortune had turned again for Kepler. Tycho had died in the autumn of 1601, and Kepler had been the heir to Tycho's position and duties in Rudolph's court. With Tycho's observational logs finally lying open before him (albeit with Tycho's relatives all too ready to snatch them away and intermittently succeeding) Kepler had spent more than half a decade beating his brains over this data and his own calculations, using all the mathematical skill he could muster and inventing new mathematics to work out the true orbit of Mars. At one desperate point he had been almost ready to relinquish his Pythagorean faith and admit that the orbit of Mars simply did not make mathematical sense at all. He had even taken issue with God about it, in words he might have used to express disappointment about a human colleague: "Heretofore we have not found such an ungeometrical conception in his other works!"

Yet eventually the universe and the Creator had turned out, in an unexpected way, to have lived up to Pythagorean standards after all. The planetary orbits, Kepler discovered, were elliptical, not circular, and with that hard-won realization everything fell into place. Kepler had been able to engineer an entire revision of astronomy in the light of Copernican theory and his own. He had found precisely how a planet's velocity changes as it moves closer to and farther away from the Sun in its orbit. He had painstakingly chronicled his "war with Mars" and stated his first two laws of planetary motion in his book *Astronomia nova*, and had sent to press the manuscript for this great work that was the fruit of his, and Tycho's, labors.* The book would win Kepler immortality. The campaign that had begun in Denmark at Uraniborg when Tycho first decided to train his fabulous instruments on Mars was over, and Kepler had awarded the victory to Copernicus, not to Tycho

* Kepler's first law of planetary motion: A planet moves in an elliptical orbit and the Sun is one focus of the ellipse. Kepler's second law of planetary moton: A straight line drawn from a planet to the Sun sweeps out equal areas in equal times as the planet travels in its elliptical orbit.

or Ptolemy. Even so, Kepler still clung to his polyhedral theory as possibly being the underlying logic of the solar system, though he now knew that it could not account for all the proportions.

That January, 1607, a letter arrived from von Hohenburg. He was trying to find a copy of Ptolemy's *Harmonics*, evidently having forgotten that he had sent a copy of that very book to Kepler six and a half years earlier. Kepler reminded him of that previous gift but asked whether he would now please find him a copy in the original Greek.

In March, the book arrived. It included commentaries by Porphyry and the fourteenth-century monk Barlaam of Seminara (the man who had tried unsuccessfully to teach Petrarch Greek). Barlaam argued that the text appearing as Ptolemy's last three chapters was not authentic. This was particularly disappointing to Kepler, who was sure that it was in those very chapters that Ptolemy must have showed how to use harmonic principles to derive the parameters of his planetary models. Kepler made plans to publish an edition of *Harmonics* in Greek with his own commentary, in which he would explain and then refute Ptolemy's theories, then compare them with his own. He also planned to undertake a new Latin translation. In the table of contents for the final volume of his *Harmonice mundi*, he listed the commentary and also an appendix of about thirty pages, translating what he felt were the most relevant parts of Ptolemy's book and re-creating Ptolemy's text in the suspect chapters. But when *Harmonice mundi* appeared in 1619, the actual appendix included, instead, only an apology that the promised material, begun ten years earlier but interrupted by a move from Prague to Linz "combined with many other troubles," was not there.

"Many other troubles" was, sadly, an understatement. Soon after New Year's, 1611, Kepler's three children had contracted smallpox. His six-year-old son, Friedrich, who had been a particular delight to Kepler, died. Troops led by a cousin of Kepler's patron, Emperor Rudolph II, overran Prague and rioted with vigilantes in the streets surrounding Kepler's house. Rudolph, always an exceedingly eccentric, reclusive ruler, and by then somewhat over the brink of madness, abdicated the throne. Kepler's wife died of a fever that July, and Rudolph himself expired the next winter. Even before the emperor's death, Kepler had foreseen the end of his usefulness in Prague and had accepted a position as teacher and district mathematician in Linz—a job on about the level of the one in which he had begun his career seventeen years before in Graz. Though he was widely known and respected, famous for his *Astronomia nova*, nothing better was available because of an earlier statement still on record that he had made and would not recant, that he believed a Calvinist also was a "brother in Christ." That opinion disqualified him from any position at a Lutheran university. His salary as imperial mathematician continued, theoretically, but in reality he was still trying to collect years of back pay from the undependable imperial treasury. If all of that had not been enough to distract him, he found it imperative, with two motherless children, to look for a new wife, and he remarried about two years after his first wife's death. In December 1615, disaster struck again. His mother was accused of witchcraft. In the next three years, while he defended her and struggled to keep his own reputation from being destroyed in the process, he and his new wife lost two infant daughters and also his much loved stepdaughter, the daughter of his first wife by a previous marriage.

In the winter of 1618, Kepler was too distracted with grief to concentrate on tedious calculations needed for the Rudolfine Tables—astronomical tables based on Tycho's observations, on which Kepler had been working intermittently for many years. "Since the *Tables* require peace," he wrote, "I have abandoned them and turned my mind to developing the *Harmony*." In *Astronomia nova*, he had completely reconstructed astronomy, and this meant not only that the harmony project had taken on much greater proportions but also that Kepler had a much stronger and more comprehensive foundation on which to work. He was now dealing with what he knew to be a real planetary system, whose mathematics and geometry he understood better than anyone else alive.

"The *Harmony*" referred to the book he had barely begun in Graz when he had first considered linking the velocities of the planets with musical harmony and had shared his thoughts with von Hohenburg and Mästlin. That period of Kepler's life had also been a time of mourning, for the death of his first and second child. Now tragedy had again decimated his family, and there must have seemed little evidence of a rational, loving God, but Kepler turned again to the effort to reveal what he believed was God's marvelous wisdom and rationality to be discovered in nature.

When Kepler began laying out the table of contents for the book that would be *Harmonice mundi*, he decided the moment had come to revive his plans to translate Ptolemy's *Harmonics*. Then the Thirty Years War broke out and the scarcity of manpower made it impossible to get material printed. Not until 1864, more than two centuries after his death, did an edition of *Harmonice mundi* appear that included his Latin translation of Ptolemy's *Harmonics*. It had survived in manuscript form.

Nevertheless, in 1618, Kepler was well acquainted with Ptolemy's *Harmonics* and had also researched what Aristotle and Pliny had written, centuries before Ptolemy, about the Pythagoreans. He decided that Ptolemy must have been trying to describe and improve on Pythagorean teachings about the harmony of the heavens but had not made it clear what those teachings had been.[4] Kepler chose to accept Pliny's opinion that Pythagoras had assigned a musical pitch to each of the eight heavenly bodies (five planets, stars, Moon, Sun) and linked the distances between them with distances (intervals) between those pitches. Kepler concluded that the Pythagorean heavenly scale must have begun with the Moon, not the Earth, because in an Earth-centered cosmos Earth would not move, and a body at rest, making no sound, has no pitch associated with it. It is something of a mystery why he thought that Pythagoras would have visualized an Earth-centered cosmos. Copernicus had used the Pythagorean concept of a central fire as a precedent for his own rearrangement of the cosmos, and Kepler too liked to point to that precedent. He had read about it in Aristotle's *De Caelo*. Perhaps he believed the central fire was an idea from later Pythagoreans, for example Philolaus, and that Pythagoras himself must have treated the Earth as the unmoving center.

Kepler based his own reconstruction of the scale Pythagoras had used on the Pythagorean reverence for the intervals of the fourth and fifth and on the intervals Pliny had chosen. In Kepler's reconstruction, the Moon was A, Mercury B flat, Venus B, Sun D, Mars E, Jupiter F, Saturn F sharp, the stars A. The first four notes (A, B flat, B, D) were separated by intervals of a half step, a half step, and a step and a half. The second four (E, F, F sharp, A) were separated by that same sequence of intervals.* The two groups were separated by a whole step (D to E, or Sun to Mars), and that whole step was between the fourth and

* A half-step is the interval between any note on the piano and the one immediately to either side of it, regardless of whether that is a white or black key.

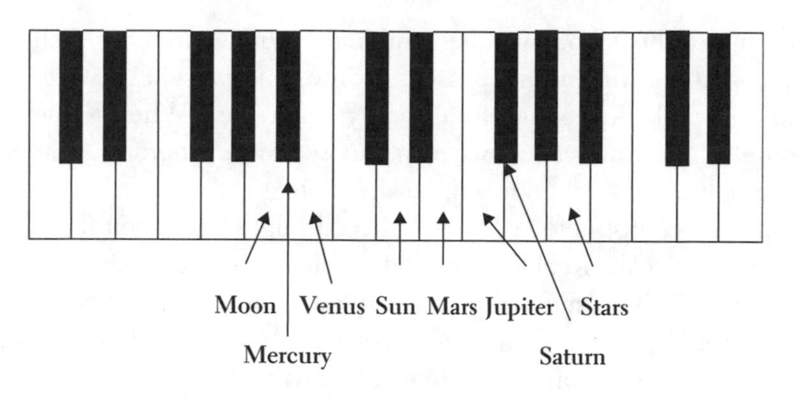

fifth of the scale, two significant Pythagorean notes because they reflected the ratios 3:4 and 2:3.

Kepler's reason for deciding that Pythagoras had associated the lowest note with the nearest heavenly body, the Moon, rather than start his scale on the most distant body, was that he thought Pythagoras, observing the sky, would have seen the higher, more distant planets appear to be moving faster than the lower ones and would not have realized that one component of that movement was the rotation of the Earth itself. The entire sky appears to rotate westward, making one complete rotation every twenty-four hours, while each planet moves in a motion contrary to that. The result of the combined motion is to fool an observer into thinking that the more distant planets (actually the slowest moving) are moving fastest.* Again, Kepler must have been trying to work out what Pythagoras himself would have thought, not later Pythagoreans, who probably did understand that the observed motion of a planet,

* Imagine you are standing across the corridor from a moving walkway in an airport. A man is walking along the walkway, from your left to your right, but he is going the wrong way and so is actually losing ground. Say he is walking at 5 miles per hour and the walkway is moving, in the opposite direction, at 10 miles per hour. From your vantage point, you see the combined movement, and the man appears to be moving 5 miles an hour toward the left. A woman is walking faster, 8 miles per hour, but also in the wrong direction. Eight miles per hour is not sufficient to avoid losing ground against the 10-mile-an-hour walkway that is moving in the opposite direction, so, again, from your vantage point, you see the combined motion, and this woman appears to be moving 2 miles per hour toward your left. You cannot be faulted for thinking that the man (who appears to be moving 5 miles per hour toward your left) is moving faster than the woman. If the walkway stopped you would find out what the true velocity of each one was, and your finding would contradict your initial impression. Likewise, Kepler concluded that if the daily rotation of the heavens had stopped, Pythagoras would have seen that Saturn is the slowest of the planets, and should be sounding the lowest tone.

the Sun, or the Moon, was the combination of two motions. Scholars still regard this understanding as one of the triumphs of early Greek astronomy, probably coming from the Pythagoreans.[5] When Kepler got around to constructing his own planetary scales and chords, he had Saturn, not the Moon, sounding the lowest note.

Having reconstructed what he believed might have been the actual Pythagorean scale, Kepler set it beside Ptolemy's and considered which he preferred. Both suffered from having an Earth-centered cosmos in mind. But Kepler was fond of Pythagoras' scale—"altogether more elegant and richer in mysteries" than Ptolemy's—because it seemed to him to give more importance to the planets' motions. On the other hand, he gave Ptolemy points for having recognized that there must be a "divine axiom" that determined the number and sizes of the spheres.

Before Kepler could decide for himself what the harmony of the heavens might be, and what ratios might underlie it, he had to determine which intervals were agreeable to the human ear. He took a great deal of trouble differentiating between types of intervals. There were the usual octaves, fourths, fifths, thirds, and sixths, all of which he called *consonantia* (they were harmonious when the two tones sounded simultaneously). Then there were several intervals that he called *concinna* that sounded pleasant following one another in a melody but not when played simultaneously. These included a "major tone" and "minor tone" (roughly equivalent to a whole step and a half step) and two other intervals that were smaller than the interval between adjacent keys on a piano. Finally there were three tiny intervals that Kepler dubbed the "doubtful *concinna*." They were not particularly pleasant to hear under any circumstances.

Kepler combined the intervals into two kinds of musical scales. One had a major third and sixth in it and was the *durus* scale, close to what we call the major scale. (The major scale beginning on C, for example, includes the intervals C to E and C to A.) The other, with a minor third and sixth, was the *mollis* scale, close to what we call the minor. (The minor scale beginning on C, for example, includes the intervals C to E flat and C to A flat.) Likewise, chords based on major thirds and sixths were *durus*; chords based on minor thirds and sixths were *mollis*.* It requires no musical training to hear the difference between the two scales

* In German, *dur* in music still means "major"; *moll* is "minor."

or chords and experience the emotional effect of this difference: the *durus* (major) is happy and the *mollis* (minor) sad. Why these sounds have any influence over human emotions is still a mystery, but the early Pythagoreans, had they known about thirds and sixths, would surely not have been surprised.

One of Kepler's goals in the research that lay behind *Harmonice mundi* was to find out whether two proposals were true: First, that certain ratios between pitches have a special "nobility" and importance and are embodied in the arrangement and movements of the solar system. Second, that the influence of music on the human soul depends on these ratios. As the Pythagoreans had known, musical intervals are the way mathematical ratios show up in sound. One usually encounters written-out music in the form of notes drawn on and between horizontal lines on a page, and seldom does one realize that it would be possible to write out the music more precisely (though less practically) as a long string of mathematical ratios. If one wrote out all the continually shifting mathematical proportions among the planets, would the result, played as music, sound harmonious and pleasing to human ears? In Book V, Chapter 9 of *Harmonice mundi*, Kepler explained why he was convinced—after a prodigious amount of study and calculation—that the details of planetary astronomy, the continually changing speeds and distances of the planets in relationship with one another, were as harmonious and pleasing as could possibly be. He also showed how this best-of-all-possible harmonious arrangements inevitably (even though God had created it) fell a bit short of perfection.

Most significant for the history of astronomy, Book V began with an ecstatic statement about discovering the relationship between the planets' orbital radiuses and their orbital periods, even though Kepler had not yet made this discovery when he began writing his book:

> At last I brought it into the light, and beyond what I had ever been able to hope, I laid hold of Truth itself: I found among the motions of the heavens the whole nature of Harmony, as large as that is, with all of its parts. It was not in the same way which I had expected—this is not the smallest part of my rejoicing—but in another way, very different and yet at the same time very excellent and perfect.[6]

Kepler felt that this discovery—his third law of planetary motion, or "harmonic law"—was so important that it was essential to go back and insert those sentences to let his readers know what was coming. It was also in Book V that he included a list giving a view of the astonishing mind of Kepler exactly where he was, beginning with a flat statement that sounds completely unremarkable to modern ears but was a blockbuster in his time. The list included his three planetary laws, which are still celebrated among the greatest discoveries in astronomy, but also—surprisingly—his old polyhedral theory.

1. The planets and Earth orbit the Sun. The Moon orbits the Earth.
2. The Sun is not at the center of a planet's orbit [in other words, planetary orbits are eccentric], meaning that each planet has a maximum and minimum distance from the Sun and also passes through all other distances between the maximum and minimum.
3. The five regular polyhedra dictate the number of planets: six.
4. The polyhedra alone cannot determine the distance from the Sun, since the orbits are eccentric (see proposition 2). Other principles are needed to establish the orbits and the diameters and the eccentricities.
5. A planet's velocity is inversely proportional to its distance from the Sun. The orbit of a planet is an ellipse, and the Sun, "the source of motion," is one focus of the ellipse. [This was Kepler's first law of planetary motion, which he had discovered while writing *Astronomia nova*.]
6. If two objects move the same actual distance, but one of them is farther away than the other, the movement of the one farther away appears smaller than the one nearer. So, if a planet never changed its speed, then, viewed from the Sun, its motions when it is farthest away would appear smaller than its motions when it is nearest. But a planet does change its speed. Its motion is not the same at its nearest and farthest points, and the difference is in proportion to the distance from the Sun. In other words, the apparent sizes of the motions are different for two reasons: The actual size of the motion is smaller. Distance makes the size of the motion look smaller. So the apparent

sizes of the motions are very nearly the *inverse square* of the proportion of their distances from the Sun.

7. When it comes to celestial harmony, it is motions as seen from the Sun that are important. Motions as seen from the Earth are irrelevant.
8. The ratio of the squares of the orbital periods of two planets is equal to the ratio of the cubes of their average distances from the Sun. [This was the great "harmonic law," one of Kepler's most significant discoveries. Kepler made the discovery as he was finishing the book and came back and inserted it in this list.]

9–13. [These have to do with applying the harmonic law. Kepler tried to spell out more clearly that the ratio of the motions of two planets as they draw closer or move apart, together with the ratio of their periodic times, determine the extreme distances they can have (closest and farthest from the Sun), and this determines how eccentric their orbits are.]

Kepler eventually came to the conclusion that celestial harmony could not possibly be audible. There were no sounds in the heavens. How could they be enjoyed? Knowing or calculating the path lengths was too complicated to give pleasure in an instinctive way. The harmony of the cosmos could be best appreciated from the Sun itself, in the visible arcs of the planetary motions as they would be seen from there.[7] (Hence number 7 in his list.)

Think of the Sun, with yourself standing on it, as being at the center of a huge clock face with the planets moving on large, nearly circular pathways near the rim of the clock face. The entire orbit of a planet is 360 degrees, all the way around the clock. The distance between one and two o'clock, viewed from the center of the clock, is 30 degrees. You, on the Sun, see Earth circling you—though "circling" is not quite the right word, since Earth's orbit is not round but slightly elliptical. Earth is at aphelion (the part of its orbit farthest from the Sun and you). You watch for a twenty-four-hour period and find that Earth has moved 57'3" (57 "minutes" and 3 "seconds"). Since there are sixty minutes in a degree, Earth has moved almost one degree. Suppose, instead, you are viewing Earth when it is at perihelion (the part of its orbit closest to the Sun). Now you find that Earth's motion is faster, 61'18" in twenty-four hours, more than 1 degree. Those two measurements—Earth's apparent diurnal motions at aphelion (57'3") and perihelion (61'18")—are not far different from one another. Earth's orbit is not very eccentric.

Kepler pondered how these two numbers might be adjusted so as to produce a harmonious interval in music. By changing 57'3" to 57'28" (a very small adjustment) he could make the interval a *concinna*, an interval that sounded pleasant in a melody though not when the two notes were played simultaneously. Kepler made similar tiny adjustments for the other planets' orbits. The most troublesome was Venus, whose motion varied so slightly that its musical interval was a *diesis*. That was small indeed but still fell into Kepler's category of *concinna*.

Having worked with each planet individually, calculating and adjusting the relationship between its motion at aphelion and at perihelion, Kepler turned to studying the motions of pairs of planets, and was pleased to find fairly good harmony. The small adjustments necessary could, Kepler wrote, easily be "swallowed" without detriment to the astronomy he had constructed using Tycho's observational data. Again Venus was a problem, and so was Mercury, but their motions were not yet well established anyway.

Satisfied with the way things were going so far, Kepler proceeded to assign actual notes to each of the planets at aphelion and perihelion and found that when he built a scale with Saturn (the lowest note) at aphelion, the result was a *durus* scale, a major scale. With Saturn at perihelion the result was a *mollis* scale, a minor scale. Planetary motion apparently did involve both types of scale. Using other planets as the starting note produced the different modes used in ancient music and church music.*

Thus far, all of these combinations had the planets at the extremes of their motions, at aphelion or perihelion. Particularly for the planets most distant from the Sun, such opportunities would actually occur only rarely. However, if the planets involved in the harmony did not have to be at those extreme positions, the harmonic opportunities were much more numerous. For example: With Saturn moving between the pitches G and B (its pitches at perihelion and aphelion) and Jupiter between B and D, Kepler found, along the way, intervals of an octave, an

* You get the same result by playing scales using only the white keys on a piano but starting on different notes. The Ionian mode (start on C) is the same as the major scale, the Dorian mode (start on D), the Phrygian mode (E), the Lydian mode (F), and the Mixolydian mode (G). The Aolian mode (start on A) is the same as the minor scale.

octave plus a major or minor third, a fourth, and a fifth. Mercury, the true coloratura in the company, offered even more opportunities because the difference between its pitches at perihelion and aphelion was greater than an octave, and it made that change in only forty-four days. The result was that Mercury as it moved along sang every harmonic interval at least once with each of the other planets.[8]

As Kepler calculated it, two-note harmonies of this sort occur almost every day, and Mercury, Earth, and Mars even sing three-part harmony fairly often. Venus, with so little eccentricity to its orbit, hardly varies its pitch at all, making it a sort of Johnny One-Note in the choir. If there is to be harmony with Venus, it must be when another planet slides into harmony with her, not the other way around. Four-note harmonies occur either because Mercury, Earth, and Mars are in adjustment with Venus' monotone, or because they have waited long enough for the slow-changing bass voice of Jupiter or Saturn to ease into the right note. "Harmonies of four planets," wrote Kepler, "begin to spread out among the centuries; those of five planets, among myriads of years."[9] As for harmony among all six planets—that grand and greatest "universal harmony"—the chord would be huge, spanning more than seven octaves. (You could not play it on most modern pianos. You would need an organ.) Kepler thought it might be possible for it to occur in the heavens only once in the entire history of the universe. Perhaps one might determine the moment of creation by calculating the past moment when all six planets joined in harmony. Kepler thought about the words of God to Job: "Where were you when I laid the Earth's foundation . . . while the morning stars sang together?"

Kepler dared to move ahead to what he felt was the true test of his theory: "Let us therefore extract, from the harmonies, the intervals of the planets from the Sun, using a method of calculation that is new and never before attempted by anyone."[10] If you did not know the astronomy of the solar system, could you deduce it correctly from the harmonic scheme he believed he had discovered? Starting with the best harmony and figuring out what planetary orbits and motions this harmony implied, what would be the observable consequences of the cosmos' adhering to this harmony? Kepler used Tycho Brahe's data for comparison and concluded that "all approach very closely to those intervals which I found from the Brahe observations. In Mercury alone there is a small difference."

Kepler proceeded to compare the solar system as dictated by his harmonic scheme with the solar system as dictated by his polyhedral theory. His conclusion was that the polyhedra, nested in the way he had earlier suggested, had been God's rather loose model for the solar system. It dictated how many planets there would be and the approximate dimensions of the spheres within which they moved. It was a sort of sketch, with the final dimensions filled out by the harmonic proportions among the planets' apparent motions as viewed from the Sun. The concept of "harmonies" was required to reflect an eternally fluid system like that found among real planets in motion, and the real solar system could not be understood apart from its motion.

Nearing the end of his book, Kepler imagined himself drifting off to sleep to the strains of the planetary harmony, "warmed by having drunk a generous draught . . . from the cup of Pythagoras."[11] He is soon dreaming about pure, simple beings who might live on the Sun, in the right position to appreciate the harmony, and of creatures on the other planets: It would be a terrible waste if there were none. They, like Earth dwellers, have no way of appreciating the harmony directly and can only learn of it, as humans had, by a combination of observation and reasoning. Kepler wrote a prayer that God would be praised by the heavens, by the Sun, Moon, and planets, by the celestial harmonies and their beholders—"by you above all, happy old Mästlin, for you used to inspire these things I have said, and you nourished them with hope"—and by his, Kepler's, own soul. He ended with a return to the old idea that was inherent from the start in the Pythagorean discovery of musical ratios: that one does not have to know about them to be moved by music. There is a mysterious inherent connection between human souls and the underlying pattern of the universe that affects us without our understanding why or how. The same was true, Tycho Brahe had thought, of the design of his palace/observatory. Kepler wrote:

> It does not suffice to say that these harmonies are for the sake of Kepler and those after him who will read his book. Nor indeed are aspects of planets on Earth for the sake of astronomers, but they insinuate themselves generally to all, even peasants, by a hidden instinct.[12]

With modern hindsight, it seems Kepler took an odd, eccentric road indeed to arrive at his great "harmonic" law. He found it twice, at first rejecting it because of a computational error on March 8, 1618, and then discovering that it was correct a few weeks later, on May 15. The comment has sometimes been made that the harmonic law was an accidental discovery in the midst of a labyrinth of worthless musical/mathematical speculation, and that Kepler hardly realized he had made an important discovery. But Kepler definitely knew it was significant. It was in response to this discovery that he fell to his knees and exclaimed, "My God, I am thinking Thy thoughts after Thee." Without the underpinning of modern mathematics and the modern scientific method, the convoluted musical path Kepler took may have been the only way he could have got there. After all, he was the one who did get there. Kepler had one of the truest ears in history for the harmony of mathematics and geometry.

CHAPTER 17

Enlightened and Illuminated

Seventeenth–Nineteenth Centuries

KEPLER'S CONTEMPORARY GALILEO wrote that "Science" was to be found "in a huge book that stands always open before our eyes—the universe." But to understand it, one needed to be able to understand the language, and "the language is mathematics."[1] Galileo was not the first in his family to win a place in history. His father, Vincenzo, appears in textbooks of music history as a prominent musician of the sixteenth century—a composer, one of the best music theorists of his time, and a fine lutenist. One of his areas of research was ancient Greek music, and there is a story that when he read Boethius' *De musica*, the account of Pythagoras hanging weights on lengths of string, plucking the strings, and discovering the ratios of musical harmony piqued his curiosity.[2] Amazingly, no records survive, from all the prior centuries during which scholars had been reading Boethius, of anyone trying this to see whether it would work. Vincenzo discovered, of course, that it did not, but he went on experimenting with the physics of vibrating strings. When his son watched a lamp swinging in the Pisa cathedral and first decided to experiment with pendulums, perhaps he had in mind his father's tests with weights and strings.

Two decades later, the younger Galileo, though largely oblivious to the work Kepler was doing, had become personally convinced that the

Copernican system was correct, and he was looking for physical evidence to support that opinion and convince other scholars. Copernicus had mentioned in *De revolutionibus* that the planet Venus might supply important evidence in the case against an Earth-centered cosmos. Venus, reflecting the Sun's light, waxes and wanes as the Moon does, but if the Ptolemaic arrangement of the cosmos were correct, Earth dwellers would never be positioned in such a way as to see the face of Venus anywhere near fully lit (the equivalent of a full Moon). As the first decade of the seventeenth century drew to a close, the newly invented telescope (Galileo did not invent it but was putting it to better use than anyone else) made it possible to observe the phases of Venus as never before, and in 1610 Galileo followed up on Copernicus' suggestion. He found that Venus had a full range of phases. How could any scholar fail to see that this was irrefutable evidence in favor of Copernicus? But Galileo's Catholic colleagues included a group of recalcitrant scholars who remind one of an unusually virulent strain of *acusmatici*.

Except in the case of Giordano Bruno, whose offenses by church standards were so flagrant and numerous that he would almost surely have been burned at the stake no matter where he thought the center of the universe was, the Catholic church hierarchy had for centuries been rather sluggishly tolerant of new astronomical theories. Not a murmur was heard when Nicholas of Cusa, in the early fifteenth century, put the Earth in motion and removed it from the center of the universe, nor when Copernicus published *De revolutionibus* in 1543. Two of Copernicus' strongest supporters were prominent Catholic clergy. But in 1616, when both Galileo and his opponents were pushing the church for a ruling on the Copernican question, a decree was issued condemning the "new" astronomy, though not actually calling it heresy—a technicality perhaps, but a victory for Galileo and the cardinals who supported him. In this decree, the Pythagoreans took an unfair hit:

> And whereas it has also come to the knowledge of the said Congregation that the Pythagorean doctrine—which is false and altogether opposed to the Holy Scripture—of the motion of the Earth and the immobility of the Sun, which is also taught by Nicolaus Copernicus . . . is now being spread abroad and accepted by many, as may be seen from a certain letter of a Carmelite Father.

The Carmelite father who had put the Pythagoreans in the range of fire was the Reverend Father Paolo Antonio Foscarini. His letter, dated the year before the decree, was titled "On the Opinion of the Pythagoreans and of Copernicus Concerning the Motion of the Earth, and the Stability of the Sun, and the New Pythagorean System of the World." Foscarini insisted this doctrine was "consonant with truth and not opposed to Holy Scripture." The church's "General Congregation of the Index," which made official judgments on such matters, felt differently. Copernicus' book *De revolutionibus*—seventy-three years after its publication—was "suspended until corrected," and Foscarini's work was "altogether prohibited and condemned." It took seventeen more years of on-and-off sparring, and Galileo's book *Dialogo*, for matters to come to a truly dangerous head in his famous trial. The Catholic church, for centuries the guardian and bastion of learning, had turned foolish to the point of malign senility and condemned herself and Italy—the ancient home of Pythagoras—to what was virtually a new scientific dark age. The center of scientific endeavor and achievement moved, irretrievably, to northern Europe and England.

As the scientific revolution continued north of the Alps in the mid-seventeenth century, Kepler's three laws of planetary motion and his Rudolfine Tables, based on Tycho Brahe's observations, rightly gave him his earthly immortality, but his polyhedral theory and most of *Harmonice mundi* were consigned to the cabinet of curiosities. No one took nested polyhedrons or cosmic chords and scales seriously or followed up on them as science. They had been the odd and unlikely midwives to Kepler's "new astronomy," helping birth the future, but in doing so had relegated themselves to the past. However, the conviction that numbers and harmony and symmetry were guides to truth because the universe was created according to a rational, orderly plan began to be treated as a given, trustworthy enough to underpin what would later be called the scientific method.

No one was using the words "science" or "scientific" yet in their modern sense, but the process for determining what was and was not true about nature and the universe was continuing to evolve, and people were discussing and beginning to agree about how this process should work. The French scientist and philosopher René Descartes, one of the first to try to establish a solid foundation for human understanding of the world, chose mathematics as the only trustworthy road to sure knowledge.[3] He tried to show that a single, united system of logical mathematical theory could account for everything that happens in the physical universe. Christiaan Huygens, Edmond Halley, and Isaac Newton all shared the conviction that when observations were inadequate, one could even with some confidence go out on a limb on the assumption that the universe is orderly, and discovering new examples of "order" was beginning to be regarded as a sign that one was on the right track. Robert Hooke, in the field of biology, suggested that crystals like those that may have alerted the Pythagoreans to the existence of the five regular solids occurred because their atoms had an orderly arrangement.[4] Robert Boyle wrote his book *The Sceptical Chymist*, which many identify as marking the beginning of modern chemistry, and cited Pythagoras, asserting that the final decisions of science must be made on the basis of both the evidence of the senses and the operation of reason. This balance, on which Kepler had performed such prodigious acrobatics as he struggled to write his *Astronomia nova*—without thinking of it as a "scientific method"—was becoming the balance of science.

Newton, born mid-century, capped off the Copernican revolution with his discovery of the laws of gravity and his 1687 book *Philosophiae Naturalis Principia Mathematica* ("Mathematical Principles of Natural Philosophy"), known as his *Principia*. A fervent believer in the harmony and order of the universe, he was convinced that the observable patterns in the cosmos were the visible manifestation of a profound, mysterious, underlying order. His theories of gravitation admirably supported the Pythagorean ideal of unity and simplicity. The same force, gravity, that kept the planets in orbit also dictated the trajectory of a ball thrown on Earth and kept human beings' feet on the ground, and its laws could be stated in a simple formula. Though he was notoriously miserly about giving credit where credit was clearly due among his contemporaries, Newton, in an extraordinary gesture, wrote that his own famous law of universal gravitation could be found in Pythagoras. Nor was this the extent of Newton's unusual attributions. He sought examples among the Greeks, the Hebrews, and other ancient thinkers, of ideas and discoveries that seemed—sometimes it was quite a stretch—to foreshadow his own. This was not modesty. Newton was by no means a modest man. It was more a way of elevating himself to the company of the greatest sages. Better than discovering something new was rediscovering knowledge that God had previously revealed only to extraordinary men of legendary wisdom. Newton thought of another link with Pythagoras when he used a prism and split the light of the Sun into seven colors. There were seven notes in the Pythagorean scale.[5]

Gottfried Leibniz, Newton's arch-rival and one of those contemporaries to whom Newton should have given considerably more credit, wrote in Pythagorean tones that "music is the pleasure the human soul experiences from counting without being aware that it is counting."[6] Leibniz tried to construct a universal language which had no words, that could express all human statements and resolve arguments in a completely unambiguous way, even, he hoped, bring into agreement all versions of Christian faith. His attempts to make good on this scheme included a use of numbers that would have pleased the Pythagoreans and annoyed Aristotle: "For example, if the term for an 'animate being' should be imagined as expressed by the number 2, and the term for 'rational' by the number 3, the term for 'man' will be expressed by the number 2×3, that is 6."[7]

NEWTON'S DISCOVERIES ABOUT gravity showed the cosmos seeming to operate like a stupendous, dependable mechanism, and, in the eighteenth century, scholars and amateur science aficionados picked up on that idea and became obsessed with mechanisms and machines. The demonstration of a new apparatus to explain or test a scientific principle was likely to cause more excitement than a lecture or a new theory at meetings of the Royal Society of London for Improving Natural Knowledge, or of the Birmingham "Lunar Men" of Charles Darwin's grandfather. It was the age of the "clockmaker's universe" and of England's industrial revolution. Careful observation and experiment became the hallmark of science, but cautious generalization was also encouraged, especially if it led to practical applications.

In other ways, in the eighteenth century, the universe was failing to live up to its promise of simplicity. The Swedish botanist Carl Linnaeus was applying two-word Latin names to more and more species that travelers and voyagers to all corners of the world were discovering. There were a greater number than anyone had ever imagined. Linnaeus saw new plants in his garden, too, and began to suspect, a century before Darwin's *Origin of Species*, that new species were emerging all the time. He decided that these had always existed in the mind of God but

Carl Linnaeus

were just now coming into material existence, a very Platonic way of assuaging his religious scruples.

No one's faith in the completeness of universal harmony and the power of numbers surpassed that of the French mathematician Pierre Simon de Laplace, whose lifetime spanned the turn of the eighteenth to the nineteenth century. For him, numbers and mathematics were an unshakably trustworthy bridge to the past and future—if one could know the exact state of everything in the universe at a given moment. His contention was that an omniscient being with that knowledge, with unlimited powers of memory and mental calculation, and with knowledge of the laws of nature, could extrapolate from that the exact state of everything in the universe at any other given moment.

Meanwhile, Pythagorean themes appeared in other than scientific settings. The Whig party praised the governmental structure which brought together king and Parliament by means of "natural" laws, with these words:

What made the planets in such Order move,
He said, was harmony and mutual Love.

The Musick of his Spheres did represent
That ancient Harmony of Government.

That was by no means an isolated allusion. The harmony of the heavens had become a beloved poetic image. William Shakespeare, a contemporary of Galileo and Kepler, had given it beautiful expression in *The Merchant of Venice*, where he had Lorenzo tell Jessica,

. . . soft stillness and the night
Become the touches of sweet harmony . . .
Look how the floor of heaven
Is thick inlaid with patines of bright gold;
There's not the smallest orb which thou behold'st
But in his motion like an angel sings.
Such harmony is in immortal souls;
But, whilst this muddy vesture of decay
Doth grossly close it in, we cannot hear it.

Shakespeare's contemporary John Davies had written a "justification of dance" titled "Orchestra" that was full of such allusions—not only to the celestial music but also to the four elements. Davies was not making a scientific or philosophical statement. He was correcting one lady's disparagement of dancing by pointing to its ancient, primordial origins.

Dancing, bright lady, then began to be
When the first seeds whereof the world did spring,
The fire, air, earth and water did agree,
By Love's persuasion, nature's mighty king,
To leave their first disordered combating
And in a dance such measure to observe
As all the world their motion should preserve.
. . . . .
The turning vault of heaven formed was,
Whose starry wheels he hath so made to pass
As that their movings do a music frame
And they themselves still dance unto the same.
. . . . .
All the world's great fortunes and affairs

Forward and backward rapt and whirled are
According to the music of the spheres.

John Milton, a later contemporary of Galileo and Kepler, like Shakespeare referred to the inability of human ears to hear this music:

But else in deep of night when drowsiness
Hath locked up mortal sense, then listen I
To the celestial Sirens' harmony . . .
Such sweet compulsion doth in music lie,
To lull the daughters of Necessity,
And keep unsteady Nature to her law,
And the low world in measured motion draw
After the heavenly tune, which none can hear
Of human mould with gross unpurged ear.

Another Englishman, John Dryden, born in 1631, the year after Kepler died, like Davies gave music a voice in creation:

From harmony, from heavenly harmony,
This universal frame began:
When Nature underneath a heap
Of jarring atoms lay
And could not heave her head,
The tuneful voice was heard from high:
Arise, ye more than dead!

Joseph Addison, born later in the century, was the author of a poem that combined the ideas expressed in Psalm 19 with the image of the music of the spheres. Christian congregations still sing it, to music by Franz Joseph Haydn. The final verse says of the planets:

What though in solemn silence all move round the dark
terrestrial ball?
What though no real voice nor sound amid their radiant
orbs be found?
In reason's ear they all rejoice, and utter forth a glorious
voice:

Forever singing as they shine, "The hand that made us is divine."[8]

Johannes Kepler (and nearly everyone who has sung this hymn) would have disagreed with the Earth-centered cosmos these lines implied, but Kepler himself—who had imagined the planets arranged in perfect harmony at the moment of creation—could not have put it better. His harmony was a harmony audible to "reason's ear." Even a century after Addison, William Wordsworth, whose lifetime spanned the turn of the century from the 1700s to the 1800s, could still be certain no explanation or footnote was required when he wrote of "harmony from Heaven's remotest spheres."

PYTHAGOREAN IDEAS AND traces of the Pythagorean tradition also showed up in more surprising contexts. One of the most bizarre examples was the reimagining, in the late eighteenth and early nineteenth centuries, of Pythagoras as the hero of intellectual revolutionaries in Europe and Russia. This use, or misuse, of Pythagorean themes was brought to light by James H. Billington in his book *Fire in the Minds of Men: Origins of the Revolutionary Faith*.[9] Billington showed that in the midst of confusion, when nothing was stable and dependable, Pythagoras became an icon of revolution, and his name and the ideals and symbols associated with him ran as leitmotifs through the decades of revolution and revolutionary thinking.

In 1776, the year of the American Declaration of Independence and eleven years before the date usually identified as the beginning of the French Revolution, a group in Bavaria founded by one Adam Weishaupt and recruited from the Masonic lodges in Munich was calling itself "Illuminist." Though "Illuminism" was difficult for anyone at the time (or today) to define, for Weishaupt it meant a "revolution of the mind," discarding and avoiding all "spiritualist distortions" and occult practices and ideas. However, the name and concepts vaguely associated with Illuminism predated Weishaupt, and so, probably, did the connection with Pythagoras. Because Illuminists were usually as secretive as Pythagoras and his earliest followers, many questions about them cannot be answered, and a danger of being a secret society is that your popular and historical image may be created not by yourself but by your most vocal and influential enemies. Some credited the Illuminists with almost single-handedly precipitating the French Revolution. Others said they never really existed at all but were a "police myth" conjured up by rightists to inspire public fear of clandestine plots, a myth half believed by the authorities themselves. Others assert that they were a fictional invention of propagandists who opposed Masonry and tried to tarnish its image by associating it with insurrection and revolution. Yet others claim that they were an extreme branch of Masonry, or something independent that "infected" Masonry. The Masons also were intensely secretive, though not necessarily for the same reasons the Illuminists were.

At the time of Columbus there were "Alumbrados" in Spain whose mysticism centered around the idea that a human soul could be subjected to inner purification leading to complete submission to God's will and direct communication with and through the Holy Spirit. Eighteenth-century Illuminists also emphasized inner perfection and purification, but with a secular stress on reason and logic. This newer Illuminist ideology either first appeared in lodges of the Freemasons and other Masonic orders, such as Weishaupt's in Bavaria, or else found fertile ground there and rapidly took over. For Masons, working toward inner perfection and purification was already central to their teaching, and it was also attractive to see themselves as re-creating an ancient brotherhood. In fact, it must have been difficult for a member of a Masonic lodge to know whether he was merely taking part in an inspiring ceremony full of ancient symbols, or dealing with something that really was supposed to have supernatural power, or fomenting revolution—or what, if any, of this made him an "Illuminist." How much more difficult for anyone looking from the outside! Not only was there "fire in the minds of men"; there was also considerable confusion. The Illuminist slant, however, does seem to have been that the road to perfection and purification could and should be taken not only by individuals but by human societies. Had not Pythagoras engineered a marvelous reconstruction of society in Croton? However, Illuminists believed that this time, in the eighteenth century, the process was going to require enormous upheaval and the violent overthrow of existing authority.

As early as 1780, seven years before the French Revolution began, the attempt to legitimize revolutionary thinking by reference to ancient ideas had ceased to be something happening only in closed lodges and secret gatherings. Intellectual revolutionaries found it inspiring and reassuring to resurrect what they regarded as primal, natural truths that had been discovered in antiquity, and much that was attributable to, or at least attributed to, the Pythagoreans entered the symbolism of the incipient revolution itself. The rhetoric and the images that began to appear openly in the 1780s featured four "Pythagorean" geometric figures: the circle, the triangle, and their solid counterparts the sphere and the pyramid. These had also been symbols for God in medieval Christianity, but that use was militantly rejected.

Pythagoras and also Prometheus seemed ideal role models. Concepts associated with Pythagoras, correctly or incorrectly—prime numbers, geometric shapes, and the harmonic ratios of music—were "truth" that was more ancient and fundamental than the doctrines of Christianity that intellectual revolutionaries had discarded. Plato had spoken of "a gift of the gods to human beings, tossed down from the gods by some Prometheus together with the most brilliant fire," and Plato's ancient readers had assumed this "Prometheus" was Pythagoras. Prometheus, according to legend, had stolen that "most brilliant fire" from the gods, and fire had long been associated with Pythagoras, the Pythagorean "central fire." So Pythagoras seemed a splendidly appropriate symbol for the hope that darkness would vanish forever, a new day was dawning, and the sun would never set. The fact that he had left Samos to avoid a tyranny also qualified him as a model intellectual turned revolutionary. In pre-revolutionary Paris, Benjamin Franklin was dubbed "the Pythagoras of the New World," when he served as Venerable Master of the Masonic Lodge of the Nine Sisters (La Loge des Neufs Soeurs), whose membership also included such noteworthy revolutionary figures as Nicolas de Bonneville, "Anarcharsis" Cloots, Georges Danton, and Sylvain Maréchal.

The French Revolution began in 1787, and the storming of the Bastille in Paris took place July 14, 1789. The execution of the French royal family, members of the nobility, and clergy began in 1792, and the guillotine was busy for several years as those who had overthrown the monarchy turned on one another. It was a time of chaos, ferment, and confusion—and not only in politics. Conflicting reinterpretations of history, religion, and science vied with one another as factions right and left sought legitimacy, and those caught in the maelstrom clutched desperately not only for safety or victory but also for new self-images. Billington pointed out that it was not insignificant that many of the musicians in Strasbourg who first played the hymn of the French Revolution,

"La Marseillaise," in 1792, the year the royal family were executed, had also played in the orchestra when Mozart's *Magic Flute* was first introduced to French audiences there a few months earlier. Illuminism had reached Mozart in its Masonic guise, and *The Magic Flute* was chock full of Masonic, "Illuminist," and Pythagorean symbols.

The opera seems, to most twenty-first-century eyes and ears, a delightful fairy tale embellished with some archaic pseudo-religious ideas. However, in the 1790s, many would have seen it differently. It spoke symbolically and eloquently for an era when traditions and assumptions were being called into question or crumbling outright, when new discoveries of science and the ideas of the Enlightenment were continuing to undermine or transform older versions of Christian faith, and, when the over-ornate, elaborate, simpering, aristocratic artificiality of the Rococo had little to offer but denial of reality. In this milieu, Mozart, Masons, Illuminists, and revolutionaries were alike in preferring simple harmonies and forms in nature that could provide a securer philosophical foothold—a new, surer, more inspiring pathway to truth.

In about 1786, a young man who would later be dubbed the "first professional revolutionary," Filippo Michele Buonarroti, had encountered Illuminism in a "Scottish Rite" Masonic lodge in Florence. This lodge had become a forum where Illuminists held sway and discussed radically revolutionary ideas. So severely did the Florentine authorities frown on Buonarroti's involvement that although he was married to a noblewoman, held a doctorate of law, and was highly regarded for his literary talents, his library was raided and Masonic and anticlerical books confiscated. Shortly thereafter, an unrepentant Buonarroti found himself banished to Corsica.[10] In 1789—the year the Bastille fell—it looked for a short time as though he would join several young Italians who were starting up of a new journal in Innsbruck (for which city they used the code word "Samos"). These men had been influenced by Weishaupt's Illuminism while studying in Bavaria. However, events in France proved too enticing to Buonarroti, and instead of going to "Samos," he was soon deeply involved in revolutionary activities there.

Weishaupt, meanwhile, had been the first in many centuries to consider what he thought were Pythagorean principles as direct guidelines for public policy. In 1787, he had published his *Pythagoras*, laying out a design for the most politicized form of Illuminism and reiterating the idea that simple principles first taught in Croton were still a splendid guide for reforming and rebuilding society. He especially approved of ending ownership of private property. Following Weishaupt's lead, when Buonarroti drew up his own blueprint for revolution, he emphasized that same practice. Others joined the Pythagorean chorus: Nicolas de Bonneville composed poetry about "the numbers of Pythagoras" and insisted that Pythagoras "brought from the Orient his system of true Masonic instruction to Illuminate the Occident." The American Thomas Paine, the famous pamphleteer of the American Revolution and author of *Common Sense*, living a liberated life in a ménage à trois with Bonneville and Bonneville's wife, worked Pythagoras into his version of the history of the Masons, though he gave the Druids primary credit for providing Masonry with an ideology that Paine thought a finer alternative to Christianity. The sun worship of the Druids—paralleling the Pythagorean belief in the central fire—had passed into Masonry, Paine wrote, in *An Essay on the Origin of Free Masonry*.

In 1799, Sylvain Maréchal wrote a six-volume biography titled *Voyages of Pythagoras* that raised its protagonist above the level of an ideal for this one revolutionary period. Kepler had dubbed Pythagoras "the grandfather of all Copernicans," but the family became considerably larger when Maréchal insisted that all revolutionaries of all times were "heirs of Pythagoras." The Pythagoras of Maréchal's biography was a great geometer who was driven from the island of Samos by the tyrant Polykrates and fled to Croton, where he founded a philosophical-religious brotherhood with the goal of transforming society. The story went on, reimagined from the point of view of those who felt themselves part of a noble, centuries-old tradition devoted to that same goal: Neo-Pythagoreans who were radical intellectual reformers had flourished in Alexandria in the second century B.C. . . . the Pythagorean Apollonius of Tyana, the itinerant wonder-worker, was not a rather ridiculous cult figure but a legitimate and important rival to Christ, since discredited by Christian writers. . . . in the Middle Ages, those attracted to Pythagorean ideas recognized that Pythagoras was a secret Jewish link between Moses and Plato. . . . Pythagoreanism had never ceased to fascinate thinkers of the Renaissance and Enlightenment but had remained only an undercurrent until the time for its new awakening had come, in the revolution that would transform France and the rest of Europe. Maréchal wrote of "the equality of nature" and a Pythagorean "republic of equals," and echoed Weishaupt and Buonarroti in advising his readers to "own everything in common, nothing for yourself." Volume VI of the *Voyages* included no fewer than 3,506 supposed "Laws of Pythagoras."

Masons, Illuminists, and intellectual revolutionaries associated Pythagoras with prime numbers, though there had been no suggestion in antiquity of such a link. Great significance was attached to what were believed to have been the central prime numbers of Pythagorean mysticism: 1, 3, 5, and 7. The most extreme uses of "Pythagorean principles" were efforts to find paths, by means of mystical numbers and numerology, to the deep truths of nature, different from the use of numbers by early Pythagoreans, and even more different from their use by scientists to reach a mathematical understanding of nature and the cosmos. In a moment of leftist paranoia about a possible Jesuit plot for a secret takeover of Masonry, there was a suggestion that 17 was the number needed to understand the Jesuit plan. A rightist pamphleteer turned that idea around and proceeded, ingeniously, to show how all of revolutionary history derived from the number 17. Other opponents on the right picked up on this same type of pseudo-Pythagorean number mysticism and produced pamphlets suggesting that the prime numbers were a code for the organization of revolution.

The obsession with Pythagoras did have something to do with the way revolutionary activities were organized, though this involved triangles and circles rather than prime numbers. The link revolutionary intellectuals made between Pythagoreans and the circle and sphere was not far-fetched. The Pythagoreans (as reflected, for instance, in the fragments of Philolaus) had been among the earliest to think in terms of a system in which the Earth and the universe were both spherical. Furthermore, Newton's laws of gravity, which Newton himself had linked with Pythagoras, revealed a "circular harmony." Another Pythagorean doctrine, the transmigration of souls, also suggested a circular movement, forever returning to begin again. Illuminist "Pythagoreans" were fond of the idea that a purification process took place within the framework of this "circular" transmigration of souls, beginning with the lowest forms of life, spiraling upward through the level of humanity to the divine spheres of pure rationality. The "rules of geometry," as they called the laws behind such schemes, were appropriate for those who thought of themselves as the "mason-architects" of a new society. The architect Pierre Patte argued that there was a superior morality about circular shapes because they were essentially more egalitarian and communal.

Accordingly, one way of organizing Illuminist groups was in a hierarchy of concentric circles. A flame "at the center" represented the central fire around which Earth, Sun, and planets moved in the Pythagorean ten-body system. As one advanced in Illuminism, one progressed from the outer circles inward, freeing oneself from physical limitations to join, or rejoin, life in the inner circle or most heavenly sphere. The same symbolism applied to societies, connecting the circles to the idea of "revolution." Like individuals, societies could revolve inward through concentric circles, freeing themselves from the limitations of old traditions and beliefs to join the inner circle of freedom and rational simplicity. According to the report of a young collaborator of Buonarroti's, Gioacchino Prati, the first organization that Buonarroti instituted, in the 1790s, the Sublime Perfect Masters, was composed of concentric circles each of which had its own secret creed. The inner circle was absolutely egalitarian and so secret that the outer circles were unaware of its existence. If the writings of some critics of the Illuminists are to be trusted, Illuminist groups organized also in another way into "circles," a code name for nine-man cells of conspiracy. Weishaupt, the Bavarian Illuminist, was particularly fond of circles as symbols and considered it symbolic to speak of "circulating" his ideas by means of "circulars."

The triangle used in revolutionary symbolism was the equilateral triangle, the *tetractus*, which had also previously been an important symbol in Masonry. On seals, stamps, placards, and banners, Liberty, Equality, and Fraternity made up sides of a triangle, colored in red, white, and blue. Hats were tricornered. In 1798 Franz Xavier von Baader wrote a book called *On the Pythagorean Square in Nature*, a strange title for a book that celebrated triangles. Three elements—fire, water, and earth (air seemed not to interest von Baader)—were given life by an "all-animating principle," a "point of sunrise," represented by a dot in the center of an equilateral triangle. This image became hugely popular.

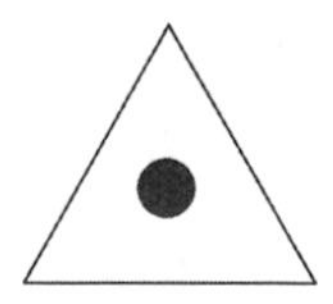

Maréchal saw triangular harmonies in the three roles of a man as father, son, and husband, three persons in one, replacing the Christian Trinity with a trinity centered in each individual.

The triangle showed up in a triangular organization of revolutionary groups. An individual from an inner group recruited two apprentices from an outer group, and eventually each of those recruited two more to form his own triangle. As Weishaupt described it,

> I have two directly under me into which I breathe my entire soul, and these two each have two others, and so forth. In this manner I am able, in the simplest way, to set thousands of people into movement and flames. In this manner the Order must be organized and operate politically.

This meant that each man knew the name and identity of only one from the inner group above him. It was a relatively secure form of organization, an interlocking system that was difficult to infiltrate effectively. The Spanish Triangle Conspiracy of 1816, a plot to kill King Ferdinand VII, was appropriately named.*

In a less potentially deadly usage, the mystic Louis-Claude de Saint-Martin wildly mixed images and cultures in his hope that Pythagorean forms and numbers could be employed to transform Paris into a new Jerusalem, with revolutionary democracy becoming a "deocracy." Others made related plans for an innovative Parisian architecture, based on the circle, triangle, pyramid, and sphere—an idea that was remarkably realized in the 1980s in I. M. Pei's controversial modern entrance to the Louvre, a glass pyramid.

The "Pythagoreans" who idealized their role model as an intellectual turned revolutionary also celebrated his association with music and were particularly fond of "songs without words." These seemed a link with the music of the spheres, expressing "the harmony of creation, or rather of the world as it should be." Antoine Fabre d'Olivet, who composed music for Napoleon's coronation, also set to music the *Golden Verses of Pythagoras*, the pseudo-Pythagorean work that had been

* The three-man cell was used again in Vietnamese communism, in Algeria in the 1950s, and in the USSR in the late 1960s.

popular in the Roman/Hellenistic era, and wrote that music was "the science of harmonic relationships of the universe."

In 1804, Napoleon, who five years earlier had installed a military dictatorship in France with himself as "First Consul"—the event usually identified as the end of the French Revolution—declared France a hereditary empire and crowned himself emperor. Thus, with the beginning of the new century, European revolutionary hopes waned seriously, but the iconic Pythagoras became important in a new way to those who opposed Napoleon. As France followed the Roman example and transformed herself from a republic to an empire, Pythagoras was viewed nostalgically as an ancient, nobler alternative to Napoleonic images of conquest, expansion, and domination. Both Paine and Maréchal envisioned themselves as still following in the footsteps of Pythagoras, as intellectuals temporarily unable to act effectively ("in exile") but devoted to constructing a brotherhood that would eventually free human society. In the words of Billington, two labels—Pythagoras and "Philadelphia" (signifying brotherly love)—"recur like leitmotifs amidst the cacophony of shifting ideals and groups during the recession of revolutionary hopes. . . . Pythagoras became a kind of patron saint for romantic revolutionaries," who more than ever were in need of "symbols of secular sanctity."

Pythagorean inspiration and iconography reached Russia the same year Napoleon became emperor in Paris, when Maréchal's biography of Pythagoras began to appear in official Russian government journals, a volume each year, and parts of it were excerpted in other Russian periodicals. A *kruzhkovshchina* (mania for circles) began in Russia and would last into the twentieth century. In 1818, in the western Ukraine, young men organized a "society of Pythagoras" with its own collection of "rules of the Pythagorean sect." A series of groups calling themselves free Pythagoreans were soon forming in other areas of the Russian empire. Groups of radicals frequently debated one another about rival sets of "laws of Pythagoras." Some preferred those that banned private ownership of property; others, those (whose Pythagorean origin was dubious) stressing that weapons and friendship could conquer all. Still others insisted that Pythagorean teachings regarding moral perfection had to be given priority over legal reform. Billington also tells of one student group in Vilnius that met at night in locations of great natural beauty to hear occult wisdom of an "arch-illuminated visitor" from an "inner circle."

A brief new tide of insurrections against monarchs in Europe that started with the Spanish Triangle Conspiracy of 1816 ebbed dramatically in 1823. The pope condemned Masonry, and several of the monarchies outlawed it. Throughout Europe, civil liberties were curtailed and organized discussions came under suspicion. Vestiges of republicanism, including Pythagorean symbols, fell out of official and public favor. The rector of the University of Kazan decreed that the Pythagorean theorem should not be taught.

The Russian revolution of December 1825 was a failed echo of the fervor that had inspired intellectuals in Europe for more than half a century. Young officers who had helped defeat Napoleon and marched into Paris in 1814 had experienced there a freer, more enlightened world. They, rather than the lower classes in Russia, had begun to organize with the hope of bringing reform to Russia—in the words of one of the Turgenev brothers, to resist being turned "back into gingerbread soldiers! And by whom? Political pygmies."

Among those whose thinking and work led up to that brief, doomed Revolution of 1825, Pythagoras was again an inspiration. F. N. Glinka, who founded a group called the Union of Salvation, one of many secret societies formed at this time, was strongly moved by a French work that he read in translation about "the institute of Pythagoras." A leading Russian periodical featured an article about the Sect of Pythagoreans that included a series of questions and answers like those favored by the *acusmatici* ("What is universal? Order. What is friendship? Equality") and a description: "not having any private property, not knowing false pride and vain praise, far from petty things that often divide, they competed with one another only in doing good . . . and learned to use things in common and forget about ownership." One of the leaders of a "circle" that helped foment the revolt, called the Green Lamp, wrote a piece that imagined St. Petersburg three hundred years in the future. In his vision, the tsar and all Orthodoxy would have given way to Pythagorean forms represented by a circular temple, music, and a phoenix with an olive branch.

When the 1825 revolution failed, five leaders were hanged and the others exiled to Siberia. Perhaps there was consolation in recalling that

Pythagoras, their iconic ancient model, had—at least in the mythology they thought they knew—been forced to flee in ignominy from a city he had tried to introduce to a better way of life.

BEYOND REVOLUTIONARY CIRCLES, other literature of the nineteenth century remembered Pythagoras. The poet Percy Bysshe Shelley wrote a piece praising the vegetarian "Pythagorean Diet," and Leo Tolstoy chose to follow it. Louisa May Alcott knew her readers would need no explanation when she wrote in *Jo's Boys* that "Grandpa March cultivated the little mind with the tender wisdom of a modern Pythagoras, not tasking it with long, hard lessons, parrot-learned, but helping it to unfold as naturally and beautifully as sun and dew help roses bloom." Honoré de Balzac attributed the saying "no man is known until he dies" to Pythagoras. Pythagoras was one of the ghosts present in Charles Dickens' *The Haunted House* and also made an appearance in *The Pickwick Papers*.

Also in the nineteenth century, the belief continued that the concept of the mathematical structure of the universe had originated with the Pythagoreans. The economist William Stanley Jevons wrote: "Not without reason did Pythagoras represent the world as ruled by number. Into almost all our acts of thought number enters, and in proportion as we can define numerically we enjoy exact and useful knowledge of the universe."[11]

One Pythagorean ideal began to come into its own in a way it could not have done earlier. The assumption that there was unity to the universe had already become one of the pillars on which science rested, but not until the nineteenth century did the knowledge and the instruments begin to be available that would allow scientists to explore the question whether this assumption was valid, or whether, like the music of the spheres, it was best relegated to the realm of poetic metaphor. The idea that there is unity to nature emerged strongly in the work of three men who had a particularly significant impact on the future of science.

When the Danish physicist and chemist Hans Christian Oersted wrote his doctoral thesis about a book by Immanuel Kant called *The Metaphysical Foundations of Knowledge*, he was already convinced that all experience could be accounted for by a correct understanding of the forces of nature, and that the forces of nature were actually not many forces but one. Kant had suggested there were two basic forces, but Oersted decided to push forward with the certainty that light, heat, chemical affinity, electricity, and magnetism were all different faces of "one primordial power." In 1820 he discovered electromagnetism, having "adhered to the opinion, that the magnetical effects are produced by the same powers as the electrical . . . not so much led to this by the reasons commonly alleged for this opinion, as by the philosophical principle, that all phenomena are produced by the same original power."

Michael Faraday was another early-nineteenth-century scientist who undertook a lifelong search for ways in which the forces of nature are unified. He began his professional life as a chemist and discovered several new organic compounds. As had been true of Linnaeus' numerous previously unknown species, those discoveries might have been taken to indicate a *lack* of unity, but instead they expanded awareness of what was out there to be unified. A tally of Faraday's most notable contributions included producing an electric current from a magnetic field, showing the relationship between chemical bonding and electricity, and discovering the effect of magnetism on light.

Michael Faraday

Faraday's work was the experimental foundation—and also a large part of the theoretical foundation—for the work of James Clerk Maxwell later in the century. Maxwell's electromagnetic field theory achieved the full unification of electricity and magnetism. The "electromagnetic force" would enter the twentieth century as one of four basic forces of nature. Maxwell's equations, based in turn on Faraday's study of electric and magnetic lines of force, would also be instrumental in setting a scientific trajectory toward the linking of mass and energy in Einstein's special theory of relativity. Science at the turn of the twentieth century was well on the way to finding the unity of nature that Pythagoreans had so fervently believed in. Paradoxically, Maxwell's work also provided a vision of reality with problems that would be resolved in the twentieth century by quantum theory. And quantum theory, in its turn, would cause a crisis of faith in the rationality of the universe, a crisis on a scale with that perhaps caused by the ancient Pythagorean discovery of incommensurability.

CHAPTER 18

Janus Face

Twentieth Century

IN THE TWENTIETH CENTURY, two major books appeared that highlighted humanity's debt to Pythagoras and the Pythagoreans. "Debt to Pythagoras" might seem to imply that there is something positive for which to thank Pythagoras and his followers, and one of the authors, Arthur Koestler, certainly believed there was. Bertrand Russell, on the other hand, insisted that most of Pythagoras' influence had been negative. Their two accounts constitute an excellent example of how taking off one pair of glasses and putting on another can change the view in astounding ways.[1]

Russell was born in 1872. In the years leading up to World War I, he tackled a question that would engage him for most of his life: whether mathematics can be, to a significant degree, reduced to logic, with one true statement implying the next. It is perhaps conventional wisdom that this is precisely the way mathematics works, but to assume so betrays a naive view. The issue is complex, and Russell knew it was. Though his place among academics was more as philosopher than mathematician, in *Principles of Mathematics* and a three-volume work that he coauthored with Alfred North Whitehead, *Principia Mathematica*, his goal was to re-found mathematics on logic alone.[2] There is nothing anti-Pythagorean about faith in mathematical logic. It was on other issues that Russell took on both Pythagoras and Plato.

Vehemently rejecting the idea that humans have any grounds for discussion of an ideal world beyond what can be extrapolated in a reasonable manner from what we experience with our five senses, Russell was convinced that "what appears as Platonism is, when analyzed, found to be in essence Pythagoreanism." It was from Pythagoras that Plato got the "Orphic elements" in his philosophy, "the religious trend, the belief in immortality, the other-worldliness, the priestly tone, all that is involved in the simile of the cave, his respect for mathematics, and his intimate intermingling of intellect and mysticism." Russell blamed Pythagoras for what he saw as Plato's view that the realm of mathematics was a realm that was an ideal, of which everyday, sense-based, empirical experience would always fall short.

Russell's chapter on Pythagoras was part of a hefty tome of nearly nine hundred pages, his 1945 *History of Western Philosophy*. He wrote it to appeal to a wide, nonacademic readership, but it was no innocent survey without an agenda. His fascination with language, with analyzing it down to its minimum requirements, transforming sentences into equations to wring from them the most trimmed-down, unmistakable message possible, had made him a master at the manipulation of language, and—it must be said—the manipulation of readers. Careless reader he sometimes was, and sometimes careless thinker, but hardly ever careless writer. His chapter about Pythagoras is peppered with tongue-in-cheek understatements, making it easy to miss the fact that he intended this clever, seductive, amusing prose to undermine not only some of the prized tenets of the mathematical sciences but also belief in God.

The book traced philosophy from Thales to himself, and Russell tried to show how this long history had culminated in, and finally found a corrective in, his own philosophy. In this context, he did not treat Pythagoras as just one more philosopher in the table of contents. The book's final paragraph, long past the chapter devoted entirely to Pythagoras, states: "I do not know of any other man who has been as influential as he was in the sphere of thought." The co-author of *Principia Mathematica*, Alfred North Whitehead, also believed Pythagoras' influence had been tremendous, the very bedrock of European philosophy and mathematics.

Russell agreed with those who thought that Pythagoras was the first to use mathematics as "demonstrative deductive argument," rather than merely a practical tool of commerce and measurement. This, he thought, made Pythagoras a founding father of the line of mathematical thinking that would lead to all of modern mathematics including his own. "Pythagoras was intellectually one of the most important men that ever lived, both when he was wise and when he was unwise," Russell wrote. "Unwise" referred to the fact that Pythagoras and Pythagoreanism seemed to Russell also to have had a mystical side, and when that encouraged Plato to introduce the Forms, the inheritance went sour.

Just as other sciences had their roots in false beliefs—astronomy in astrology; chemistry in alchemy—mathematics, wrote Russell, had begun with "a more refined type of error," the belief that although mathematics is certain, exact, and applicable to the real world, it nevertheless can be done by thought alone with no need to observe the real world. He had a point. Think of the ten-body cosmos. Even though the Pythagoreans discovered the ratios of musical harmony by listening (one of the senses) and observing where they were putting their fingers on the strings of the lyre (involving both sight and touch), they proceeded in an unfortunate way that involved trusting thought, not checked by observation. What Russell insisted had emerged as a result was a view of the realm of mathematics as an ideal from which sense-based, empirical knowledge would always fall short. Once that was in the air, lamented Russell, goodbye to the idea that observation of the real world was a useful guide to truth.

Plato, as interpreted by Russell, had believed that anyone on a quest for truth had to reject all empirical knowledge and regard the five senses as untrustworthy, even false witnesses. Absolute justice, absolute beauty, absolute good, absolute greatness, absolute health, "the essence and true nature of everything"—the only way to reach that level of knowledge was, Plato had Socrates say, by means of "the mind gathered into itself."[3] Actually, there is no record of Pythagoras, or pre-Platonic Pythagoreans, insisting that truth about the universe must be discovered by thought alone, but, to Russell's mind—although it was Plato who articulated the idea—its source was the Pythagoreans; it was implicit in the way they thought and the conclusions they reached. Russell was convinced that the idea of the superiority of thought and intellect over

Bertrand Russell

direct sense observation of the world would not have emerged at all had it not been for the combination of the Pythagorean view of numbers and Plato's idea of Forms, which together created an unfortunate legacy that endures to the present and that has motivated people to look for ways of coming closer to what they saw as the mathematician's ideal. "The resulting suggestions were the source of much that was mistaken in metaphysics and theory of knowledge. This form of philosophy begins with Pythagoras."

Having read Plato, one must take issue. He did not think of numbers and mathematics as Forms or "ideals" at all—not even as a sure path to discovering them. In his creation of the world-soul in his *Timaeus*, for example, and when Socrates taught about "recollection" in the *Meno* by drawing the square and the isosceles triangle for the untutored slave boy, mathematics for Plato was a way of reaching out toward the ultimate level of knowledge, toward the Forms, of trying to get there. It does not appear, in these passages, that Plato thought he *was* there or that numbers and mathematics were going to get him there. His pupils later thought of numbers as on the level of Forms, but even they did not necessarily believe human thinkers could reach that level of mathematics.

Russell had another objection to Pythagoras. The Pythagorean insight that numbers and number relationships underlie all of nature—not created or invented by humans but discovered by them—was, he believed, a false vision and an enormous and tragic misstep in the history of human thought. Following that Pythagorean fantasy, mathematics was doomed always to have in it "an element of ecstatic revelation." "Revelation" was, for Russell, an impossible concept. He wrote that those mathematicians who have "experienced the intoxicating delight of sudden understanding that mathematics gives, from time to time," find the Pythagorean view "completely natural even if untrue." In this he was ignoring the fact that neither the Pythagoreans nor any major mathematician from the late sixteenth century on, not even the ecstatically religious Kepler, ever claimed to have received a mathematical "revelation." But Russell equated "discovery" of truth with "revelation," and "revelation" with "illusion." With that equation in mind, what seemed to be the discovery of the underlying level of mathematical reality equaled a leap of faith to a false "ideal world." And, according to Russell, that idea had been foisted off on a gullible future.

Russell nailed all this down by attributing to the "delighted mathematicians" a different idea (though many mathematicians would disagree with it): that mathematics is something created by mathematicians in the same way that music is something created by composers. This could have been an insightful parallel, had Russell followed up on it: From a background having to do with which tones and meters are possible, which sounds are pleasant and which not—and much else that one might *discover* about hearing, sounds, and their effect on human emotions—a composer is still left with a vast number of choices. The result depends on the composer's creativity and inventiveness in using basic, unchangeable material. Perhaps from a background of true mathematical possibilities, a mathematician likewise has a vast number of choices. Even if the uncharted territory one is exploring is not subject to choice or invention, the trails leading into it and across it are a matter of choice and creativity.

Russell had something else in mind. He was opting for a different philosophy of mathematics, that mathematics is a human construction to impose logical order on the universe or draw a map through territory that is not inherently mathematical at all. He laid twofold blame on Pythagoras: first, for the Platonic idea that there is a realm not perceptible to human senses but perhaps to human intelligence, and, second, for the belief that mathematicians were discovering mathematical truth, not inventing it. Because numbers are eternal, not existing in time, it was possible to conceive of numbers and mathematics as "God's thoughts," and just there, said Russell, rooted in Pythagoreanism, was Plato's idea that God is "a geometer." A sort of "rational" religion had come to dominate mathematics and mathematical method.

Russell was willing to concede one positive outcome from the Pythagorean doctrine of a universe undergirded with rationality and mathematical order: It had led people to be dissatisfied with movements in the heavens that were irregular and complicated, as they appear to a naive observer. Such a messy situation was not "what a Pythagorean creator would have chosen," and that puzzle had led astronomers like Ptolemy, and later Copernicus and Kepler, to propose systems that an orderly designer would have preferred.

Russell wrote *The History of Western Philosophy* before the discovery of the scribal tablets that showed that the "Pythagorean" theorem was known long before Pythagoras. Justifiably, he was confident in calling the Pythagorean theorem the "greatest discovery of Pythagoras." He sympathized with the misfortune of the Pythagoreans, the discovery of incommensurability. He had reason to be sympathetic, for during his lifetime several discoveries occurred that seemed to undermine his own efforts, in the same way that the discovery of incommensurability had traditionally undermined Pythagorean faith that the world was based on rational numerical relationships. One of the discoveries was "Russell's paradox." He was trying to set mathematics on a better track by seeking to found it on logic, with one true mathematical statement implying the next. However, a true statement sometimes implies more than one next statement. Sometimes it implies two statements that contradict one another.* That paradox was no trivial snag. Russell wrote a letter about it to the German mathematician and logician Gottlob Frege, who received it as he was completing the second volume of a treatise on the logical

* Picture a collection of coins. Call it Set A. A collection of coins is an example of a "set" that cannot be a member of its own set. In other words, a collection of coins is not a coin. Picture, then, another set (call it Set B) that contains all things that are not coins. This Set B itself is not a coin, so it must be a member of itself. In other words, Set B is a member of Set B. Now picture a third set—Set C. This one contains all the sets that are *not* members of themselves. Is Set C a member of itself or not? You will find that it *is* if, and only if, it *is not*.

foundations of arithmetic that had taken twelve years of painstaking work. Frege responded by adding the following sad words to his book:

> A scientist can hardly meet with anything more undesirable than to have the foundation give way just as the work is finished. In this position I was put by a letter from Mr. Bertrand Russell as the work was nearly through the press.[4]

Russell spent some time in his chapter on Pythagoras considering the problem of incommensurability. He thought that the square root of 2, being the simplest form of the problem, was the "first irrational number to be discovered" and that it was known to early Pythagoreans who had found the following ingenious method for approximating its value.* Suppose you have drawn an isosceles triangle, the one Plato used in his *Meno*, which contains the problem of incommensurability. Russell thought it was while studying this triangle that the Pythagoreans came upon the problem, so let us follow his thinking.

First, review the problem. The Pythagorean theorem says that the square of Side A plus the square of Side B will equal the square of Side C. Say that Side A measures 1 inch. Side B also measures 1 inch. The square of 1 is 1. So the square of Side A plus the square of Side B (1+1) equals 2. If the Pythagorean theorem is correct, the square of Side C

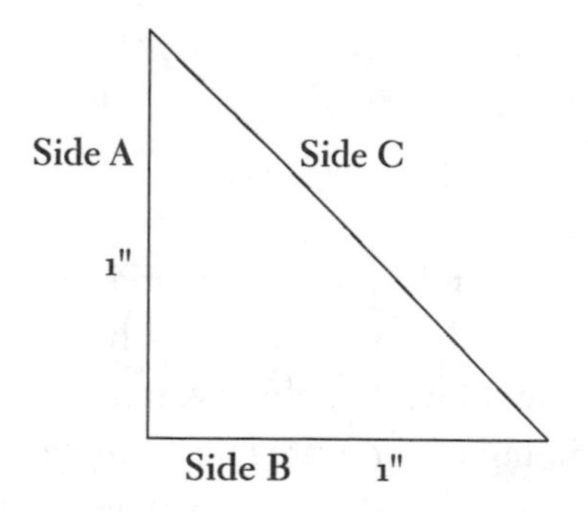

must likewise be 2, but what is the *length* of Side C? You cannot find out if you cannot calculate the square root of 2. Here is how Russell suggested the Pythagoreans might have approximated it:

* However, the simplest form of a problem is not always the form in which it is first encountered. If it were, the history of mathematical and scientific discovery would have gone much more smoothly!

Make two columns: Column A and B, and let each begin with the number 1.

A	B
1	1

To get the next pair of numbers:

> For Column A, add the first A and B (1+1).
> For Column B, double the first A and add the first B (2+1)

A	B
1	1
2	3

Continue using the same method of getting the next pair of numbers, always using the two previous numbers as your "former A and B," and you soon have:

A	B
1	1
2	3
5	7
12	17
29	41
70	99

For each pair the following is true: 2A squared minus B squared equals either 1 or minus 1. In each case, B divided by A is close to the square root of 2, and the farther down the chart you move, the closer it is to the square root of 2, though it never quite gets there because the square root of 2 is not a rational number. Would this have satisfied the Pythagoreans? One cannot help thinking that for people who believed they had found complete rationality and simplicity in the universe, it would have been poor consolation.

Russell in great part credited Pythagoras with linking philosophy with geometry and mathematics, with the result that geometry and mathematics had been an influence on philosophy and theology ever since—an influence Russell regarded as "both profound and unfortunate." In geometry, as Euclid and other Greeks established it, and as it is still taught today, one does not begin in a void, thinking nothing true unless proved. There are statements that are not proved but are "self-evident" (or at least seem to be), called axioms. Some bit of self-evident truth must be there as the starting place. That may seem a shaky foundation to build on, but many generations have managed to accept it and proceed. Beginning with the axioms, the next step is to use deductive reasoning to arrive at things that may not be at all self-evident, called theorems. Axioms and theorems are supposed to be true about actual space; they are something that could be experienced. In other words, by taking something self-evident and using deductive thinking it is possible to discover things that are true of the actual world.

Russell had no argument with this line of thinking in geometry. His regret was that it been applied to other areas. The American Declaration of Independence, for example, declared, "We hold these truths to be self-evident," on the assumption that there are, indeed, things having nothing to do with geometry or mathematics that are so clearly true that no sane person would question them. The words "self-evident" were one of Benjamin Franklin's changes in the draft of the Declaration. Thomas Jefferson had written, "We hold these truths to be sacred and undeniable," a less down-to-earth version of the same idea. The point was that everyone could proceed from there without looking back. But could they?

Russell was not really trying to undermine Franklin, but he was disgruntled that the process by which geometry is done had been co-opted not only by brilliant rebels but by theologians. Thomas Aquinas had used it in arguments for the existence of God. His arguments did not start from nothing, but rather from "first principles." In fact, what Aquinas meant by "science" was a body of knowledge that has "first principles" or "givens." Again, Russell blamed the Pythagoreans: "Personal religion is derived from ecstasy, theology from mathematics; and both are to be found in Pythagoras." The Pythagorean marriage of mathematics and theology had polluted the religious philosophy of Greece, then the Middle Ages, and so on through Immanuel Kant and beyond. In his essay "How to Read and Understand History," Russell lamented,

> There was a serpent in the philosophic paradise, and his name was Pythagoras. From Pythagoras this outlook descended to

> Plato, from Plato to Christian theologians, from them, in a new form, to Rousseau and the romantics and the myriad purveyors of nonsense who flourish wherever men and women are tired of the truth.[5]

Russell identified some characteristics of what he saw as a blending of religion and reasoning, of "moral aspiration with logical admiration of what is timeless," in Plato, Augustine, Thomas Aquinas, Descartes, Spinoza, and Kant. Their offenses were belief in insight or intuition as a valid route to knowledge, a route distinct from analytic intellectual processes; denial of the reality of time and the passage of time in the ultimate scheme of things; belief in a unity of all things and a resistance to any fragmentation of our knowledge of the world. This "philosophical mysticism"—a term used not by Russell but coined by the physicist John Barrow—according to Russell "distinguished the intellectualized theology of Europe from the more straightforward mysticism of Asia."[6] However, he believed it was a much earlier form of Eastern mysticism that had entered, through Orphism, into Pythagoreanism, in which fertile ground it had taken root to develop into the intellectualized but still partly mystical theology of Europe.

Russell was not a lone voice. He was one of the founders of a school of thought called logical analysis, an effort "to eliminate Pythagoreanism from the principles of mathematics," ridding it of "mysticism" and "metaphysical muddles." He and those who joined him in this movement refused to indulge in what they saw as "falsification of logic to make mathematics appear mystical, and the practice of passing off, as authentic intuitions of reality, prejudices about what is real." Russell also tried to put logic to work in an attempt to clarify issues in philosophy, making "logical analysis the main business of philosophy," rejecting any notion that moral considerations have a place in philosophy or that philosophy might either prove or disprove the truth of religious doctrine. Philosophy, stripped of its "dogmatic pretensions," would nevertheless "not cease to suggest and inspire a way of life."

While Russell and his colleagues recognized there were questions they could not answer, they preferred to leave them unanswered rather than cling to what they felt were foolish and misleading "answers," or believe there are "higher" sources of answers:

> The pursuit of truth, when it is profound and genuine, requires also a kind of humility which has some affinity to submission to the will of God. The universe is what it is, not what I choose that it should be. Towards facts, submission is the only rational attitude, but in the realm of ideals there is nothing to which to submit.[7]

Reading that, one cannot avoid the conclusion that Russell was far more ambivalent about the issue of "discovery" versus "invention" than he was willing to admit.

Though he deplored the way mathematics had been "misused" in other areas, Russell believed that what he was insisting philosophy do—utilize logical analysis, adopt methods of science, and try to base its conclusions on impersonal, disinterested observations and inferences—should be applied in all spheres of human activity. This would bring about a decrease of fanaticism and an increase in sympathy and mutual understanding. He attempted, with scant success, to apply logical analysis to fields such as metaphysics, epistemology, ethics, and political theory, making (ironically) what was arguably a "Pythagorean" leap of faith that what seemed to be a good idea in one area of experience would be a good idea in all.

Russell decried yet another aspect of the Pythagorean legacy: The Pythagoreans lived by an ethic that held the contemplative life in high esteem and had bequeathed to the future something he called "the contemplative ideal." In the fable about the people at the Olympic Games, Pythagoras and his followers were in the third group, those who had come to watch. These "onlookers" celebrated not practical but "disinterested" science—in other words, they were disengaged from the world of buying, selling, and competing, able to view the whole scene with greater objectivity—thinking that their roles as independent observers placed them in a better position on the path of escape from the eternal circle of the transmigration of souls. Russell contrasted this view with a modern set of values that sees the players on the field as superior to mere spectators, and that admires politicians, financiers, and those who govern the state, the "competitors in the game," above those who keep to the sidelines and watch and make wise observations.

Nevertheless, said Russell, the elevated status of the "gentlemanly on-looker" who does not dirty his hands has endured, and this began in ancient Croton, was carried forward with the Greek idea of genius, then with the monks and scholars of the church, and later with the academic university life. He criticized all these, including "saints and sages," who, except for a few activists, had lived on "slave labor," "or at any rate upon the labor of men whose inferiority is unquestioned." It is these "gentlemen," these "spectators at the Games," he lamented, who have given us pure mathematics, and that contribution has meant, for them, prestige and success in theology, ethics, and philosophy, because pure mathematics is generally regarded as a "useful activity." Russell did not mention that he himself was one of these gentlemen he was criticizing—literally so, for he was born into the British nobility, studied at the University of Cambridge and became a fellow of Trinity College there, and spent most of his life as an academic and writer. However, he did, certainly, become one of the activists as well.

Parodoxically, Russell believed passionately in some ideals that he could not have arrived at by confining himself to strict empiricism, deductive thinking, and the scientific method. Perhaps his intuitions about what is right and what is wrong were, indeed, self-evident. To judge from his writings, these ideals became for him a higher priority than his logical analysis. He was a pacifist during World War I, and this unpopular stand cost him his Cambridge fellowship and landed him in prison for a while, but in 1939, in the face of the Nazi threat, he renounced pacifism. He was a lifelong, outspoken opponent of Nazism, Soviet communism, and belief in God. He campaigned vigorously for nuclear disarmament and against the Vietnam War. In these causes he was superbly able to write essays for popular readers that often seemed to begin as polemics but ended with reasoned arguments.

Russell was an impassioned and influential man, who recognized that there was a directionality built into human beings that makes us at least seem to be existing somewhere on a continuum from evil to good, ugliness to beauty, unfairness to justice, mediocrity to greatness, weakness to strength, with the ultimate in every case being off the scale, over the horizon of human comprehension or imagination. In his espousal of a mathematical philosophy that would soon be outdated, but still more in the positions he took against nuclear weapons, war, and what he saw as cruel dogmatisms, Russell, paradoxically, lived by this Platonic, perhaps Pythagorean-based view of the world. Ironically, it was the decisions he made on that foundation that ultimately made him memorable.

ARTHUR KOESTLER'S PICTURE of Pythagoras was far more positive. Born in Budapest in 1905, Koestler was an author and journalist and probably the most widely read political novelist in the world in the 1940s and early 1950s. His writing highlighted the moral dilemmas caused by the rise of communism and the two world wars. Koestler, like Russell, spent time in prison. While serving as a foreign correspondent in Spain, he was captured by Franco's troops and sentenced to death. The British government intervened and Koestler was able to return to London. As he aged, he took an increasing interest in science and the history of ideas and knowledge. His 1959 *The Sleepwalkers* was a masterpiece when it came to splendid writing and an ability to convey Koestler's passion for science, scientists, and scientific ideas. It was the first in a trilogy that continued in 1964 with *The Act of Creation* and in 1967 with *The Ghost in the Machine*. Koestler died in 1983. He was suffering from leukemia and Parkinson's disease, and he and his wife, both advocates of voluntary euthanasia, together took their own lives.

In *The Sleepwalkers*, Koestler wrote of Pythagoras: "His influence on the ideas, and thereby on the destiny, of the human race was probably greater than that of any single man before or after him." Koestler called the sixth century B.C. a "turning point for the human species," a "miraculous century." It was also the century of Buddha, Confucius, and Lao-tzu, and in the Greek world, Thales and Anaximander. Still, it was, in a sense, like an orchestra tuning up,

> each player absorbed in his own instrument only, deaf to the caterwaulings of the others. Then there is a dramatic silence, the conductor enters the stage, raps three times with his baton, and harmony emerges from the chaos. The maestro is Pythagoras of Samos.

For Koestler, the power of the Pythagorean vision came from its "all-embracing, unifying character; it unites religion and science, mathematics and music, medicine and cosmology, body, mind, and spirit in an inspired and luminous synthesis." "Cosmic wonder and aesthetic delight no longer live apart from the exercise of reason," and the intuitions of religion had also been joined to the whole in the concept of a scientific/philosophical search for God. Religious fervor had been channeled into intellectual fervor, "religious ecstasy into the ecstasy of discovery." Koestler concluded that although one cannot know which specific discoveries to attribute to what person or to what date, it is clear that the "basic features were conceived by a single mind," making Pythagoras the founder of "a new religious philosophy and of science as the word is understood today." In fact, the transmigration of souls itself was not a new religious philosophy, and Koestler gave a long description of Orphic religion. As for founding science, the discovery of the ratios of musical harmony was, Koestler said, the "first successful reduction of quality to quantity, the first step towards the mathematization of human experience."

According to Koestler, the reduction of experience to a straitjacket of numbers rightly arouses misgivings in the modern world, but for the Pythagoreans it did not diminish or impoverish anything. It enriched them. Because numbers were sacred to the Pythagoreans, reduction to numbers did not mean a loss of "color, warmth, meaning, and value." Instead, marrying music to numbers ennobled music. Koestler may be correct, but one could also reasonably believe that the Pythagoreans did not think numbers were sacred until they had made the discovery of their connection with music. Possibly only after that did numbers seem to them to have the marvelous, immortal qualities that Koestler ecstatically described, and come to be regarded as a link between humans and the divine mind. Koestler probably would have liked either interpretation equally well.

Koestler also singled out the idea that "disinterested science leads to purification of the soul and its ultimate liberation" as a major contribution of the Pythagoreans, and wrote about the enormous historical importance of this idea. "Harnessing science to the contemplation of the eternal, entered, via Plato and Aristotle, into the spirit of Christianity and became a decisive factor in the making of the Western world." Indeed it did keep the feeble flame of something resembling science alive during the Middle Ages and caused scholar-clerics to welcome with immense thirst and enthusiasm the rediscovery of ancient knowledge. Through the time of Kepler and Galileo, the scientific quest and the quest for the knowledge of God were considered to be the same quest.

As for Pythagorean secrecy, Koestler wrote that "even a lesser genius

Arthur Koestler

than Pythagoras might have realized that Science may become a hymn to the creator or a Pandora's box, and that it should be trusted only to saints."

Koestler seductively clothed the bare skeletal outline of Pythagoras with the garments of creative hindsight and beautiful prose, and fashioned a legend for the twentieth century. But, remembering the little ancient community, trapped in many ways in the thinking of that time, able—except for the great discovery of rationality in the ratios of musical harmony—to make only feeble attempts to link numbers with nature and the cosmos and creation, believing in a unity of all being that there was no way to demonstrate, one is forced to conclude that he was looking through the glasses of his own ideals. Nevertheless, his interpretation makes wonderful reading. He was truly the master of the magnificent overstatement that sounds so beautiful and convincing that we long for it to be correct. His is an "ode to Pythagoras," or an orchestral variation on a brief, sketchy "theme of Pythagoras," but it resonates better than any other existing account with the awe with which the modern world—hardly knowing Pythagoras at all—nevertheless regards his name. Koestler's retelling is not quite the truth about Pythagoras, and it is also more than the truth. In any case, it is Koestler's truth.

At the end of Koestler's chapter about Pythagoras, there are two statements with which not even the most skeptical scholars would disagree. The first is that Pythagoreanism had the "elastic" quality of all truly great systems of ideas, the "self-regenerating power of a growing crystal or a living organism." The second is that the Pythagoreans were probably the first to believe that mathematical relations hold the secrets of the universe. The world, concluded Koestler, "is still blessed and cursed with this heritage." By "cursed" he meant that the modern age should rightly have misgivings about the reduction of experience to a straitjacket of numbers. The Pythagorean conviction that numbers hold the secrets of the universe had carried us magnificently to the edges of time and space, but "our hypnotic enslavement to the numerical aspects of reality has dulled our perception of non-quantitative moral values; the resultant end-justifies-the-means ethics may be a major factor in our undoing."

CHAPTER 19

The Labyrinths of Simplicity

Twentieth and Twenty-first Centuries

THE "SCIENTIFIC METHOD" as it is taught in science classrooms and practiced by scientists all over the world is not very old when compared with the spans of time covered in this book. It emerged in the seventeenth century. No committee put it together, not even one so august as the Royal Society of London for Improving Natural Knowledge or the French Academy. "Emerged" is the correct word. In their day-to-day labors, Tycho Brahe, Galileo, and Johannes Kepler knew no "scientific method." They were working out, by trial and error, employing common sense and genius, how science from their time forward was going to operate. But their procedure for systematically separating what is true from what is not had not yet been assigned a name or analyzed precisely. Little if any consideration was given to the fact that it incorporated and rested on unproved articles of faith that are not even self-evident—principles that were much older and already so embedded in the European worldview that no one thought to debate whether they were valid. Bertrand Russell might lament that the practice of building on some truths without questioning them was being employed in other areas besides geometry, but with regard to science, G. K. Chesterton was on target when he wrote, "You can only find truth with logic if you have already found truth without it."

From a twentieth- or twenty-first-century vantage point—with hindsight and knowledge of what has happened since the seventeenth century—it is easier to recognize what an essential role the Pythagorean legacy played in providing this basic foothold for the scientific method and how much it came into its own in that method. The conviction that the universe is rational, the belief in underlying order and harmony, the confidence that truth is accessible by way of numbers, and the assumption that there is unity to the universe have become the pillars undergirding science. In the twentieth century, challenge after challenge was hurled at this list, by investigators and by nature itself, but the scientist who gets up and goes to work in the morning does so largely assuming that these articles of faith do hold true. An essentially Pythagorean faith remains as instrumental in driving science as the Aristotelian insistence on observation and experiment. Indeed, if the universe is not rational and ordered, if numbers are not a reliable guide, if there is no unity to the universe, observation and experiment are shortsighted and futile and there is little possibility of doing science at all. The conclusion is inevitable: Either the Pythagoreans in the sixth century B.C. brilliantly and prophetically uncovered truths that have not failed to hold in two thousand, five hundred years . . . or their persuasive philosophy has for all these centuries pulled the wool over our eyes so effectively that we are incapable of recognizing and following up on evidence that would expose their worldview as a mirage . . . or (a third possibility) when Arthur Koestler wrote of a truly great system of ideas, with the "self-regenerating power of a growing crystal or a living organism," he was only clothing a group of self-evident ideas, erroneously traced to an ancient cult, in beautiful language.

It was not only the ancient assumptions underlying modern cutting-edge science that made the twentieth century a Pythagorean century. There were also discoveries that caused crises of faith in the power of numbers and the rationality of the universe.

One of the most dramatic, successful stories of trusting numbers and mathematics as guides into the unknown in a scientific search was the discovery of black holes. The physicist Stephen Hawking commented, "I do not know any other example in science where such a great extrapolation was successfully made solely on the basis of thought."[1] The "thought" was mathematical thought. By the mid-1960s physicists had discovered solutions to Albert Einstein's equations that made it difficult not to conclude that there must be black holes in the universe, even though there was no observational evidence for them. By the mid-1980s, confidence ran high that black holes did indeed exist, and there were several "candidates," but still no unequivocal evidence. It was not until the 1990s that there was convincing observational evidence of the presence of several black holes and reason to conclude that there are many, many more. Still, the evidence was indirect, circumstantial. The discovery of a black hole was an ingenious collaboration of theory, mathematics, and observational astronomy. But there is now little question that black holes do exist, and old candidates and new ones are not difficult to evaluate.

The nonexpert public, though intrigued by such discoveries as black holes and eager to read about Stephen Hawking, has not been so entirely convinced by the power of mathematical thinking as the scientists, nor by the travelogues into the wilds of physics theory that these experts have provided for those who cannot follow the equations. In 1988, Hawking's first wife, Jane Hawking, told an interviewer, "There's one aspect of his thought that I find increasingly upsetting and difficult to live with. It's the feeling that, because everything is reduced to a rational, mathematical formula, that must be the truth."[2] One could well imagine the wife of Pythagoras saying something like that. Jane Hawking was not the only one who had trouble sharing the faith in mathematics that leads the thinking of theoretical physicists. Arthur Koestler deplored "our hypnotic enslavement to the numerical aspects of reality."

Writers like myself who explain science for nonexpert readers are often approached by intelligent people who have read of such things as the extra dimensions of physics theory—sometimes more dimensions, sometimes fewer, but hardly ever just the three of space and one of time that humans experience—and who say, "I can picture it easily enough the way you describe it, the dimensions rolled up into little hose-like tubes, but how does it actually link with reality? Is it only a mathematical reality?" That "only" betrays a suspicion that mathematicians and physicists immersed in their own Pythagorean universe are at a loss to explain away. In the sixth century B.C., no one could see ten bodies in the heavens. In the twenty-first century, not only can no one see the extra dimensions, no one can even *imagine* them. Hawking has admitted that anyone who thinks he or she can imagine what the extra dimensions would be like has either made a large evolutionary leap in mental capacity or is mistaken. But that has not kept theoretical physicists from following eagerly the paths of the equations in which such things do make sense.

Scientists are not the only ones who adopt a Pythagorean view of numbers as the strongest vehicles on the avenue to truth and progress. Pythagorean faith in mathematics shows up at nearly all school curriculum meetings. Though no one proposes resurrecting the quadrivium, educators seem to have decided that a child who can talk and read and calculate holds the essential keys to all knowledge, and many would argue that the third—"calculate"—is potentially the most powerful by far. Music has, however, tended to fall by the wayside.

When Hawking wrote in the late twentieth century about his high hopes that he and others would find the Theory of Everything that would unify all of physics, and when he brought that quest into the public mind in his *Brief History of Time*—even for those who only read *about* that book—he was expressing another Pythagorean theme. Many physicists were hoping, indeed expecting, complete knowledge of the universe to turn out, ultimately, to be unified, harmonious, and simple. This hope was not based only on wishful thinking. Listen, for example, to the way the physicist Richard Feynman traced its history.

There was a time, wrote Feynman, when we had something we called motion and something else called heat and something else again called sound,

> but it was soon discovered, after Sir Isaac Newton explained the laws of motion, that some of these apparently different things were aspects of the same thing. For example, the phenomena of sound could be completely understood as the motion of atoms in the air. So sound was no longer considered something in addition to motion. It was also discovered that heat phenomena are easily understandable from the laws of motion. In this way, great globs of physics theory were synthesized into a simplified theory.[3]

In the early twentieth century, physics seemed to be coming together in a thoroughly Pythagorean unity. Einstein unified space and time and explained gravity in a way that the physicist John Archibald Wheeler could encapsulate in one short sentence: "Spacetime grips mass, telling it how to move; mass grips spacetime, telling it how to curve."[4] Einstein's theory of special relativity could be summarized in an equation on a T-shirt: $E = mc^2$. The Russian mathematician Alexander Friedmann predicted that anywhere we might stand in the universe we would see the other galaxies receding from us, just as we do from Earth, and better understanding of the expansion of the universe has shown he was undoubtedly right, although no one has been able to try it yet. Just as Nicholas of Cusa thought in the fifteenth century, the universe is homogenous.

Two of four forces of nature known to underlie everything that happens in the universe—the electromagnetic force (already a unification) and the weak nuclear force—were combined by the "electroweak theory" in the early 1980s. There was also work going on that promised to show that, if we could observe the extremely early universe, it would be obvious that all four forces were originally united and that nature was composed of symmetries well concealed in our own era of the universe's history. James Watson and Francis Crick and their colleagues discovered the simple pattern of the structure of DNA, the double helix. Those who were insisting that Darwin's nineteenth-century theory of evolution was no threat to religious faith were pointing out that it was difficult to imagine anything that could more eloquently support the conviction that there was a brilliant and unified (and some would add, pitiless) rationality behind the universe. John Archibald Wheeler wrote his essentially Pythagorean poem:

Behind it all
is surely an idea so simple,
so beautiful,
so compelling that when—
in a decade, a century,
or a millennium—
we grasp it,
we will all say to each other,
how could it have been otherwise?
How could we have been so stupid
for so long?

All was not, however, a story of undiluted success for the Pythagorean vision of a "unity of all being." Einstein, a firm believer in the unity of nature, spent thirty years trying to construct a theory that would explain electromagnetism in terms of space-time, as he had explained gravity. He never succeeded, and many physicists would blame his failure in part on the fact that he so stubbornly refused to admit quantum mechanics into the picture. But a new theory, called string theory, that saw the elementary particles as tiny strings or loops of string and that certainly had no qualms about accepting quantum mechanics, was gaining supporters in the 1980s. It offered hope of doing what Einstein had failed to do: gathering into the fold the most rebellious of the four forces (when it came to unification)—gravity. As the first decade of the twenty-first century progressed, however, physicists were becoming impatient with string theory. It had been able to come up with no prediction that could be tested in a way that would show whether the theory was correct. Aristotle would have been happier with this development than Pythagoras or Plato, not because Aristotle wanted to tear down theories, but because twenty-first-century mathematical physicists were clearly not out of touch with the need for truth to be linked with the perceptible world. However, even with string theory looking less promising than it had, no one really questioned the essential unity of the universe.

Such faith is hard to lose, especially when no evidence definitively shows that it is wrong. However, some serious mathematical and scientific blows to Pythagorean convictions have occurred during the past one hundred years. Humans seem fated to discover again and again that the universe is not so rational after all—at least, not by the best current human standards of rationality. Such discoveries have challenged and stretched scientists to dig deeper in search of a level of reality where the Pythagorean principles still hold. One of the greatest manifestations of symmetry, harmony, unity and rationality in the universe is the fact that, although drastic changes do occur over time and from situation to situation, and although things can look dramatically different in different parts of the universe—and act in what even seem contradictory ways—the underlying laws that govern how change occurs apparently do not change. Maybe this is convincing evidence that our Pythagorean assumption of unity is correct, or it might be that our assumption is leading us to a false impression. We can only answer by pointing to past experience.

The search for a more fundamental law often begins with the discovery that something that has seemed fundamental and unchanging fails to hold under some circumstances. When that happens, the Pythagorean assumption of unity and symmetry kicks in and compels everyone to conclude that whatever it is they have been regarding as bedrock is not that at all. It is merely an approximation. Researchers put their noses back to the grindstone and explore for a deeper underlying law that does not change.

There have been many examples of this process of discovery. Newton's laws of gravity hold true except when movement approaches the speed of light or when gravity becomes enormously strong, as it does near a black hole. Einstein's newer, more fundamental description in terms of space-time does not break down, as Newton's laws do, in these extreme circumstances. But Einstein's description also presents problems that challenge the assumption of unity and harmony. They predict that there will be singularities—points of infinite density—at the origin of the universe and at the center of black holes. At a singularity, all the laws of physics break down. And so the search must go on for a more fundamental set of laws, on the Pythagorean assumption that at absolute bedrock there are laws that break down in no situations whatsoever. The underlying unchanging laws, whatever they are, and the nearest approaches to them that have been found, do obviously allow a vast range of changes and events to occur, a vast range of behavior and experience. How far we have come from the early Pythagoreans, as they hurriedly and superficially applied this same faith in numbers! How unfathomably deep beyond their imagination the true connections lie! Beyond ours, too, perhaps.

The first challenge to the Pythagorean assumption of rationality in the universe to occur in the twentieth century was Russell's paradox, the discovery of Bertrand Russell that was discussed in Chapter 18. That happened early, in 1901. Another, in 1931, was Austrian Kurt Gödel's "incompleteness theorem." Gödel was then a young man working in Vienna; he would later join Einstein at the Institute for Advanced Study in Princeton. Gödel's discovery was that in any mathematical system complex enough to include the addition and multiplication of whole numbers—hardly fringe territory; any schoolchild is familiar with that—there are propositions that can be stated, that we can see are true, but that cannot be proved or disproved mathematically within the system. This means that all significant mathematical systems are open and incomplete. Truth goes beyond the ability to prove that it is true. Gödel also showed that it is not possible to prove whether or not any system rich enough to include addition and multiplication of whole numbers is self-consistent.

These discoveries constituted a serious reversal of hopes for some, and a serious undermining of assumptions for others. The great mathematician David Hilbert and his colleagues had previously been able to demonstrate that logical systems less complex than arithmetic were consistent, and it seemed certain that they would be able to go on to demonstrate the same for all of arithmetic. Not so. With Gödel, the soaring Pythagorean staircase to sure knowledge, built of numbers, became something more resembling a staircase in an Escher drawing, and it is no wonder that the most famous book about Gödel is Douglas R. Hofstadter's *Gödel, Escher, Bach*. The Bach is Johann Sebastian Bach. Bertrand Russell was one of those who were badly shaken by Gödel's theorems—particularly so because he misread Gödel and thought he had proved that arithmetic was not incomplete but *inconsistent*. Instead, Gödel had demonstrated that no one ever would be able to prove whether it was consistent or not. David Hilbert was not so discouraged as Russell: Until his death in 1943, he refused to recognize that Gödel had put paid to his hopes. The influence of Gödel's discoveries was profound, and yet, on one level, rather inconsequential. As John Barrow wrote in 1992, "It loomed over the subject of mathematics in an ambiguous fashion, casting a shadow over the whole enterprise, but never emerging to make the slightest difference to any truly practical application of mathematics."[5]

Though Gödel's discoveries may have undermined some forms of faith in mathematics, in a manner that seemed to resemble the Pythagorean discovery of incommensurability, Gödel's view of mathematics was, in fact, Pythagorean. He believed that mathematical truth is something that actually exists apart from any invention by human minds—that his theorems were "discoveries" about objective truth, not his own creations.

This was not a popular idea in the 1930s. Many mathematicians disagreed. In fact, the concept of anything existing in an objective sense—waiting out there to be discovered and not in any way influenced by the actions of the investigator—had been called into question by a development in physics. A far more dramatic and far-reaching crisis than the one caused by Gödel's incompleteness theorem had occurred in the 1920s and was having a profound effect on the way scientists and others viewed the world. It was the discovery of the uncertainty principle of quantum mechanics.

The way cause and effect work had long seemed good evidence that the universe is rational. It also seemed that if cause and effect operate as they do on levels humans can perceive, they surely must operate with equal dependability in regions of the universe, or at levels of the universe, that are more difficult— or even impossible—to observe directly. Cause and effect could be used as a guide in deciding what happened in the very early universe and what conditions will be like in the far distant future. No one was thinking of belief in cause and effect as a "belief" at all, though, in fact, there was nothing to prove that cause and effect would not cease to operate in an hour or so, or somewhere else in the universe. Then, in the 1920s, came developments that required reconsideration of the assumption that every event has an unbroken history of cause and effect leading up to it.

The quantum level of the universe is the level of the very small: molecules, atoms, and elementary particles. It is on that level that a commonsense description breaks down. Here there are uncaused events, happenings without a history of the sort it is normally assumed any event must have. Atoms are not miniature solar systems. You cannot observe the position of an electron orbiting the nucleus and predict where it will be at a later given moment and what path it will take to get there or say where it was an hour ago—as you could with fair accuracy for the planet Mars in the solar system. An electron never has a definite position and a definite momentum at the same time. If you measure precisely the position of a particle, you cannot at the same time measure its momentum precisely. The reverse is also true. It is as though the two measurements—position and momentum—are sitting at opposite ends of a seesaw. The more precisely you pin down one, the more up-in-the-air and imprecise the other becomes. This is the Heisenberg uncertainty principle of quantum physics—the twentieth century's "incommensurability." It was first articulated by Werner Heisenberg in 1927. Not only did it undermine faith in a rational universe, it also seemed to undermine the notion that truth was something objective, something waiting out there to be discovered. On the quantum level, your measurement affects what you find.

On the other hand, the existence of quantum uncertainty itself was apparently a very unwelcome piece of objective truth waiting out there that no physicist could change, as much as he or she might wish to, no matter what observational methods he or she used. Einstein in particular rebelled at the notion that no future advance in science and no improvement in measuring equipment was ever going to resolve this uncertainty. Until his death, he went on trying to devise thought experiments to get around it. He never succeeded, nor has he succeeded posthumously as others have found ways to carry out experiments he invented in his head. "God does not play dice!" Einstein wrote on one occasion to Niels Bohr, who was far more ready to accept quantum uncertainty than Einstein. "Albert, don't tell God what he can do!" Bohr answered. The Bohr-Einstein debate about how to interpret the quantum level of the universe continued and became famous.

It is easy to sympathize with Einstein. The quantum world and the paradoxes implicit in it did not seem to be the work of a rational mind. Einstein might have rephrased the complaint Kepler registered when faced with a similar problem: "Heretofore we have not found such an ungeometrical conception in His other works!" How could what happened to one particle affect another over time and space with no link between them? How could a cat be both dead and alive at the same time—as one had to accept in the famous example of "Schrödinger's cat"? How could something be a wave some times and a particle at others, depending on the experimental situation? It was a *Through the Looking Glass* world—and still is, in spite of the reassurance that it *is* possible to predict things on the quantum level of the universe, if one can be satisfied with probabilities. It does seem that the staircase to knowledge about the universe can have a firm footing on the quantum level, with probabilities forming a sort of superstructure above the quagmire. All is far from lost for the Pythagorean climb.

The dawning awareness of a new aspect of the universe, in chaos and complexity theories developed later in the twentieth century, was not nearly so great a shock as quantum uncertainty. However, it did seem to hint that science had been discovering one orderly, predictable system after another only because it was impossible or at least terribly discouraging to try to study any other kind of system in a meaningful fashion. The relatively easy to study predictable systems actually turned out to be the exception rather than the rule. But for those of a Pythagorean cast of mind, it was the discoveries of the repeating patterns in chaos—the pictures deep in the Mandelbrot and Julia sets, and also in nature itself—that gloriously seemed to uphold, as never before, the ancient conviction that beauty and harmony are hidden everywhere in the universe and have nothing to do with any invention of humans. Less immediately mind-boggling, but no less impressive, was the realization in the study of chaos and complexity that there seem to be mysterious organizing principles at work. There are probabilities, but by some calculations they are vanishingly low, that the universe would have organized itself into galaxies, stars, and planets; that life on this earth would have been organized into ecosystems and animal and human societies. Yet that is what has happened. Thus, as with the other challenges to faith in the Pythagorean assumptions underlying science, when scientists began to get a handle on chaos and complexity, the theories having to do with them became not threats but new avenues in the search for better understanding of nature and the universe.

Twentieth-century "postmodern" thinking, combined with suspicions raised by the discovery of quantum uncertainty and our inability to examine the quantum world without affecting it, led to fresh doubts about other Pythagorean pillars of science. Is there really such a thing as objective reality? Is anything real, waiting to be discovered? Does the fact that science continues to discover things that make sense, and suspects or dismisses anything that does not, mean that we are finding out more and more about a rational universe . . . or only that we are selecting the information and discoveries that fit our very Pythagorean expectations?

The assumption of rationality lies at the root of modern arguments about "intelligent design." It is true that the world's design, as the Pythagoreans found out, is intelligent to a degree that would send any discoverer of a new manifestation to his or her knees—but before what, or whom? Does discovering rationality necessarily mean one has glimpsed the Mind of God? On the other hand, does a good scientist have to repress the strong impression that it does? Those who attack belief in God do so from several directions. One is rather old-fashioned now, but still heard: Everything is so perfectly laid out, in so tight and orderly a design, that there is no room for God to act at any point. It all goes like clockwork. Or, a newer argument: Everything happens—and has always happened—entirely by chance. The impression of any underlying rationality in nature is an illusion. The "anthropic principle" says that if things had not fallen out just the way they have, we could not be here to observe them—and that is the *only* reason we find a universe that is amenable to our existence. Or . . . our entire picture of the universe is created, by us, in the self-centered image of our own minds, and we are discovering something not far different from the ten heavenly bodies of the Pythagoreans. Plato might have enjoyed the late-twentieth-century discussions about whether mathematical rationality might be powerful enough to create the universe, without any need for God. Quantum theory made possible the suggestion that "nothingness" might have been unstable in a way that made it statistically probable that "nothingness" would decay into "something."

Pythagorean principles and issues also showed up in other ways in twentieth-century culture. Peter Shaffer's trilogy of plays *The Royal Hunt of the Sun*, *Equus*, and *Amadeus* were all profound explorations of the theme of rationality and irrationality and reflected the sort of love/hate humanity has for both: Is there a Mind behind the universe? Is that Mind sane or mad? Tennessee Williams dubbed the so-called "rationality" of God the rationality of a senile delinquent. In music, "twelve tone" compositions were the most mathematically bound compositions ever written, but this form of music was also clear evidence that the Pythagorean insight had been correct that certain combinations of tones—and *only* certain combinations—have a deep link with what the human ear recognizes as harmonious and beautiful. On *Sesame Street*, numbers came to life and danced and sang in a way that probably would have delighted the Pythagoreans—if they did not find it irreverent—but probably would have annoyed Aristotle.

The music of the spheres remained a popular metaphor, but in the second half of the century it moved beyond the "spheres."[6] As Richard Kerr has put it, "the idea of heavenly harmonics is now making a comeback among astronomers. Instead of listening to the revolutions of the spheres, modern astronomers are tuning in to the vibrations within stars."[7]

In 1962, astronomers studying the Sun discovered that sound waves traveling through the Sun cause a bubbling of its visible surface, the photosphere.[8] They described it as a "solar symphony" that is somewhat like a "quivering gong," or "a large spherical organ pipe," or a "ringing bell," for the Sun has millions of different overtones.[9] Ours is not, of course, the only star that vibrates in this way. The giant star XiHydrae is a "sub-ultra-bass instrument," with oscillations of several hours.

In a book entitled *Einstein's Unfinished Symphony: Listening to the Sounds of Space-time*, Marcia Bartusiak described the possibility of detecting a black hole "by the melody of its gravity wave 'song.'"[10] Black holes have now indeed joined the heavenly choir. When material falls toward a supermassive black hole, that produces a jet of high-energy particles that blasts away from the black hole at nearly the speed of light. This jet plows into the gas around the black hole, creating a magnetized bubble of high-energy particles. An intense sound wave rushes ahead of the expanding bubble.[11] The NASA satellite *Chandra*, named for Subrahmanyan Chandrasekhar, the first scientist to see that, given Einstein's theories, black holes were inevitable, has found evidence of acoustic waves like this in the gaseous regions around two supermassive black holes. One of them, at the center of the Perseus galaxy cluster, plays the deepest note discovered so far in the universe, B flat fifty-seven octaves below middle C.[12]

Mark Whittle of the University of Virginia has produced a tape of "Sounds from the Infant Universe" which reproduces the power spectrum of the Cosmic Background Radiation—radiation that is still reaching us from the early universe—as an audible sound, covering the first million years of the cosmos in ten seconds.[13] In order to make the acoustic waves hearable by the human ear, he had to shift them upward approximately fifty octaves. The tape begins in silence, as the universe did, because there were no acoustic waves as long as the infant universe was symmetrical. Eventually there arose acoustic waves of deeper and deeper tone. The expansion of the universe stretched the wavelengths, making for an overall drop in pitch as the tape continues. The largest variations compare to "rock concert volume."[14]

The prediction was that a "ripple" in the distribution of galaxies in the universe would reflect the acoustic waves in the Cosmic Background Radiation. At the January 2005 meeting of the American Astronomical Society, the report came that this evidence had been found.[15] Those who announced it likened the discovery to "detecting the surviving notes of a cosmic symphony" and the difficulties of the observations to trying to hear the "last ring" of a bell that "gets forever quieter and deeper in tone as the Universe expands."[16] One cannot help thinking that Kepler would have been intensely interested in projects like these.

Kent Cullers, who works at SETI, the Search for Extraterrestrial Intelligence, and on whom Carl Sagan based one of his characters in the novel and film *Contact*, is blind and claims this is an advantage as he listens to signals from outer space. "When I hear signals from distance regions, my mind goes out there. I try to ride those waves, extend my senses to a realm where they've never been, hear songs from a cloud of gas."[17] In the 1970s, it was proposed that the Pythagorean theorem, or "Pythagorean triples" of numbers that make right triangles, be beamed as messages into space, in the hope that rational life in other star systems might receive the signals and realize that there was rational life on Earth. It is a signal like that that Cullers is hoping to hear, coming to us from deep space—evidence of how truly primordial this knowledge is.

EPILOGUE

Music or Silence

N GENERATION AFTER GENERATION, men and women have recognized the essential truth of the ancient insight that rationality and order underlie the variety and confusion of nature. The image of Pythagoras himself has shifted and occasionally become distorted, but through all the centuries and all the paradigm shifts, this Pythagorean vision has never been extinguished or forgotten, and it has almost always been cherished. He and his first followers could not begin to conceive how vast a landscape lay beyond the door they opened. From unimaginably tiny flickering wisps of uncertainty to the uncountable galaxies, into multiple dimensions, and maybe even to an infinity of other universes. Yet numbers and number relationships seem to have guided the way through this labyrinth of the physical universe as effectively as Pythagoras himself could ever have hoped.

If civilization as we know it were wiped out and only a remnant were left to start over, would someone make that same discovery? Break the code again? Surely they would! Is it not basic *truth*? Or . . . maybe they wouldn't. Maybe the Pythagoreans got it wrong, and we have been living in a dream. Maybe the world really never got beyond a formless "unlimited," and we are only imagining the pattern, or creating it ourselves. The human soul has not proved so easy to map with numbers . . . and yet we are the "rational beings" on the Earth, presumably reflecting the rationality of the universe. How can it be that we are the most difficult of all territory? We do not yet know. Meanwhile most of us are too intoxicated by the music of Pythagoras to suffer a crisis of faith.

We send our tiny beeps into the far distant reaches of space, certain that any intelligent beings out there, no matter how "other" they may be in some respects, could not have failed to discover what our world did . . . sure that our little signaled evidence of rationality will look familiar to them. In spite of the still unsolved mysteries—and the possibility that they may never be solved—our Pythagorean ideal of the unity and kinship of all being tells us this must be so.

Pythagoras . . . are you there?

Appendix

The proof for the Pythagorean theorem that Jacob Bronowski thought may have been used by Pythagoras.[1]

Start with a right triangle.

Create a square using four triangles identical to that one, but rotated, so that the "leading points" of the triangles point to the four points of the compass (north, south, east, and west), and the long side of each triangle ends at the leading point of its neighbor:

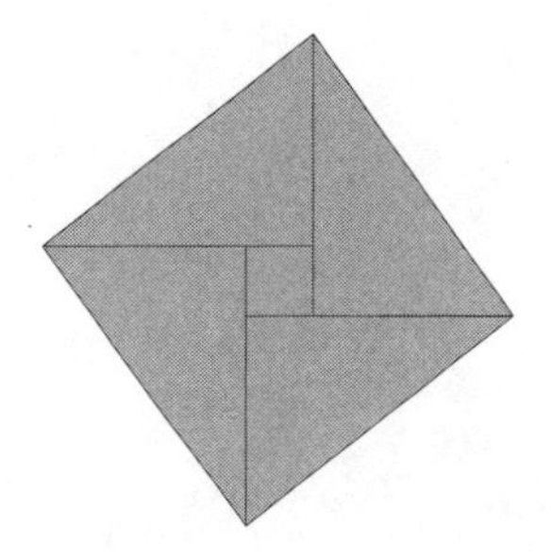

What you now have is a square based on the long side of the original triangle—the "square on the hypotenuse." It is this total area that must equal the sums of the squares of the other two sides, if the Pythagorean theorem is correct. As you proceed, remember that however you rearrange these five shapes, their total area stays the same. So, rearrange them into the following shape. Place a rod across your design and look at it carefully. You will see that you have two squares, and they are the squares on the other two sides of the triangle. Using no numbers, you have proved the Pythagorean theorem.

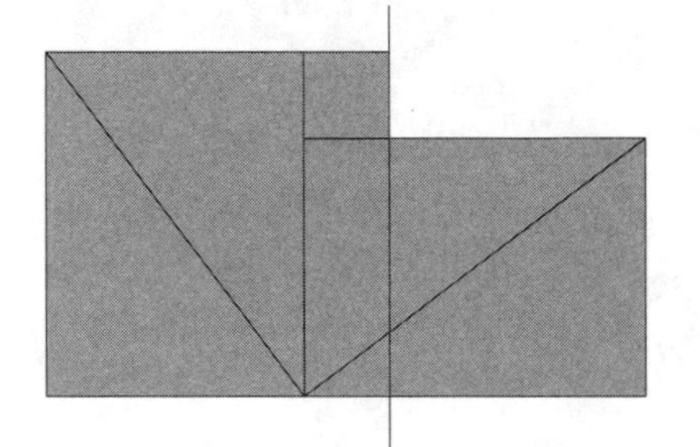

Notes

Chapter 1: The Long-haired Samian

1. Iamblichus' *Pythagorean Life* or *Life of Pythagoras* is available in translation by Thomas Taylor: *Iamblichus Life of Pythagoras* (Rochester, Vt.: Inner Traditions International, 1986). It and the biographical treatments by Porphyry and Diogenes Laertius are available in translation by Kenneth Sylvan Guthrie: *The Pythagorean Sourcebook and Library: An Anthology of Ancient Writings Which Relate to Pythagoras and Pythagorean Philosophy* (Grand Rapids: Phanes Press, 1987). The Guthrie anthology also contains some of the pseudo-Pythagorean works.
2. Diogenes Laertius' and Porphyry's "lives" of Pythagoras are reprinted in K. S. Guthrie.
3. Jacob Bronowski, *The Ascent of Man* (Boston: Little, Brown, 1973), p. 156.
4. Quoted in Richard Buxton, ed., *From Myth to Reason: Studies in the Development of Greek Thought* (Oxford, U.K.: Oxford University Press, 1999), p. 74.
5. Kurt A. Raaflaub, "Poets, Lawgivers, and the Beginnings of Political Reflection in Archaic Greece," in Christopher Rowe and Malcolm Schofield, eds., *The Cambridge History of Greek and Roman Political Thought* (Cambridge, U.K.: Cambridge University Press, 2000), p. 51.
6. Plato, *Thaetetus*, 174 A., quoted by Thomas L. Heath, *Greek Astronomy* (London: J. M. Dent and Sons, 1932), p. 1, and Arthur Koestler, *The Sleepwalkers: A History of Man's Changing Vision of the Universe* (London: Hutchinson, 1959), p. 22.
7. The story of Thales and the river Halys was one of those collected by Herodotus and included in his *Histories* I 75.3–5. Reprinted in Barnes, p. 10.

8. Ian Shaw, *Ancient Egypt: A Very Short Introduction* (Oxford, U.K.: Oxford University Press, 2004). p. 12.
9. Porphyry's biography is reprinted in K. S. Guthrie, p. 124.

Chapter 2: "Entirely different from the institutions of the Greeks"

1. For the information about what Pythagoras might have learned in Egypt, I have relied on David P. Silverman, ed., *Ancient Egypt* (New York: Oxford University Press, 1997).
2. For information about Babylon in this era, I have relied on H. W. F. Saggs, *Everyday Life in Babylonia and Assyria* (Assyrian International News Agency, 1965); and Joan Oates, *Babylon* (London: Thames & Hudson, 1979). Speculation about the historical timing of Pythagoras' abduction from Egypt is based on Saggs, p. 25. Modern scholarly knowledge about the city of Babylon during this period comes from a variety of sources: the biblical and Greek tradition, Nebuchadnezzar's building inscriptions, business, legal and administrative records, and the excavation of the city, which together give a fairly clear picture of life in the Babylonian capital under Nebuchadnezzar II, though there are many details which we do not yet know and may never know.

Chapter 3: "Among them was a man of immense knowledge"

1. Exhibits in the Museo Archeologico Nazionale di Croton suggest the appearance of the ancient city.
2. Information about Achaea comes from N. G. L. Hammond, *A History of Greece to 322 B.C.* (Oxford, U.K.: Oxford University Press, 1986), pp. 13 and 118.
3. Porphyry's *The Life of Pythagoras*, reprinted in K. S. Guthrie, 1987, p. 135.
4. Acts 17:21.
5. Kurt A. Raaflaub, "Poets, Lawgivers, and the Beginnings of Political Reflection in Archaic Greece," in Christopher Rowe and Malcolm Schofield, eds., *The Cambridge History of Greek and Roman Political Thought* (Cambridge, U.K.: Cambridge University Press. 2000), p. 57.
6. Guthrie, William Keith Chambers, *The Earlier Presocratics and the Pythagoreans*, Vol.1 of A *History of Greek Philosophy*. (Cambridge, U.K.: Cambridge University Press, 2003), pp. 176–177. Guthrie refers to the historian C. T. Seltman.
7. Ibid., pp. 176–77.

Chapter 4: "My true race is of Heaven"

1. This overview of Greek beliefs about immortality and the manner in which Pythagorean doctrine fits into this context is based on the discussion in Guthrie (2003), beginning on p. 196, and on W. K. C. Guthrie, *The Greeks and Their Gods* (Cambridge, U.K.: Cambridge University Press, 1951).
2. See Betrand Russell, *The History of Western Philosophy* (London: George Allen & Unwin, 1945).
3. This is the way the verse in Homer was translated by Alexander Pope.
4. The story was told by Diogenes Laertius and also by Diodorus in his *Universal History* X, quoted in Barnes, p. 34.
5. See W. K. C. Guthrie (2003), pp. 201–202.
6. Quoted in ibid, p. 199.
7. Ibid, p. 199.
8. Ibid.
9. Eudemus, *Physics*, fragment quoted by Simplicius in *Commentary on the* "Physics" 732.23–33. Quoted in Barnes, p. 35.
10. See Barnes, pp. 167–68.
11. Material is from Aulus Gellius, *Attic Nights*, Book IV, xi 1–13. *Attic Nights* was twenty volumes long, a compendium of miscellaneous knowledge. It and parts of it are available in a number of editions.
12. For the discussion around the suggestion that the miraculous stories were meant to discredit Pythagoras, see Walter Burkert, *Lore and Science in Ancient Pythagoreanism* (Cambridge, Mass.: Harvard University Press, 1972), p. 146

Chapter 5: "All things known have number"

1. Ibid, p. 377.
2. W. K. C. Guthrie (2003), p. 178.

Chapter 6: "The famous figure of Pythagoras"

1. Bronowski, p. 156.
2. From "Surveying" article in the *Encyclopaedia Britannica*, 2007, Online, 3 Mar. 2007 http://www.britannica.com/eb/article-51763, p. 2.
3. This information comes from a conversation with John Barrow and from his book *Pi in the Sky: Counting, Thinking, and Being* (Oxford, U.K.: Clarendon Press, 1992), pp. 73–75. The information about Indian women and their doorstep paintings comes from personal experience in Kothapallimitta, South India, in 2000, and trying to do it myself.

4. W. K. C. Guthrie (2003), p. 187.
5. Except where otherwise footnoted, and except for some information about Tell Harmal, the information in these paragraphs about Babylonian mathematics comes primarily from Eleanor Robson, "Three Old Babylonian Methods for Dealing with Pythagorean Triangles," *Journal of Cuneiform Studies* (1997) 49, pp. 51–72.
6. Robson, "Mesopotamian Mathematics: Some Historical Background," in Victor Katz, ed., *Using History to Teach Mathematics: An International Perspective* (Cambridge, U.K.: Cambridge University Press, 2000), p. 154
7. Plimpton 322 is now in the collection of Columbia University, in New York City.
8. See Taha Baqir, "An Important Mathematical Problem Text from Tell Harmal," *Sumer 6* (1950), pp. 39–55. Taha Baqir was curator of the Iraq Museum.
9. Diagram and text reconstruction are from Robson, "Three Old Babylonian Methods," p. 57.
10. See, for example, Ross King, *Brunelleschi's Dome: How a Renaissance Genius Reinvented Architecture* (London: Penguin, 2000).
11. John Noble Wilford, "Early Astronomical Computer Found to Be Technically Complex," *New York Times*, November 30, 2006.
12. For discussion, see Robson, "Mesopotamian Mathematics: Some Historical Background," pp. 154–55. The quotation is from Robson, "Influence, Ignorance, or Indifference? Rethinking the Relationship Between Babylonian and Greek Mathematics," *The British Society for the History of Mathematics*, Bulletin 4 (Spring 2005), pp. 2, 3.
13. Ibid., p. 14.
14. Ibid., p. 10.
15. See discussion in ibid, pp. 2, 3.
16. Charles H. Kahn, *Pythagoras and the Pythagoreans*: A *Brief History* (Indianapolis: Hackett, 2001), p. 134.
17. Marcus Vitruvius Pollio, *De Architectura, Book IX*. Vitruvius' work is reprinted as *Vitruvius: Ten Books of Architecture* (Cambridge, U.K.: Cambridge University Press, 2001).
18. Bronowski, p. 160.

Chapter 7: A Book by Philolaus the Pythagorean

1. For the discussion of the *acusmatici* and the *mathematici* and the question about which were truer to the original teachings of Pythagoras, I have relied

on Burkert, particularly the section entitled "*Acusmatici* and *Mathematici*."

2. The names of some *mathematici* have survived. One was Archytas of Tarentum, and he mentioned Eurytus of Tarentum as one of his predecessors. This was the same Eurytus linked with Philolaus in Plato's *Phaedo*. Eurytus and Philolaus had students whose names Aristoxenus listed. They were from Chalcidice in Thrace and from Phlius.
3. I have mostly followed Burkert regarding the authenticity of fragments of Philolaus; I have also relied on W. K. C. Guthrie (2003) in the discussion of Philolaus' book and on Guthrie and Jonathan Barnes, *Early Greek Philosophy* (London: Penguin Books, 1987), for translations of quotations.
4. Plutarch, *On the Face in the Moon* 929AB, quoted in Barnes, p. 89.
5. From Plutarch, *Pericles*; passage quoted in full in Barnes, p. 92.
6. Quoted from Aristotle's *Metaphysics*, in W. K. C. Guthrie (2003), pp. 287–88.
7. See W. K. C. Guthrie (2003) p. 248, for the seed of this idea.
8. W. K. C. Guthrie (2003) wrote that Philolaus was Aristotle's "favorite author" (p. 260). The quotation is in W. K. C. Guthrie (2003), pp. 307–308.
9. Quoted in ibid., p. 309.
10. See ibid., p. 233, for the arguments each way concerning Alcmaeon's dates. The quotation appears in W. K. C. Guthrie (2003), p. 313.
11. Plato, *Phaedo*, quoted in ibid., p. 310.
12. Quoted in ibid., p. 311.
13. Quoted in ibid., p. 312.

Chapter 8: Plato's Search for Pythagoras

1. Information about Plato's visits to Megale Hellas (as he would have called it) and Syracuse can be found in many sources. I have used W. K. C. Guthrie (2003), and Malcolm Schofield, "Plato and Practical Politics," in Christopher Rowe and Malcolm Schofield, eds., *The Cambridge History of Greek and Roman Political Thought* (Cambridge, U.K.: Cambridge University Press, 2000).
2. The discussion of Archytas' work is based on Kahn, p. 40 ff. Most of the quotations from Archytas are drawn from Kahn's book.
3. Burkert, p. 68.
4. Modern scholars such as Kahn have made a distinction that sets Archytas a little more apart from the earlier Pythagoreans but still keeps him in the tradition. Kahn described Archytas' harmonic theory as "work of original genius . . . working in the Pythagorean musical tradition that is represented for us by the earlier theory of Philolaus" (pp. 32–43).

5. From Eudemus (also mentioned by Aristotle), quoted in Kahn, p. 43.
6. Archytas' description of a bull-roarer, or *rhomboi*, is in W. K. C. Guthrie (2003), p. 227.
7. Aristotle, quoted in ibid., p. 335.
8. For this discussion and the question about which were truer to the original teachings of Pythagoras, I have relied on Burkert, particularly the section entitled "*Acusmatici* and *Mathematici*."
9. The original is Aristophon, fragment 12; see Burkert, p. 199, for the quotation. Burkert was not sure the Greek word used actually referred to the Pythagoreans.
10. Quotations are from the musician Stratonicus and from Sosicrates, in Burkert, p. 202.

Chapter 9: "The ancients, our superiors . . ."

1. Kahn, p. 50 ff, is especially helpful in interpreting Plato's thought as it related to the Pythagoreans, and he pinpoints these two themes.
2. The quotations from Plato's *Timaeus*, unless otherwise noted, are taken from the translation by Desmond Lee (London: Penguin, 1977).
3. From Plato's *Philebus*, in Kahn, p. 14.
4. Ibid., p. 58.
5. From Plato's *Gorgias*, Quoted in W. K. C. Guthrie (2003), p. 305.
6. Book 7 of Plato's *Republic*, quoted in ibid., p. 162.
7. Plato, *Timaeus*, in the Stephanus edition (1578), p. 52.

Chapter 10: From Aristotle to Euclid

1. W. K. C. Guthrie (2003), p. 331. I have closely followed Guthrie in his discussion of Aristotle's reactions to the Pythagoreans. Where not otherwise noted, the quotations from Aristotle are drawn from Guthrie's book.
2. Burkert lists the writers in whose work fragments from Aristotle appear (pp. 28, 29).
3. "What the sky encloses" is a quotation from Burkert (p. 31), but he was paraphrasing Aristotle.
4. It is no longer generally accepted that, as Burkert states, "like all pre-Socratics they take everything that exists in the same way, as something material" (Burkert, p. 32). That does not apply correctly either to the Pythagoreans or to the other pre-Socratics.
5. Burkert, pp. 45–46.

6. W. K. C. Guthrie (2003), p. 259.
7. The quotation is from Burkert (p. 431), in his discussion of these different possibilities, but he did not favor this choice.
8. This discussion draws on W. K. C. Guthrie (2003), p. 266 ff, and Burkert, p. 68 ff.
9. Ibid., p. 226n.
10. Burkert's paraphrase of Aëtius, in which he seems to have given only the final four words in direct quotation (Burkert, p. 70).
11. Quoted in W. K. C. Guthrie (2003), pp. 263–64.
12. Historical information about this era comes in part from Greg Woolf, ed., *Cambridge Illustrated History of the Roman World* (Cambridge, U.K.: Cambridge University Press, 2003); and from Paul Cartledge, ed., *Cambridge Illustrated History of Ancient Greece* (Cambridge, U.K.: Cambridge University Press, 1998).
13. Specifically in the *Elements*, Book VII.
14. One scholar, the German B. L. van der Waerden, insisted that there were writings before Archytas and long before Euclid that dealt with this same material. Kahn calls some of his claims "excessive" (Kahn, p. 41n.). Propositions in Book II of the *Elements* have very early Babylonian precursors, as the Pythagorean theorem did, that Euclid probably was not aware of (Robson [2005], p. 4).
15. For information and discussion, see Burkert, p. 432.
16. This quotation is from Iamblichus, *On Common Mathematical Knowledge* 91.3–11, translation in I. Mueller, "Mathematics and Philosophy in Proclus's Commentary on Book I of Euclid's *Elements*," in J. Pépin and H. D. Saffrey, eds., *Proclus, Lecteur et Interprète des Anciens* (Paris: CNRS, 1987). Quoted in S. Cuomo, *Ancient Mathematics* (London: Routledge, 2001), pp. 236–37.

Chapter 11: The Roman Pythagoras

1. Marcus Tullius Cicero, *On The Republic*, Book 2, 28, 29. Reprinted in translation by Niall Rudd, in *Cicero: The Republic and The Laws* (Oxford, U.K.: Oxford University Press, 1998), pp. 43–44.
2. Quoted in Kahn, pp. 89–90.
3. From Pliny, *Natural History* 34.26. As retold in Kahn, p. 86.
4. From the *Pythagorean Notebooks*, excerpt quoted in Diogenes Laertius' *The Life of Pythagoras*, reprinted in K. S. Guthrie, pp. 148–49.
5. Quoted in Simon Hornblower and Antony Spawforth, eds., *The Oxford Companion to Classical Civilization* (Oxford, U.K.: Oxford University Press, 1998), p. 245.

6. Thomas Wiedemann, "Reflections of Roman Political Thought in Latin Historical Writing," in Christopher Rowe and Malcolm Schofield, eds., *The Cambridge History of Greek and Roman Political Thought* (Cambridge, U.K.: Cambridge University Press, 2000), p. 526–27.
7. Sextus Empiricus, quoted in Kahn, p. 84.
8. Elizabeth Rawson, *Intellectual Life in the Late Roman Republic* (Baltimore: Gerald Duckworth & Co., 1985), p. 310.
9. Cicero, *Timaeus*, "Introduction." Quoted in Kahn, p. 73.
10. Kahn, p. 88.
11. Cicero, *Vatinium* 6. Quoted in Kahn, p. 91.
12. Cicero, *On the Commonwealth*, quoted from the edition translated by George Holland Sabine and Stanley Barney Smith (New York: The Liberal Arts Press, 1929), pp. 114–15.
13. Ibid., p. 206.
14. Cicero, *On Divination*, quoted in Barnes, p. 165–66.
15. Cicero, "Scipio's Dream," from *On the Republic*, translated by Cyrus R. Edmonds and Moses Hadas, in *The Basic Works of Cicero* (New York: Random House, 1951), p. 165.
16. Ibid., p. 166.
17. Ibid., p. 166.
18. Vitruvius (Marcus Vitruvius Pollio). *De architectura*, Book VI. Vitruvius' work has been reprinted as *Vitruvius: Ten Books of Architecture* (Cambridge, U.K.: Cambridge University Press, 2001).
19. Cesariano's drawing for Vitruvius: ccat.sas.upenn.edu/george/vitruvius.html. The quotation is from Book IX of Vitruvius.
20. Vitruvius, Book I.
21. Ibid.
22. The information about King Juba is from a footnote in Kahn (p. 90) to E. Zeller, *Die Philosophie der Griechen in ihrer geschichtlichen Entwicklung* I–III (Leipzig: Reisland 1880–92), p. 97.
23. *Lysis's Letter to Hipparchus*, quoted in Diogenes Laertius, *The Life of Pythagoras*, in K. S. Guthrie, pp. 141–55.
24. Burkert dates the letter to the third century B.C. A. Staedele dates it to the first. Kahn, p. 75, mentions contemporaneity as Burkert's suggestion, citing Walter Burkert, "Hellenistische Pseudopythagorica," *Philologus* 105 (1961).
25. Introduction to the Occelus piece in K. S. Guthrie, p. 203.
26. Mentioned in Kahn (p. 79), where the footnote refers to a quotation in J. Dillon, *The Middle Platonists* (Ithaca, NY: Cornell University Press, 1977), p. 156n.

27. See Bruno Centrone, "Platonism and Pythagoreanism in the Early Empire," in *Cambridge History of Greek and Roman Political Thought* (Cambridge, U.K.: Cambridge University Press, 2000). In Centrone's words (p. 568): "Here it is an artificial language, which only reproduces the commonest features of Doric."
28. Kahn, p. 76.

Chapter 12: Through Neo-Pythagorean and Ptolemaic Eyes

1. For the discussion of the neo-Pythagorean philosophers and cults, I have relied on Kahn and Centrone.
2. From the *Pythagorean Golden Verses*, reprinted in K. S. Guthrie, p. 164.
3. Seneca, in a "Letter to Lucilius" (108.17–21). Quoted in Kahn, p. 151.
4. Philostratus, *Life of Apollonius of Tyana*, quoted in Robin Lane Fox, *Pagans and Christians in the Mediterranean World from the Second Century A.D. to the Conversion of Constantine* (London: Penguin Books, 1986), p. 245.
5. Philostratus, in Fox, p. 248.
6. Eudorus, quoted or paraphrased in Arius Didymus. Quoted in Kahn, p. 96.
7. Information about the Alexandrian Jewish community is from Greg Woolf, ed., *Cambridge Illustrated History of the Roman World* (Cambridge, U.K.: Cambridge University Press, 2003), p. 277.
8. Quoted in Kahn, p. 101.
9. Centrone, p. 561n.
10. Philo of Alexandria, *De opific*, quoted in Kahn, p. 100n.
11. The two descriptions come from Harry Austryn Wolfson and Valentin Nikiprowetsky, as reported by Centrone, p. 561.
12. Centrone, p. 561.
13. Number 41 in *The Catalogue of Lamprias*, a list of Plutarch's works that probably was compiled in the fourth century A.D.
14. This was how Porphyry reported Moderatus' views: see Kahn, p. 105.
15. As quoted and/or paraphrased by Porphyry: see Kahn, p. 106.
16. Theon of Smyrna, *Mathematics Useful for Understanding Plato*, excerpted as Appendix 1 in K. S. Guthrie.
17. W. K. C. Guthrie (2003), p. 406.
18. Dodds (1963), p. 259; quoted in Kahn, p. 118.
19. Numenius fragment #2; quoted in Kahn, p. 122.
20. Numenius fragment, no number; quoted in Kahn, p. 122.
21. Numenius fragment #52; quoted in Kahn, p. 132.

22. Johannes Kepler, *Johannes Kepler Gesammelte Werke*, Max Caspar et al., eds. (Munich: Deutsche Forschungsgemeinschaft, and the Bavarian Academy of Sciences, 1937–), vol. 6, p. 289.
23. Pliny, *Natural History*, 2:20, translated by Bruce Stephenson, p. 24, in Stephenson *The Music of the Heavens: Kepler's Harmonic Astronomy* (Princeton, N.J.: Princeton University Press, 1994).
24. Stephenson, p. 29. Stephenson cites Von Jan, "Die Harmonie der Sphären," *Philologus* 52 (1893).
25. Simplified by the author from Stephenson, p. 31.
26. Stephenson, p. 37.

Chapter 13: The Wrap-up of Antiquity

1. E. R. Dodds, *The Greeks and the Irrational* (Berkeley, Calif.: University of California Press, 1951), p. 296.
2. Ibid., p. 285.
3. Description of Rome in this era is based on Michael Grant, *History of Rome* (London: Faber & Faber, 1978), p. 284 ff.
4. Edward Gibbon, *Decline and Fall of the Roman Empire*, Chapter X. Quoted in Russell (1945), p. 287.
5. Description taken from Grant, p. 294.
6. Ibid.
7. Plotinus quoted in Dodds, pp. 285–86.
8. Ibid., pp. 286–87.
9. Kahn, p. 134.
10. Dodds, p. 287.
11. Ibid.
12. Fox, p. 190.
13. John 1:1–5, 9–12, 14. Paraphrased from the Oxford New International Version translation.
14. Augustine, *City of God*, translated by Gerald Walsh et al. (Garden City, N.Y.: Doubleday, 1958), pp. 241–42.
15. H. J. Blumenthal and R. A. Markus, eds., *Neo-Platonism and Early Christian Thought: Essays in Honour of* A. H. *Armstrong* (London: Variorum Publications, 1981), p. 90. This concept is the "*soma-sema* formula."
16. Information about this period comes from H. G. Koenigsberger, *Medieval Europe: 400–1500* (London: Longman, 1987).
17. Most of the information about Macrobius is from Stephenson, pp. 38–41.
18. Quoted in Richard E. Rubenstein, *Aristotle's Children: How Christians,*

Muslims, and Jews Rediscovered Ancient Wisdom and Illuminated the Dark Ages (New York: Harcourt, 2003), p. 62, from Josef Pieper, *Scholasticism: Personalities and Problems of Medieval Philosophy*, translated by Richard and Clara Winston (New York: McGraw-Hill, 1964), p. 30.

Chapter 14: "Dwarfs on the Shoulders of Giants"

1. Joscelyn Godwin, *The Harmony of the Spheres: A Sourcebook of the Pythagorean Tradition in Music* (Rochester, Vt.: Inner Traditions International, 1993).
2. Quotations from Hunayn's *Nawadir al-Falasifa* are from excerpts translated by Isaiah Sonne and reprinted in Godwin, pp. 92–98.
3. Hunayn, quoted in Godwin, p. 92.
4. Quoted in David C. Lindberg, *The Beginnings of Western Science: The European Scientific Tradition in Philosophical, Religious, and Institutional Context, 600 B.C. to A.D. 1450* (Chicago: University of Chicago Press, 1992), p. 176.
5. From *The Epistle on Music of the Ikhwan al-Safa'*, translated by Amnon Shiloah; quoted from an excerpt reprinted in Godwin, p. 113.
6. Ibid., p. 113.
7. Ibid., p. 115.
8. Ibid.
9. From Al-Hasan Al-Katib, *Kitah Kamal Adal Al-Gina'*, translated by Amnon Shiloah; quoted from an excerpt reprinted in Godwin, p. 122.
10. The information about Aurelian comes from Godwin, p. 99.
11. Aurelian of Réôme, *Musica Disciplina*, translated by Joseph Ponte; quoted from an excerpt reprinted in Godwin, pp. 101–102.
12. Information about Eriugena comes from ibid., pp. 104–105.
13. John Scotus Eriugena, *Commentary on Martianus Capella*, translated by Joscelyn Godwin; quoted from an excerpt reprinted in ibid., p. 105.
14. Ibid., p. 106.
15. John Scotus Eriugena, *Periphyseon* or *De Divisione naturae*, quoted in ibid., p. 104.
16. Regino of Prüm, "Epistola de harmonica institutione," the introduction to his book about plainsong melodies, *Tonarius*. This excerpt translated by Sister Mary Protase LeRoux and reprinted in Godwin, p. 110.
17. Ibid., p. 111.
18. For the emergence of the universities, see Thomas Kuhn, *The Structure of Scientific Revolutions* (Chicago: University of Chicago Press, 1962), p. 102.
19. The archbishop's translation project is described at length in Rubenstein.

20. Information about the Seven Liberal Arts is from Koenigsberger, p. 199 ff.
21. As discussed by Burkert, beginning on p. 386.
22. Information from Burkert, p. 406, including footnote 31.
23. For the *Ars Geometrae* supposedly composed by Boethius, see Burkert, p. 406.
24. Koenigsberger, p. 202.
25. Quoted in Koenigsberger, p. 201.
26. This information comes in part from a website of the University of Notre Dame Jacques Maritain Center: (http://maritain-nd.edu) Ralph McInerny, *A History of Western Philosophy*, vol. 2, part III, chapter IV.
27. John Hedley Brooke, *Science and Religion: Some Historical Perspectives* (Cambridge, U.K.: Cambridge University Press, 1991), p. 45.
28. Ibid., p. 25.

Chapter 15: "Wherein Nature shows herself most excellent and complete"

1. Letter from Petrarch to Francesco Bruni, October 25, 1362, reprinted in Ernst Cassirer, Paul Oskar Kristeller, and John Herman Randall, Jr., eds., *The Renaissance Philosophy of Man: Selections in Translation* (Chicago and London, University of Chicago Press, 1948, 1969), p. 34.
2. From the introduction to the excerpts from Petrarch in ibid., p. 25.
3. Petrarch, *On His Own Ignorance*, reprinted in ibid., p. 92.
4. Ibid., p. 94.
5. Ibid., p. 24.
6. Marsilio Ficino, *Five Questions Concerning the Mind*, reprinted in ibid., pp. 209–210.
7. Pico della Mirandola, *On the Dignity of Man*, reprinted in ibid., pp. 232–33.
8. G. Pico della Mirandola, *Conclusiones sive Theses*, edited and translated by Bohdan Kieszowski, reprinted in Godwin, p. 176. For an attempt to make some sense out of this list, and connections with Plato, Nicomachus, Ptolemy, and, indeed, Oscar Wilde, see Godwin, p. 447.
9. See Cassirer et al., p. 245.
10. "Letter to Leo X," quoted in Kahn, p. 158, from A. E. Chaignet, *Pythagore et la philosophie pythagoricienne*, vol. II (Paris, 1873), p. 330.
11. Leon Battista Alberti, *The Ten Books of Architecture* (Mineola, N.Y.: Dover replica Edition, 1987), Chapter 5 of Book 9.
12. Nicholas of Cusa, *Of Learned Ignorance* (1440). Quoted in Koenigsberger, p. 367.
13. Prefatory letter to *De revolutionibus, Gesamtausgabe*. Vol. II: *De revolu-*

tionibus. Kritischer Text, eds. H. M. Nobis and B. Sticker (Hildesheim, Germany: 1984), p. 4, as quoted in T. S. Kuhn, *The Copernican Revolution: Planetary Astronomy in the Development of Western Thought* (Cambridge, Mass.: Harvard University Press, 1957), p. 137.

14. Prefatory letter to *De revolutionibus, Gesamtausgabe*. Vol. II, *De revolutionibus*, p. 4, as quoted in Kuhn, *Copernican Revolution*, p. 142.
15. Mentioned in Brian L. Silver, *The Ascent of Science* (Oxford, U.K.: Oxford University Press, 1998), p. 177.
16. Book 1, Chapter 10 of *De revolutionibus, Gesamtausgabe. Vol. II, De revolutionibus*, p. 4, as quoted in Kuhn (1957), pp. 179–80.
17. Andrea Palladio, *I quattro libri dell' architettura*. In a reproduction of the Isaac Ware 1738 English edition: *The Four Books on Architecture* (New York: Dover, 1978).
18. See, for example, Rudolph Wittkower, *Architectural Principles in the Age of Humanism* (New York: Norton, 1971).
19. It was Victor Thoren who called attention to these specifics about the way Tycho carried out Pythagorean/Palladian ideals in the design of Uraniborg; see his *The Lord of Uraniborg: A Biography of Tycho Brahe* (Cambridge, Mass.: Cambridge University Press, 1990).

Chapter 16: "While the morning stars sang together"

1. Johannes Kepler, letter to Michael Mästlin, June 11, 1598, *Johannes Kepler Gesammelte Werke*, Max Caspar, Salther von Dyck, Franz Hammer and Volker Bialas, eds. (Munich: Deutsche Forschungsgemeinschaft and the Bavarian Academy of Sciences, 1937–), vol. XIII, p. 219.
2. Godwin, pp. 104–105.
3. Kepler, *Harmonice mundi*, Book V, in *Gesammelte Werke*, vol. 6, p. 289.
4. Stephenson is an extraordinarily thorough and invaluable guide through the labyrinth of Kepler's *Harmonice mundi*.
5. Plato, *Timaeus* (London: Penguin, 1965), p. 15, p. 50n.
6. Kepler, *Harmonice mundi*, in *Gesammelte Werke*, vol. 6, p. 289.
7. For Kepler's complete table, see p. 150 in Stephenson.
8. For a much more detailed explanation, see Stephenson, p. 171.
9. Kepler, *Harmonice mundi*, in *Gesammelte Werke*, vol. 6, p. 323.
10. Kepler, *Harmonice mundi*, in *Gesammelte Werke*, vol. 6, p. 356.
11. Stephenson points out that Kepler had read Proclus's Platonic/neo-Pythagorean hymns.
12. Kepler, in a letter to Vincenzo Bianchi, February 17. Letter number 827 in *Gesammelte Werke* 17.326.213–19. Quoted in Stephenson, p. 241.

Chapter 17: Enlightened and Illuminated

1. Galileo Galilei, *Il Saggiatore*, 1623. Quoted and translated in Daniel T. Max, *The Family That Couldn't Sleep* (New York: Random House, 2006), p. 5.
2. The episode having to do with Galileo's father is retold in Silver, p. 176.
3. Barrow, p. 127.
4. Silver, p. 158.
5. Ibid., p. 177; Bronowski (p. 234) also mentions Newton's attribution to Pythagoras.
6. Quoted in Barrow, p. 127.
7. Quoted in ibid., p. 128, from G. Leibniz, *The Philosophical Works of Leibniz*, translated by G. Duncan (New Haven, Conn.: Tuttle, Morehouse and Taylor, 1916).
8. Joseph Addison, paraphrase of Psalm 19:1–6. Hymn 409 in *The Hymnal 1982, according to the use of the Episcopal Church*.
9. The paragraphs about how the image of Pythagoras was used by revolutionaries are based on James H. Billington, *Fire in the Minds of Men: Origins of the Revolutionary Faith* (New York: Basic Books, 1980). All quotations, unless otherwise noted, also come from quotations in his book.
10. Information about Buonarroti comes from Elizabeth L. Eisenstein, *The First Professional Revolutionist: Filippo Michele Buonarroti (1761–1837)*, A *Biographical Essay* (Cambridge, Mass.: Harvard University Press, 1959).
11. From Jevons, *Principles of Science*, quoted in Lindberg, pp. 371–72, n. 15.

Chapter 18: Janus Face

1. The two books discussed in this chapter are Bertrand Russell, *The History of Western Philosophy* (London: George Allen & Unwin, 1945); and Arthur Koestler, *The Sleepwalkers: A History of Man's Changing Vision of the Universe* (London: Hutchinson, 1959). All quotations are from these works except where otherwise footnoted.
2. Whitehead, Alfred North, and Bertrand Russell, *Principia Mathematica*, 3 vols. (Cambridge, U.K.: Cambridge University Press, 1910, 1912, 1913).
3. Aristotle, quoted in Russell (1945), p. 136.
4. Frege's labors were not wasted; his book is considered a classic. It is *The Foundations of Arithmetic: A Logico-mathematical Enquiry into the Concept of Number*, available in an edition translated by J. R. Austin (Oxford, U.K.: Blackwell, 1980).
5. Bertrand Russell, "How to Read and Understand History," in *Understanding History and Other Essays* (New York: Philosophic Library, 1957).

6. Barrow, p. 293.
7. Bertrand Russell, "The Value of Free Thought,"in *Understanding History*.

Chapter 19: The Labyrinths of Simplicity

1. Quoted in Kitty Ferguson, *Prisons of Light: Black Holes* (Cambridge, U.K.: Cambridge University Press, 1996), p. 114.
2. Bryan Appleyard, "Master of the Universe: Will Stephen Hawking Live to Find the Secret?" *Sunday Times* (London)
3. Richard Feynman, *QED: The Strange Theory of Light and Matter* (Princeton, N.J.: Princeton University Press, 1985), p. 4.
4. John Archibald Wheeler, *Journey into Gravity and Spacetime* (New York: Scientific American Library, 1990), p. xi.
5. Barrow, p. 129.
6. These paragraphs about the new music of the spheres rely on information from Kristine Larsen, "From Pythagoras to WMAP: The 'Music of the Spheres' Revisited," paper presented to the Society of Literature, Science, and the Arts (November 13, 2005), and published on the Internet (www.physics.ccsu.edu/larsen/wmap.html). The articles and papers cited below are all cited in Larsen's paper.
7. Richard A. Kerr, "Listening to the Music of the Spheres," *Science 1991*, 253: 1207–1208.
8. P. Demarque and D. B. Guenther (1999) "Helioseismology: Probing the Interior of a Star," *Proceedings of the National Academy of Sciences 96*: 5356–69.
9. ESO (May 15, 2002), "Ultrabass Sounds of the Giant Star Xi Hya." http://www.eso.org/outreach/press-rel/pr-2002/pr-10-02.html. The ESO is the European Organization for Astronomical Research in the Southern Hemisphere, or European Southern Observatory.
10. Marcia Bartusiak, *Einstein's Unfinished Symphony: Listening to the Sounds of Space-time* (Washington, D.C.: National Academies Press, 2000).
11. Steve Roy and Megan Watzke, "Giant Galaxy's Violent Past Comes into Focus," Harvard University press release, May 10, 2004. http://chandra.harvard.edu/press/04_releases/press_051004.html
12. Don Savage, Steve Roy, and Megan Watzke, "Chandra 'Hears' a Black Hole for the First Time," Harvard University press release, September 9, 2003. http://chandra.harvard.edu/press/03_releases/press_090903.html
13. Mark Whittle, "Sounds from the Infant Universe." Abstract for American Astronomical Society talk, June 3, 2004. http://www.astro.virginia.edu/-dmw8f/sounds/aas/aas_abs.pdf

14. Mark Whittle, "Primordial Sounds: Big Bang Acoustics," press release: American Astronomical Society Meeting, June 1, 2004. http://www.astro.virginia.edu/-dmw8f/sounds/aas/press_release.pdf
15. Shaun Cole et al. (August 5, 2005), "The 2dF Galaxy Redshift Survey: Power-Spectrum Analysis of the Final Dataset and Cosmological Implications," arXiv: astro-ph/0501174; Daniel J. Eisenstein et al. (January 10, 2005), "Detection of the Baryon Acoustic Peak in the Large-scale Correlation Function of SDSS Luminous Red Galaxies." arXiv: astro-ph/0501171.
16. Ron Cohen, "Ultimate Retro: Modern Echoes of the Early Universe." *Science News Online* 167(3), Jan. 15, 2005. http://www.sciencenews.org/articles/20050115/fob1.asp
17. Diane Richards, "Listening to Northern Lights," *Astronomy*, Dec. 2001, p. 63.

Appendix

1.Bronowski, pp. 158–160.

Bibliography

Books I have used and that appear in the endnotes.

Al-Hasan Al-Katib. *Kitah Kamal Adal Al-Gina'*. Translated by Joscelyn Godwin, from a previous French translation by Amnon Shiloah and excerpted in Godwin, Joscelyn, ed. *The Harmony of the Spheres: A Sourcebook of the Pythagorean Tradition in Music*. Rochester, Vermont: Inner Traditions International, 1993. P. 122.

Alberti, Leon Battista. *The Ten Books of Architecture*. Dover Replica Facsimile Edition. Mineola, N.Y.: Dover, 1987.

Appleyard, Bryan. "Master of the Universe: Will Stephen Hawking Live to Find the Secret?" *Sunday Times* (London), 1988.

Augustine. *City of God*. Translated by Gerald Walsh et al. Garden City, N.Y.: Doubleday, 1958.

Aulus Gellius. *Attic Nights*. Books I–V. Translator J. C. Rolfe. Cambridge, Mass.: Harvard University Press, 1927.

Aurelian of Réôme. *Musica Disciplina*. Translated by Joseph Ponte. Colorado Springs: Colorado College Music Press, 1968. Pp. 20–23.

Baqir, Taha. "An Important Mathematical Problem Text from Tell Harmal." *Sumer* 6 (1950).

Barnes, Jonathan. *Early Greek Philosophy*. London: Penguin, 1987.

Barrow, John. *Pi in the Sky: Counting, Thinking, and Being*. Oxford, U.K.: Clarendon Press, 1992.

Bartusiak, Marcia. *Einstein's Unfinished Symphony: Listening to the Sounds of Space-time*. Washington, D.C.: National Academies Press, 2000.

Billington, James H. *Fire in the Minds of Men: Origins of the Revolutionary Faith*. New York: Basic Books, 1980.

Blumenthal, H. J., and R. A. Markus, eds. *Neo-Platonism and Early Christian Thought: Essays in Honour of A. H. Armstrong*. London: Variorum Publications, 1981.

Bronowski, Jacob. *The Ascent of Man*. Boston: Little, Brown, & Co., 1973.

Brooke, John Hedley. *Science and Religion: Some Historical Perspectives.* Cambridge, U.K.: Cambridge University Press, 1991.

Burkert, Walter. *Lore and Science in Ancient Pythagoreanism*. Cambridge, Mass.: Harvard University Press, 1972.

Buxton, Richard, ed. *From Myth to Reason: Studies in the Development of Greek Thought*. Oxford, U.K.: Oxford University Press, 1999.

Capparelli, Vincenzo. *La sapienza di Pitagora*. Padua: CEDAM, 1941.

Cartledge, Paul, ed. *Cambridge Illustrated History of Ancient Greece*. Cambridge, U.K.: Cambridge University Press, 1998.

Caspar, Max. *Kepler*. Translated and edited by C. Doris Hellman. Reissue, with references by Owen Gingerich and bibliographical citations by Gingerich and Alain Segonds. New York: Dover, 1993.

Cassirer, Ernst, Paul Oskar Kristeller, and John Herman Randall, Jr., eds. *The Renaissance Philosophy of Man: Selections in Translation*. Chicago and London, University of Chicago Press, 1948, 1969.

Centrone, Bruno. "Platonism and Pythagoreanism in the Early Empire." In Christopher Rowe and Malcolm Schofield, eds. *Cambridge History of Greek and Roman Political Thought*. Cambridge, U.K.: Cambridge University Press, 2000.

Cesariano. Drawing for Vitruvius: ccat.sas.upenn.edu/george/vitruvius.html.

Cicero, Marcus Tullius. *On the Commonwealth*. Translated by George Holland Sabine and Stanley Barney Smith. New York: The Liberal Arts Press, 1929.

———. *Cicero: The Republic and The Laws*. Niall Rudd, translator. Oxford, U.K.: Oxford University Press, 1998.

———. "Scipio's Dream." In *On the Republic*. Translated by Cyrus R. Edmonds and Moses Hadas. In *The Basic Works of Cicero*. New York: Random House, Modern Library, 1951.

Cohen, Ron. "Ultimate Retro: Modern Echoes of the Early Universe." *Science News Online* 167 (3), Jan. 15, 2005. http://www.sciencenews.org/articles/20050115/fob1.asp

Cole, Shaun, et al. August 5, 2005. "The 2dF Galaxy Redshift Survey: Power-spectrum Analysis of the Final Dataset and Cosmological Implications." arXiv: astro-ph/0501174.

Copernicus, Nicolaus. *De Revolutionibus*. *Gesamtausgabe*. Vol. II, *De Revolutionibus. Kritischer Text*. Eds. H. M. Nobis and B. Sticker. Hildesheim, Germany, 1984.

Cuomo, S. *Ancient Mathematics*. London: Routledge, 2001.

Demarque, P., and D. B. Guenther (1999) "Helioseismology: Probing the Interior of a Star." *Proceedings of the National Academy of Sciences* 96: 5356–69.

Diogenes Laertius. *The Life of Pythagoras*. Translated and reprinted in Kenneth Sylvan Guthrie. *The Pythagorean Sourcebook and Library: An Anthology of Ancient Writings Which Relate to Pythagoras and Pythagorean Philosophy*. Grand Rapids: Phanes Press, 1987.

Dodds, E. R. *The Greeks and the Irrational*. Berkeley, Calif.: University of California Press, 1951.

Eisenstein, Daniel J. et al. January 10, 2005. "Detection of the Baryon Acoustic Peak in the Large-scale Correlation Function of SDSS Luminous Red Galaxies." arXiv: astro-ph/0501171

Eisenstein, Elizabeth L. *The First Professional Revolutionist: Filippo Michele Buonarroti (1761–1837), a Biographical Essay*. Cambridge, Mass.: Harvard University Press, 1959.

The Epistle on Music of the Ikhwan al-Safa'. Translated by Amnon Shiloah, in Joscelyn Godwin, ed., *The Harmony of the Spheres: A Sourcebook of the Pythagorean Tradition in Music*. Rochester, Vt.: Inner Traditions International, 1993, p. 113.

ESO (European Organization for Astronomical Research in the Southern Hemisphere, or European Southern Observatory) (May 15, 2002) "Ultrabass Sounds of the Giant Star Xi Hya." http://www.eso.org/outreach/press-rel/pr-2002/pr-10-02.html

Ferguson, Kitty. *Prisons of Light: Black Holes*. Cambridge, U.K.: Cambridge University Press, 1996.

_____. *Tycho and Kepler: The Unlikely Partnership That Forever Changed Our Understanding of the Heavens*. New York: Walker, 2002.

Feynman, Richard. *QED: The Strange Theory of Light and Matter*. Princeton, N.J.: Princeton University Press, 1985.

Fox, Robin Lane. *Pagans and Christians in the Mediterranean World from the Second Century A.D. to the Conversion of Constantine*. London: Penguin Books, 1986.

Frege, Gottlob. *The Foundations of Arithmetic: A Logico-mathematical Enquiry into the Concept of Number*. Translated by J. R. Austin. Oxford, U.K.: Blackwell, 1980.

Godwin, Joscelyn, ed. *The Harmony of the Spheres: A Sourcebook of the Pythagorean Tradition in Music*. Rochester, Vt.: Inner Traditions International, 1993.

Grant, Michael. *History of Rome*. London: Faber & Faber, 1978.

Guthrie, Kenneth Sylvan. *The Pythagorean Sourcebook and Library: An Anthology of Ancient Writings Which Relate to Pythagoras and Pythagorean Philosophy*. York Town, ME: Phanes Press, 1987. Contains Iamblichus', Porphyry's and Diogenes Laertius' biographies of Pythagoras as well as some of the pseudo-Pythagorean works.

Guthrie, William Keith Chambers. *The Earlier Presocratics and the Pythagoreans*. Vol 1 of A *History of Greek Philosophy*. Cambridge, U.K.: Cambridge University Press, 1962, 2003.

———. *The Greeks and Their Gods*. Cambridge, U.K.: Cambridge University Press, 1951.

Hammond, N. G. L. A *History of Greece to 322 B.C.* Oxford, U.K.: Oxford University Press, 1986.

Heath, Thomas L. *The Thirteen Books of Euclid's Elements* I. Cambridge, U.K.: Cambridge University Press, 1926.

Heath, Thomas L. *Greek Astronomy*. London: J. M. Dent & Sons, 1932.

Hornblower, Simon, and Antony Spawforth, eds. *The Oxford Companion to Classical Civilization*. Oxford, U.K.: Oxford University Press, 1998.

Hunayn. *Nawadir al-Falasifa*. Excerpts translated by Isaiah Sonne, in Joscelyn Godwin, *The Harmony of the Spheres: A Sourcebook of the Pythagorean Tradition in Music*. Rochester, Vt.: Inner Traditions International, 1993. Pp. 92–98.

Iamblichus of Chalcis. *Life of Pythagoras*. Translated by Thomas Taylor. In Thomas Taylor. *Iamblichus' Life of Pythagoras*. Rochester, Vt.: Inner Traditions International, 1986.

———. *On Common Mathematical Knowledge* 91.3–11. Translation in I. Mueller, "Mathematics and Philosophy in Proclus's Commentary on Book I of Euclid's *Elements*." In J. Pépin and H. D. Saffrey, eds. *Proclus, Lecteur et Interprète des Anciens*. Paris: CNRS, 1987.

John Scotus Eriugena. *Commentary on Martianus Capella*. Translated by Joscelyn Godwin from a ninth-century Bodleian Library manuscript and excerpted in Godwin, Joscelyn, ed. *The Harmony of the Spheres: A Sourcebook of the Pythagorean Tradition in Music*. Rochester, Vt.: Inner Traditions International, 1993. P. 105.

Kahn, Charles H. *Pythagoras and the Pythagoreans: A Brief History*. Indianapolis: Hackett, 2001.

Kepler, Johannes. *Harmonice mundi*. Translated by Eric J. Aiton, A. M. Duncan, and J. V. Field as *Five Books of the Harmony of the World*. Philadelphia: 1993.

———. *Johannes Kepler Gesammelte Werke*. Max Caspar, Salther von Dyck, Franz Hammer, and Volker Bialas, eds. Munich: Deutsche Forschungsgemeinschaft and the Bavarian Academy of Sciences, 1937– .

Kerr, Richard A. "Listening to the Music of the Spheres." *Science 1991*, 253: 1207–1208.

King, Ross. *Brunelleschi's Dome: How a Renaissance Genius Reinvented Architecture*. London: Penguin, 2000.

Koenigsberger, H. G. *Medieval Europe: 400–1500*. London: Longman, 1987.

Koestler, Arthur. *The Sleepwalkers: A History of Man's Changing Vision of the Universe*. London: Hutchinson, 1959.

Kuhn, Thomas S. *The Copernican Revolution: Planetary Astronomy in the Development of Western Thought*. Cambridge, Mass.: Harvard University Press, 1957.

———. *The Structure of Scientific Revolutions*. Chicago: University of Chicago Press, 1962.

Larsen, Kristine. "From Pythagoras to WMAP: The 'Music of the Spheres' Revisited." Paper presented to the Society of Literature, Science, and the Arts, November 13, 2005, and published on the Internet. www.physics.ccsu.edu/larsen/wmap.html

Lee, Desmond. Introduction to Plato's *Timaeus*. London: Penguin, 1965.

Leibniz, G. *The Philosophical Works of Leibniz*. Translated by G. Duncan. New Haven, Conn.: Tuttle, Morehouse and Taylor, 1916.

Lindberg, David C. *The Beginnings of Western Science: The European Scientific Tradition in Philosophical, Religious, and Institutional Context, 600 B.C. to A.D. 1450*. Chicago: University of Chicago Press, 1992.

Max, Daniel T. *The Family That Couldn't Sleep*. New York: Random House, 2006.

McInery, Ralph. *A History of Western Philosophy*. Vol. 2. From the website of the University of Notre Dame Jacques Maritain Center. http://maritain.nd.edu

Oates, Joan. *Babylon*. London: Thames & Hudson, 1979.

Palladio, Andrea. *The Four Books on Architecture*. New York: Dover, 1978. Replica of the Isaac Ware English edition (1738) of *I quattro libri dell' architettura*.

Philostratus. *Life of Apollonius of Tyana*. Translated by F. C. Conybeare. Cambridge, Mass.: Harvard University Press, 1912.

Pieper, Josef. *Scholasticism: Personalities and Problems of Medieval Philosophy*. Translated by Richard and Clara Winston. New York: McGraw-Hill, 1964.

Plato. *Gorgias*. Translated by Chris Emlyn-Jones and Walter Hamilton. London: Penguin, 2004.

———. *Protagoras and Meno*. Translated by Robert C. Bartlett. London: Penguin, 1956.

———. *The Republic*. Translated by Desmond Lee. London: Penguin, 2003.

———. *Timaeus*. Stephanus edition. 1578.

———. *Timaeus*. Translated by Desmond Lee. London: Penguin, 1965.

Raaflaub, Kurt A. "Poets, Lawgivers, and the Beginnings of Political Reflection in Archaic Greece," In Christopher Rowe and Malcolm Schofield, eds. *The Cambridge History of Greek and Roman Political Thought*. Cambridge, U.K.: Cambridge University Press, 2000.

Rawson, Elizabeth. *Intellectual Life in the Late Roman Republic*. Baltimore: Johns Hopkins University Press, 1985.

Regino of Prüm. *Tonarius*. Introduction, "Epistola de harmonica institutione." Translated by Sister Mary Protase LeRoux. In *The "De harmonica Institutione" and "Tonarius" of Regino of Prüm*. Ph.D. diss., Catholic University of America, 1965. Excerpted in Godwin, p. 110.

Richards, Diane. "Listening to Northern Lights." *Astronomy*, Dec. 2001, p. 63.

Robson, Eleanor. "Influence, Ignorance, or Indifference? Rethinking the Relationship Between Babylonian and Greek Mathematics." *The British Society for the History of Mathematics*, Bulletin 4 (Spring 2005).

———. "Mesopotamian Mathematics: Some Historical Background." In Victor J. Katz, ed. *Using History to Teach Mathematics: An International Perspective*. Cambridge, U.K.: Cambridge University Press. 2000.

———. "Three Old Babylonian Methods for Dealing with 'Pythagorean' Triangles." *Journal of Cuneiform Studies* (1997) 49, 51–72.

Roy, Steve, and Megan Watzke. "Giant Galaxy's Violent Past Comes into Focus." Harvard Press Release, May 10, 2004. http://chandra.harvard.edu/press/04_releases/press_051004.html

Rubenstein, Richard E. *Aristotle's Children: How Christians, Muslims, and Jews Rediscovered Ancient Wisdom and Illuminated the Dark Ages*. New York: Harcourt, 2003.

Russell, Bertrand. *The History of Western Philosophy*. London: George Allen & Unwin, 1945.

———. "How to Read and Understand History" and "The Value of Free Thought." In *Understanding History and Other Essays*. New York: Philosophic Library, 1957.

Saggs, H. W. F. *Everyday Life in Babylonia and Assyria*. Assyrian International News Agency, 1965.

Savage, Don, Steve Roy, and Megan Watzke. "Chandra 'Hears' a Black Hole for the First Time." Harvard Press Release, September 9, 2003. http://chandra.harvard.edu/press/03_releases/press_090903.html

Schofield, Malcolm. "Plato and Practical Politics." In Christopher Rowe and Malcolm Schofield, eds. *The Cambridge History of Greek and Roman Political Thought*. Cambridge, U.K.: Cambridge University Press, 2000.

Shaw, Ian. *Ancient Egypt: A Very Short Introduction*. Oxford, U.K.: Oxford University Press, 2004.

Silver, Brian L. *The Ascent of Science*. Oxford, U.K.: Oxford University Press, 1998.

Silverman, David P., ed. *Ancient Egypt*. New York: Oxford University Press, 1997.

Stephenson, Bruce. *The Music of the Heavens: Kepler's Harmonic Astronomy*. Princeton, N.J.: Princeton University Press, 1994.

Thoren, Victor. *The Lord of Uraniborg: A Biography of Tycho Brahe*. Cambridge, U.K.: Cambridge University Press, 1990.

Vitruvius [Marcus Vitruvius Pollio]. *De architectura*. Reprinted as *Vitruvius: Ten Books of Architecture*. Cambridge, U.K.: Cambridge University Press, 2001.

Whitehead, Alfred North, and Bertrand Russell. *Principia Mathematica*. 3 vols. Cambridge, U.K.: Cambridge University Press, 1910, 1912, 1913.

Wheeler, John Archibald. *Journey into Gravity and Spacetime*. New York: Scientific American Library, 1990.

Whittle, Mark. "Primordial Sounds: Big Bang Acoustics." Press release: American Astronomical Society meeting, June 1, 2004. http://www.astro.virginia.edu/-dmw8f/sounds/aas/press_release.pdf

_____. "Sounds from the Infant Universe." Abstract of American Astronomical Society talk, June 3, 2004. http://www.astro.virginia.edu/-dmw8f/sounds/aas/aas_abs.pdf

Wiedemann, Thomas. "Reflections of Roman Political Thought in Latin Historical Writing." In Christopher Rowe and Malcolm Schofield, eds. *The Cambridge History of Greek and Roman Political Thought*. Cambridge, U.K.: Cambridge University Press, 2000.

Wilford, John Noble. "Early Astronomical Computer Found to be Technically Complex." *New York Times*. November 30, 2006.

Wittkower, Rudolph. *Architectural Principles in the Age of Humanism*. New York: Norton, 1971.

Woolf, Greg, ed. *Cambridge Illustrated History of the Roman World*. Cambridge, U.K.: Cambridge University Press, 2003.

Wright, J. Robert., ed. *Ancient Christian Commentary on Scripture, Old Testament IX*. Downers Grove, Ill.: Intervarsity Press, 2005.

Index

Page numbers in *italics* refer to illustrations.

A Note on the Author

Kitty Ferguson is the author of the highly acclaimed *Tycho & Kepler: The Unlikely Partnership That Forever Changed Our Understanding of the Heavens*; *Measuring the Universe: Our Historic Quest to Chart the Horizons of Space and Time*; *The Fire in the Equations: Science, Religion, and the Search for God*; *Prisons of Light: Black Holes*; and *Stephen Hawking: Quest for a Theory of Everything*. She is also a Juilliard-trained professional musician.